December 1988

To Cliff

With best regards

Hiroshi

Toxicity of Hormones in Perinatal Life

Editors

Takao Mori, Ph.D.
Assistant Professor
Zoological Institute
Faculty of Science
University of Tokyo
Tokyo, Japan

and

Hiroshi Nagasawa, Ph.D.
Professor
Experimental Animal Research Laboratory
Meiji University
Kawasaki, Japan

CRC Press, Inc.
Boca Raton, Florida

Library of Congress Cataloging-in-Publication Data

Toxicity of hormones in perinatal life.

Includes bibliographies and index.
1. Fetus—Effect of drugs on. 2. Hormones, Sex—Toxicology. 3. Teratogenesis. I. Mori, Takao, 1928- . II. Nagasawa, Hiroshi. [DNLM: 1. Abnormalities, Drug-Induced. 2. Prenatal Exposure Delayed Effects. 3. Sex Hormones—adverse effects. WK 900 T755]
RG627.6.D79T68 1988 618.3'2 87-35536
ISBN 0-8493-6862-6

Direct all inquiries to CRC Press, Inc., 2000 Corporate Blvd., N.W., Boca Raton, Florida, 33431.

International Standard Book Number 0-8493-6862-6

Library of Congress Card Number 87-35536
Printed in the United States

PREFACE

Sex steroid hormones are essential in mammals for sex differentiation of the brain and development of the reproductive organs and their postnatal growth, morphogenesis, and function. Experimental research on the effects of sex steroids and related substances on the newborn or fetus is an effective method of analyzing abnormal development of brain function and the pathological abnormalities in the reproductive and accessory organs. Since the findings of the relationship between intrauterine exposure of human fetuses to diethylstilbestrol and various types of reproductive tract abnormalities in the female offspring, perinatal treatment of animals with sex steroids and related substances has been extensively used as a valuable model for human consequence.

All contributors invited, who are among the world's leading investigators in this field, reviewed the most recent advances in the specific fields on perinatal hormone treatment and experimental and clinical consequence. This book will be of much benefit not only to basic and clinical researchers but also to students of biomedicine.

Takao Mori
Hiroshi Nagasawa

CONTRIBUTORS

Yasumasa Arai, Ph.D.
Professor
School of Medicine
Department of Anatomy
Juntendo University
Tokyo, Japan

Howard A. Bern, Ph.D.
Professor of Zoology
Cancer Research Laboratory
University of California
Berkeley, California

Robert M. Bigsby, Ph.D.
Assistant Professor
Department of Obstetrics and Gynecology
University of Indiana
Indianapolis, Indiana

Paul S. Cooke
Department of Veterinary Biosciences
University of Illinois
Urbana, Illinois

Gerald R. Cunha, Ph.D.
Professor of Anatomy
Department of Anatomy
University of California
San Francisco, California

Annemarie A. Donjacour, Ph.D.
Assistant Research Anatomist
Department of Anatomy
University of California
San Francisco, California

John-Gunnar Forsberg, M.D.
Professor and Chairman
Department of Anatomy
University of Lund
Lund, Sweden

Keith W. Frey, B.A.
Project Coordinator
Department of Obstetrics and Gynecology
Chicago Lying-In Hospital
University of Chicago
Chicago, Illinois

William B. Gill, Ph.D.
Associate Professor
Department of Urology and Surgery
University of Chicago
Chicago, Illinois

Robert W. Goy, Ph.D.
Wisconsin Regional Primate Research Center
University of Wisconsin
Madison, Wisconsin

Arthur L. Herbst, M.D.
Professor and Chairman
Department of Obstetrics and Gynecology
Chicago Lying-In Hospital
University of Chicago
Chicago, Illinois

Rikard Holmdahl, Ph.D.
Associate Professor
Department of Medical and Physiological Chemistry
Uppsala University
Uppsala, Sweden

Taisen Iguchi, Ph.D.
Associate Professor
Department of Biology
Yokohama City University
Yokohama, Japan

Terje Kalland, M.D., Ph.D.
Associate Professor
Department of Anatomy
University of Lund
Lund, Sweden

Akira Matsumoto, Ph.D.
Assistant Professor
Department of Anatomy
Juntendo University
Tokyo, Japan

John A. McLachlan
Health and Human Services Department
National Institute of Environmental Health Services
National Institutes of Health
Research Triangle Park, North Carolina

Takao Mori, Ph.D.
Assistant Professor
Zoological Institute
Faculty of Science
University of Tokyo
Tokyo, Japan

Hiroshi Nagasawa, Ph.D.
Professor
Experimental Animal Research Laboratory
Meiji University
Kawasaki, Japan

Retha R. Newbold
Laboratory for Reproduction and Development Toxicology
National Institute of Environmental Health Services
National Institutes of Health
Research Triangle Park, North Carolina

Masako Nishizuka, Ph.D.
Assistant Professor
Department of Anatomy
Juntendo University
Tokyo, Japan

Jacob Rotmensch, M.D.
Assistant Professor
Department of Obstetrics and Gynecology
Chicago Lying-In Hospital
University of Chicago
Chicago, Illinois

Samuel A. Sholl, Ph.D.
Associate Scientist
Wisconsin Regional Primate Research Center
University of Wisconsin
Madison, Wisconsin

Noboru Takasugi, D.Sc.
Professor
Department of Biology
Yokohama City University
Yokohama, Japan

Hideo Uno, Ph.D.
Senior Scientist
Wisconsin Regional Primate Research Center
University of Wisconsin
Madison, Wisconsin

Korehito Yamanouchi
Associate Professor
Department of Basic Human Sciences
School of Human Sciences
Waseda University
Tokorozawa, Saitama-Ken, Japan

TABLE OF CONTENTS

Chapter 1
Introduction: Abnormal Genital Tract Development in Mammals Following Early Exposure to Sex Hormones 1
N. Takasugi and H. A. Bern

Chapter 2
Perinatal Sex Steroid Exposure, Brain Morphology, and Neuroendocrine and Behavioral Functions 9
Y. Arai, A. Matsumoto, K. Yamanouchi, and M. Nishizuka

Chapter 3
Mesenchymal-Epithelial Interactions as a Mechanism for Regulating Hormonally Induced Epithelial Differentiation and Growth 21
G. R. Cunha, R. M. Bigsby, P. S. Cooke, and A. A. Donjacour

Chapter 4
Histogenesis of Irreversible Changes in the Female Genital Tract after Perinatal Exposure to Hormones and Related Substances 39
J.-G. Forsberg

Chapter 5
Long-Term Effects of Perinatal Treatment with Sex Steroids and Related Substances on Reproductive Organs of Female Mice 63
T. Mori and T. Iguchi

Chapter 6
Long-Term Effects of Perinatal Exposure to Hormones and Related Substances on Normal and Neoplastic Growth on Murine Mammary Glands 81
H. Nagasawa and T. Mori

Chapter 7
Neoplastic and Non-neoplastic Lesions in Male Reproductive Organs Following Perinatal Exposure to Hormones and Related Substances 89
R. R. Newbold and J. A. McLachlan

Chapter 8
Estrogens and Immunity: Long-Term Consequences of Neonatal Imprinting of the Immune System by Diethylstilbestrol 111
T. Kalland and R. Holmdahl

Chapter 9
Psychological and Anatomical Consequences of Prenatal Exposure to Androgens in Female Rhesus 127
R. W. Goy, H. Uno, and S. A. Sholl

Chapter 10
Effects on Female Offspring and Mothers after Exposure to Diethylstilbestrol 143
J. Rotmensch, K. Frey, and A. L. Herbst

Chapter 11
Effects on Human Males of *In Utero* Exposure to Exogenous Sex Hormones..........161
W. B. Gill

Index..........179

Chapter 1

INTRODUCTION: ABNORMAL GENITAL TRACT DEVELOPMENT IN MAMMALS FOLLOWING EARLY EXPOSURE TO SEX HORMONES

N. Takasugi and H. A. Bern

As early as 1921 Stockard[1] claimed that there are certain stages in vertebrate ontogeny during which toxic substances have more injurious effects than in adulthood. These stages were referred to as ''critical periods'', being evident during organogenesis in mammals. At that time, however, it was not known whether or not there were critical periods of sensitivity to hormone action during development. Also in 1921, Evans and Long[2] found that 7 out of 800 of their rats possessed anovulatory ovaries along with persistent vaginal cornification (''persistent estrus'', PE). Wolfe et al.[3] also found some ''persistent-estrous'' old rats in their colony; later, Bloch[4] described spontaneous PE as an aging phenomenon in rats. In the meantime, Pfeiffer[4a] induced PE experimentally in female rats by transplanting a littermate testis during early postnatal life. Neonatal injections of androgen or estrogen also induced PE in adult rats.[5] The term PE is recognized as a misnomer, inasmuch as the affected animals are in fact anovulatory and noncycling and not generally behaviorally estrous. The syndrome resulting from early exposure to sex hormones has been extensively studied in rats, revealing a mechanism of persistent secretion of ovarian estrogen owing to permanent functional alteration of the hypothalamo-hypophysial system.[5] Takewaki[6] contributed much to our understanding of the biology of this syndrome. PE can be induced by sex hormone in rats only when it is administered within 10 days after birth, indicating the presence of critical period. Male rats and mice treated neonatally with estrogen or androgen show testes with persistent suppression of spermatogenesis together with atrophic seminal vesicles and prostates.[7] It is not yet fully elucidated whether the suppressed spermatogenesis is based on depressed hypothalamo-hypophysial activity, since neonatally administered sex hormones resulted in varying degrees of injury to the cells of the testis.[7]

Long-term administration of estrogen in adulthood is known to cause hyperplastic, dysplastic, and neoplastic lesions in the mammary and pituitary glands, kidney, gonad, and accessory organs in male and female rats, mice, and hamsters.[34] What happens in these same organs in animals treated with various hormones during the perinatal period? In studies on neonatally androgen- or estrogen-treated *rats*, investigations were focused largely on the permanent functional alteration of the hypothalamo-hypophysial system, with little attention paid to the possibility of permanent changes in peripheral tissues. However, studies of *mice* neonatally treated with estrogen revealed permanent changes in peripheral target organs induced by sex hormones given during the critical perinatal period.[11-13] The vaginal epithelium of mice treated neonatally with large amounts of 17β-estradiol (E_2) showed persistent proliferation and cornification, which were not abolished by adrenalectomy or hypophysectomy following ovariectomy. Vaginas of these mice maintained these epithelial changes even after they had been transplanted into untreated ovariectomized hosts, revealing an irreversible estrogen-independent phenomenon.[9,10] These findings became the origin of further studies on the occurrence of permanent vaginal changes in mice exposed perinatally to natural and synthetic estrogens. Similar changes occurred with lower doses of estrogen but these could be abolished by ovariectomy. Thus, it became necessary to recognize two syndromes: one involving permanent changes in the genital organs themselves and the other occurring secondarily to a permanently altered hypothalamo-hypophysio-ovarian axis — ovary independent as opposed to ovary dependent. In both cases, the permanent vaginal cornification was induced by estrogen or androgen only when the treatment was started within 3 days after birth, indicating the presence of a critical period for both syndromes.[11,12] The permanently proliferated vaginal epithelium of neonatally estrogen-treated mice frequently resulted in precancerous or cancerous lesions after more than 10 months.[8,11-13] Thus, short-term neonatal estrogen treatment in the critical period causes severe lesions in the target organs, similar to the changes induced by continued long-term administration of the hormone initiated after puberty. In the case of higher doses of estrogen, the neonatally induced effect was directly upon the vagina.

At the same time as these studies were conducted, Forsberg[14] was extensively analyzing the differentiation of the vaginal epithelium in various mammals, including humans, exposed and unexposed to estrogen. He demonstrated that heterotopic columnar epithelium, identified as adenosis, occurs in the fornical areas of the vagina of NMRI stock of mice exposed neonatally to estrogen, resulting in preneoplastic and neoplastic lesions of a glandular nature.[14] Adenosis-like lesions and adenocarcinomatous and/or squamous-cell carcinomatous lesions (some lesions are clearly "mixed") were observed in the cervicofornical region of the vagina in various strains of mice.[8,14] The morphological, histochemical, and biochemical nature of estrogen-independent persistent vaginal cornification and histogenesis has been examined by many investigators (e.g., Dunn, Takasugi, Kimura, Mori, Bern, Jones, Kohrman, Forsberg, Cunha, McLachlan, Walker, Iguchi, Taguchi, Nomura, Vorherr, Yasuda, Warner and their collaborators; see References 8 and 11 to 16). It is of interest that neonatally injected E_2 inhibits the normal differentiation of the Müllerian epithelium of the mouse vagina;[14] however, Juillard and Delost[17] claim that the Müllerian epithelium is replaced by proliferating urogenital sinus epithelium in these mice.

Crystals and/or concretions also appear in the vaginal lumen in neonatally E_2-treated adult mice.[12] The concretions are composed of magnesium, phosphorus, calcium, and potassium salts with about 45% organic matter. Bern et al.[18] recently demonstrated that implantation of the concretions into the vaginal lumen of adult ovariectomized mice exposed neonatally to E_2 at what was regarded as a near-threshold dose resulted in an increase in the incidence of vaginal epithelial pegs and downgrowths over that seen in the concretion-free vaginas of similarly treated mice. It is possible that the physicochemical nature of the concretions promotes the development of abnormal epithelial lesions in the vaginal epithelium of mice exposed perinatally to estrogen and suggests that early hormone exposure may increase the sensitivity of genital organs to later stimuli.

In 1970, Herbst and Scully[19] reported the occurrence during the previous 2-year period of vaginal tumors in eight adolescent women at Massachusetts General Hospital; prior to this time, only a few vaginal tumors had been recorded in young women. Seven of these cases were born of mothers who had received a synthetic estrogen, diethylstilbestrol (DES), for the supposed prevention of premature birth. Association of DES exposure during pregnancy with the occurrence of vaginal cancer became established in later reports by Herbst and his associates.[8,20] The incidence of vaginal cancer, however, is low in the DES-exposed women, being calculated at 0.14 to 1.4 per thousand through age 24, although structural changes in the vagina including adenosis are present in at least one third of the exposed females.[8,20] A study from the Boston Collaborative Drug Surveillance Program estimates that 100,000 to 160,000 women per annum were prenatally exposed to DES in the U.S. during the period 1960 to 1970, reaching more than one million in the 10 years.[8]

The embryology of the human vagina is different from that of the mouse vagina. Forsberg and his associates[8,14] demonstrated that at the level of the dorsal wall of the urogenital sinus, a pair of Müllerian ducts fuse to form an epithelial tube with a single lumen (uterovaginal canal). In contrast, the mouse vagina has a dual origin: the fused posterior part of Müllerian ducts is connected with a solid cord of urogenital sinus; the cranial three fifths is derived from Müllerian duct and the caudal two fifths from urogenital sinus. Adenosis in human females exposed to DES *in utero* frequently appeared in the anterior wall of the upper vagina.[8] The columnar epithelial cells in adenosis are similar to those lining the endocervical region. Clear-cell adenocarcinomas are the common type of tumor appearing in the cervicovaginal area of prenatally DES-exposed human females, although there are some reports of the occurrence of squamous-cell carcinomas.[8] In prenatally DES-exposed mice, adenosis-like lesions were observed mainly in the cervicofornical area of the vagina.[8,14] The columnar cells in adenosis resemble those in the pseudostratified epithelium of the Müllerian vagina of neonatal mice. Squamous-cell carcinomatous lesions were commonly observed in all parts of the vagina of old DES-exposed mice, whereas adenocarcinomatous lesions, frequently mixed with squamous-cell carcinomatous lesions, were encountered in the upper part of the vagina.[8,11-14]

Neonatally castrated 9-month-old male rats which had received daily injections for the first 30 days after birth exhibited epithelial metaplasia and downgrowths of ejaculatory ducts, coagulating glands, and dorsal urethral wall situated near the opening of the ducts or glands.[7] Neonatal E_2 treatment of male rats also resulted in hyperplastic epithelial lesions of the coagulating gland including squamous metaplasia, the incidence of which was higher in castrated mice than in intact mice.[7] McLachlan et al. injected high doses of DES into mice on days 9 through 16 of gestation. The prenatally DES-exposed males at ages of 9 to 10 months possessed cryptorchid testes with or without epididymal cysts, in which small nodules and sheets of interstitial cells were frequently formed.[21] Their seminal vesicles and/or coagulating glands had nodular enlargements adjacent to the seminal colliculus, in which squamous metaplasia was frequently observed. In DES-exposed males aged 20 to 26 months, neoplastic changes were found in the coagulating glands, seminal vesicles, and testes.[15,21]

Prior to the report by Herbst and Scully[19] of the relationship between prenatal exposure to DES and cervicovaginal tumorigenesis, it was experimentally demonstrated that cervicovaginal tumors occurred in adult mice given neonatal injections of DES or E_2.[8,11-13] Inasmuch as mice are born in a development state corresponding with an early stage of human fetal life, neonatal mice given estrogens can be considered as a model of human fetuses exposed to DES.

In an approach to elucidating the developmental mechanism of permanently proliferated vaginal epithelium, Cunha and his associates[22] recombined trypsin-separated epithelium and stroma from normal and neonatally estrogen-treated mice and suggested that permanent proliferation resulted from the cooperative expression of permanent effects on both the developing epithelium and stroma when exposed to estrogen neonatally.

Although most attention has been paid to the vagina and cervix of perinatally sex hormone-exposed experimental rodents (mice, rats, hamsters), other regions of the female reproductive tract also show permanent changes: the uterus,[23] the oviduct,[24] and the ovary[25] itself. Squamous metaplasia and fibromuscular wall pathology in the uterus, extensive hyperplasia of the oviductal epithelium, and polyovular follicles in the ovary are the most prominent changes that have been reported. Mammary gland hyperplasia, dysplasia, and neoplasia may also result but these changes are ovary dependent. Effects on the male genital tract can also be prominent (see above). There is evidence for abnormal gonadal[26] and hypophysial (including prolactin[27]) secretory patterns. Changes in the nervous system[28] (hypothalamus and its input) underlie some of the endocrine changes, but other components of the nervous system may also be affected. Changes in hormone sensitivity (receptor levels) of target organs also occur.[28a] Important alterations in the immune system have been reported,[8] and diffuse effects on metabolism as a consequence of altered enzyme patterns in the liver,[8] for example, are also potential consequences of exposure to abnormal levels of sex hormones at critical periods (see Figure 1).

There have been few attempts to search for effective factor(s) which can prevent the occurrence of hormone-induced lesions other than a series of experiments with retinoids (such as vitamin A). Injections of vitamin A resulted in prevention of the occurrence of

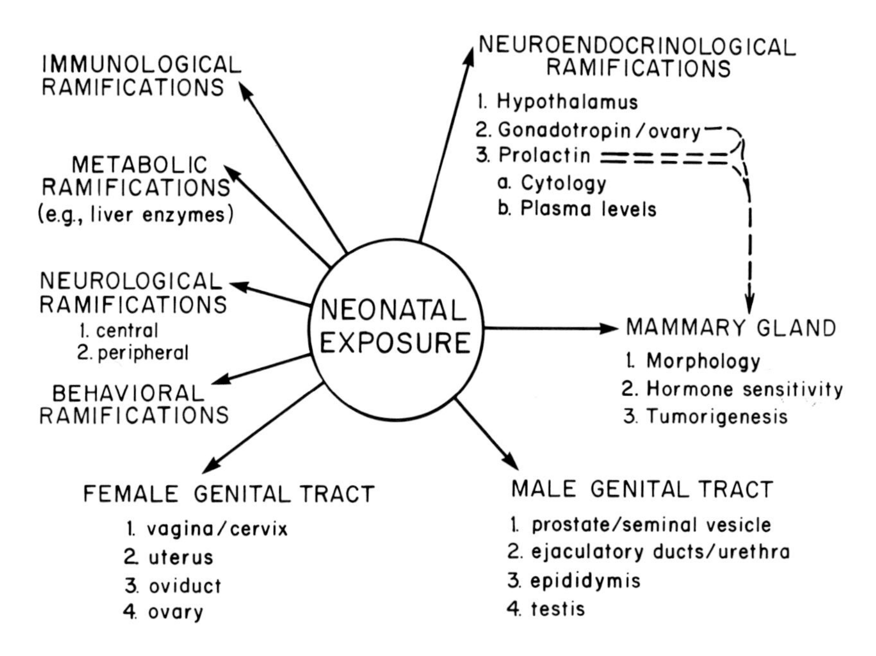

FIGURE 1. Ramifications of consequences of neonatal exposure of mice to natural and synthetic estrogens including DES. The mouse model indicates possible areas for investigation in human offspring exposed *in utero* to DES or other estrogenic agents. (Reprinted with considerable modification by permission from Herbst, A. L. and Bern, H. A., *Developmental Effects of DES in Pregnancy*, Thieme Medical Publishers, New York, 1981, 130.)

vaginal lesions in ovariectomized mice treated perinatally with E_2, DES, or dihydrotestosterone, when given simultaneously with these hormones.[11-13,29] Progesterone given postnatally reduced the severity of lesions.[30] Further studies on suppression of hyperplastic lesions in the genital tract of rodents given DES perinatally are needed.

In the cervicovaginal area of human females exposed to DES *in utero,* clear-cell adenocarcinomas were found at a median age of 18.9 years. Prior to these cases, clear-cell adenocarcinomas had only been encountered in older women.[8] A second peak in clear-cell carcinoma occurrence, therefore, remains a possibility in DES-exposed offspring postmenopausally. In addition, concern should continue in regard to the occurrence of squamous-cell carcinomas with advancing age;[31] the mouse model certainly indicates this possibility. A case of endometrial adenocarcinoma in a DES daughter has recently been reported.[32] Whether there will prove to be an increased risk of breast cancer depends on the applicability of the mouse model, where a mammary tumor virus is involved.[8] However, Boylan and Calhoon's[33] experiments on mammary sensitivity to chemical carcinogens in the rat suggest that early exposure to DES may increase tumor risk. In DES-exposed men, tumors of the genitourinary tract have not yet been reported, other than case histories of testicular cancer possibly associated with an increased incidence of cryptorchidism; a variety of structural abnormalities has been encountered in the testis (including abnormal spermatozoa), epididymis, and penis.[8,34] However, whether it is epidemiologically valid to ascribe these lesions to DES exposure is still being debated.[35] Again, however, mouse and rat studies[7,8,15,21] raise the possibility that dysplastic and even neoplastic lesions in the prostate of men exposed to DES *in utero* may arise at an advanced age.

The work with rodent models suggests the need for a much broader concern with the consequences of intrauterine exposure to DES for humans. In view of the predictive value of many of the observations made on mice in particular, attention should be paid to effects on systems other than the reproductive tract and the neuroendocrine axes governing its development and function. Immunological involvement could be visualized in increased susceptibility to other disease conditions, indicating the value of thorough epidemiological studies of the DES-exposed population. Neurological involvement suggests that further studies of psychological features of the DES-exposed population could be valuable. Metabolic involvement raises the possibility of distantly related consequences of DES exposure, which could be intensified with aging of the population.

Finally, the arguments for the need of further clinical and experimental analyses presented herein are not limited to DES exposure, although DES residues in food are still a matter of concern.[36,37] Exogenous factors other than DES, to which the developing fetus and the neonate could be exposed, include plant estrogens such as coumestrol[38] or zearalenone[39] (which are constituents of animal feeds and may occur in the human dietary intake) and progestins (which may be prescribed in an attempt to reduce the risk of spontaneous abortion). Although neonatal mouse experiments with progestins indicate effects not unlike those seen with estrogens,[40] the clinical literature to data indicates little evidence for the occurrence of fetal anomalies in humans.[41] ''Environmental estrogens'' must continue to be a matter of concern.[42,43] The fetus is also exposed to endogenous sex hormones during development, and the long-term effects of exposure to abnormal levels of endogenous hormones need to be evaluated. The central point is that the developing organism is subject to a variety of stimuli, some of which are still undefined, the influence of which during a critical period in development may result not only in teratological alterations that are readily discernible but also in subtle changes expressed much later in life. One expression of these delayed effects could be an alteration in tumor risk in tissues affected directly or indirectly by the initial stimulus.

REFERENCES

1. **Stockard, C. R.,** Developmental rate and structural expression: an experimental study of twins, '' double monsters'' and single deformities, and the interaction among embryonic organs during their origin and development, *Am. J. Anat.,* 28, 115, 1921.
2. **Evans, H. M. and Long, J. A.,** On the association of continued cornification of the vaginal mucosa with the presence of large vesicles in the ovary and the absence of corpus luteum, *Anat. Rec.,* 21, 60, 1921.
3. **Wolfe, J. M., Bryan, W. R., and Wright, A. W.,** Histologic observations on the anterior pituitaries of old rats with particular reference to the spontaneous appearance of pituitary adenomata, *Am. J. Cancer,* 34, 352, 1938.
4. **Bloch, S.,** Untersuchungen über das funktionelle Altern tierischer Genitalorgane, *Gynaecologia,* 144, 313, 1957.

4a. **Pfeiffer, C. A.,** Sexual differences of the hypophyses and their determination by the gonads, *Am. J. Anat.,* 58, 195, 1936.

5. **Barraclough, C. A.,** Modifications in reproductive function after exposure to hormones during the prenatal and early postnatal period, in *Neuroendocrinology,* Vol. 2, Martini, L. and Ganong, W. F., Eds., Academic Press, New York, 1967, chap. 19.
6. **Takewaki, K.,** Some aspects of hormonal mechanism involved in persistent estrus, *Experientia,* 18, 1, 1962.
7. **Arai, Y., Mori, T., Suzuki, Y., and Bern, H. A.,** Long-term effect of perinatal exposure to sex steroids and diethylstilbestrol on the reproductive system of male mammals, *Int. Rev. Cytol.,* 84, 235, 1983.
8. **Herbst, A. L. and Bern, H. A., Eds.,** *Developmental Effects of Diethylstilbestrol (DES) in Pregnancy,* Thieme-Stratton, New York, 1981, chaps. 1 to 15.
9. **Takasugi, N., Bern, H. A., and DeOme, K. B.,** Persistent vaginal cornification in mice, *Science,* 138, 438, 1962.
10. **Takasugi, N.,** Vaginal cornification in persistent-estrous mice, *Endocrinology,* 72, 607, 1963.
11. **Takasugi, N., Kimura, T., and Mori, T.,** Irreversible changes in mouse vaginal epithelium induced by early postnatal treatment with steroid hormones, in *Postnatal Development of Phenotype,* Kadzda, S. and Denenberg, V. H., Eds., Butterworths, London, 1970, chap. 19.

12. **Takasugi, N.,** Cytological basis for permanent vaginal changes in mice treated neonatally with steroid hormones, *Int. Rev. Cytol.,* 44, 193, 1976.
13. **Takasugi, N.,** Development of permanently proliferated and cornified vaginal epithelium in mice treated neonatally with steroid hormones and the implication in tumorigenesis, *Natl. Inst. Cancer Monogr.,* 51, 57, 1979.
14. **Forsberg, J.-G.,** Developmental mechanism of estrogen-induced irreversible changes in the mouse cervicovaginal epithelium, *Natl. Inst. Cancer Monogr.,* 51, 41, 1979.
15. **McLachlan, J. A.,** Prenatal exposure to diethylstilbestrol in mice: toxicological studies, *J. Toxicol. Environ. Health,* 2, 527, 1977.
16. **Kohrman, A. F.,** The newborn mouse as a model for study of the effects of hormonal steroids in the young, *Pediatrics,* 62, 1143, 1978.
17. **Juillard, M. T. and Delost, P.,** Les effets de l'estradiol injecté au cours de l'ontogénèse sur la différentiation du vagin et de l'utérus chez la Souris, *C.R. Soc. Biol.,* 159, 1541, 1965.
18. **Bern, H. A., Mills, K. T., and Mori, T.,** Effects of long-term implantation of vaginal concretions on the cervicovaginal epithelium of mice, *Proc. Soc. Exp. Biol. Med.,* 177, 303, 1984.
19. **Herbst, A. L. and Scully, R. E.,** Adenocarcinoma of the vagina in adolescence: a report of 7 cases including 6 clear cell carcinomas (so-called mesonephromas), *Cancer,* 25, 745, 1970.
20. **Herbst, A. L., Anderson, S., Hubby, M. M., Haenszel, W. M., Kaufman, R. H., and Noller, K. L.,** Risk factors for the development of diethylstilbestrol-associated clear cell adenocarcinoma: a case-control study, *Am. J. Obstet. Gynecol.,* 154, 814, 1986.
21. **McLachlan, J. A., Newbold, R. R., and Bullock, B.,** Reproductive tract lesions in male mice exposed prenatally to diethylstilbestrol, *Science,* 190, 991, 1975.
22. **Cunha, G. R., Lung, B., and Kato, K.,** Role of the epithelial-stromal interaction during the development and expression of ovary-independent vaginal hyperplasia, *Dev. Biol.,* 56, 52, 1977.
23. **Ostrander, P. L., Mills, K. T., and Bern, H. A.,** Long-term responses of the mouse uterus to neonatal diethylstilbestrol treatment and later sex hormone exposure, *J. Natl. Cancer Inst.,* 74, 121, 1985.
24. **Newbold, R. R., Bullock, B. C., and McLachlan, J. A.,** Progressive proliferative changes in the oviduct of mice following developmental exposure to diethylstilbestrol, *Teratol. Carcin. Mutat.,* 5, 473, 1985.
25. **Iguchi, T.,** Occurrence of polyovular follicles in ovaries of mice treated neonatally with diethylstilbestrol, *Proc. Jpn. Acad. Ser. B,* 61, 288, 1985.
26. **Tenenbaum, A., Sernvi, C., and Forsberg, J.-G.,** Ovarian progesterone synthesis and content and plasma progesterone levels in adult mice treated with diethylstilbestrol neonatally, *Biol. Res. Preg.,* 6, 143, 1985.
27. **Vaticón, M. D., Fernández Galaz, M. C., Tejero, A., and Aguilar, E.,** Alteration of prolactin control in adult rats treated neonatally with sex steroids, *J. Endocrinol.,* 105, 429, 1985.
28. **Arai, Y. and Matsumoto, A.,** Synapse formation of the hypothalamic arcuate nucleus during post-natal development in the female rat and its modification by neonatal estrogen treatment, *Psychoneuroendocrinology,* 3, 31, 1978.

28a. **Bern, H. A., Edery, M., Mills, K. T., Kohrman, A. F., Mori, T., and Larson, L.,** Long-term alterations in histology and steriod receptor levels of the genital tract and mammary gland following neonatal exposure of female BALB/cCrgl mice to various doses of diethylstilbestrol, *Cancer Res.,* p. 47, 1987.

29. **Iguchi, T., Iwase, Y., Kato, H., and Takasugi, N.,** Prevention by vitamin A of the occurrence of permanent vaginal and uterine changes in ovariectomized adult mice treated neonatally with diethylstilbestrol and its nullification in the presence of ovaries, *Exp. Clin. Endocrinol.,* 85, 129, 1985.
30. **Jones, L. A., Verjan, R. P., Mills, K. T., and Bern, H. A.,** Prevention by progesterone of cervicovaginal lesions in neonatally estrogenized BALB/c mice, *Cancer Lett.,* 23, 123, 1984.
31. **Anon.,** Report of the 1985 DES Task Force, National Institutes of Health-National Cancer Institute, U.S. Department of Health and Human Services, Washington, D.C., 1985.
32. **Barter, J. F., Austin, J. M., and Shingleton, H. M.,** Endometrial adenocarcinoma after in utero diethylstilbestrol exposure, *Obstet. Gynecol.,* 67, 84S, 1986.
33. **Boylan, E. S. and Calhoon, R. E.,** Transplacental action of diethylstilbestrol on mammary carcinogenesis in female rats given one or two doses of 7,12-dimethylbenz(*a*) anthracene, *Cancer Res.,* 43, 4879, 1983.
34. **Ward, E., Ed.,** Sex Hormones (II), Vol. 21, IARC Monogr. on the Evaluation of the Carcinogenic Risk of Chemicals to Humans, International Agency for Research on Cancer, Lyon, 1979, 205.
35. **Leary, F. J., Resseguie, L. J., Kuriand, L. T., O'Brien, P. C., Emslander, R. F., and Noller, K. L.,** Males exposed in utero to diethylstilbestrol, *JAMA,* 252, 2984, 1984.
36. **Loizzo, A., Gatti, G. L., Macri, A., Moretti, G., Ortolani, E., and Palazzesi, S.,** The case of diethylstilbestrol treated veal contained in homogenized baby foods in Italy. Methodological and toxicological aspects, *Ann. Ist. Sup. Sanità,* 20, 215, 1984.
37. **McLachlan, J. A., Korach, K. S., Newbold, R. R., and Degen, G. H.,** Diethylstilbestrol and other estrogens in the environment, *Fundam. Appl. Toxicol.,* 4, 686, 1984.

38. **Burroughs, C. D., Bern, H. A., and Stokstad, E. L. R.,** Prolonged vaginal cornification and other changes in mice treated neonatally with coumestrol, a plant estrogen, *J. Toxicol. Environ. Health,* 15, 51, 1985.
39. **Burroughs, C. D., Williams, B. A., Mills, K. T., and Bern, H. A.,** Genital tract abnormalities in female C57BL/Crgl mice exposed neonatally to phytoestrogens (coumestrol and zearalenone), *Proc. Am. Assoc. Cancer Res.,* 27, 220, 1986.
40. **Jones, L. A. and Bern, H. A.,** Cervicovaginal and mammary gland abnormalities in BALB/cCrgl mice treated neonatally with progesterone and estrogen, alone or in combination, *Cancer Res.,* 39, 2560, 1979.
41. **Check, J. H., Rankin, A., and Teichman, M.,** The risk of fetal anomalies as a result of progesterone therapy during pregnancy, *Fertil. Steril.,* 45, 575, 1986.
42. **McLachlan, J. A., Ed.,** *Estrogens in the Environment,* Elsevier/North Holland, New York, 1980.
43. **McLachlan, J. A., Ed.,** *Estrogens in the Environment II, Influence on Development,* Elsevier/North Holland, New York, 1985.

Chapter 2

PERINATAL SEX STEROID EXPOSURE, BRAIN MORPHOLOGY, AND NEUROENDOCRINE AND BEHAVIORAL FUNCTIONS

Y. Arai, A. Matsumoto, K. Yamanouchi, and M. Nishizuka

TABLE OF CONTENTS

I. Introduction 10

II. Gonadal Steroids and Brain Sexual Differentiation 10

III. Gonadal Steroids and Structural Sex Difference in the Neuroendocrine Brain ... 11

IV. Synaptic Parameters of the Sexually Dimorphic Brain 11

V. Gonadal Steroids and Synaptogenesis in the Developing Neuroendocrine Brain 16

VI. Consequences of Perinatal Exposure to Normal or Abnormal Levels of Gonadal Steroids in the Neuroendocrine Brain 16

Acknowledgments 18

References 18

I. INTRODUCTION

Evidence is now available indicating sexually dimorphic functions of the brain. The most functional differences between male and female animals are those in reproductive physiology and reproductive behavior. Male-female differences in the brain have also been demonstrated in morphologic parameters such as nuclear volume, dendritic field, and synaptic organization in certain brain regions.[1,2] Many of these sexually dimorphic brain morphology and functions are brought about by the exposure of undifferentiated brain to perinatal gonadal steroids, especially aromatizable androgen or estrogen, because intraneuronal aromatization of testicular androgen to estrogen is prerequisite for the exertion of these organizational effects.[3-5] Sex steroids have been reported to modulate maturation and neuritic growth.[6] Synaptogenesis can also be facilitated by gonadal steroids in certain neuronal groups in the developing neuroendocrine brain.[1,7] These organizational effects of gonadal steroids appear to be regionally specific and correlated with the topographic localization of sex steroid receptor-concentrating neurons. These findings may provide evidence for the mechanisms underlying sexual differentiation of the brain.

In addition to the normal process of brain sexual differentiation, abnormal levels of systemic gonadal hormones when occurring during brain differentiation might act as teratogens. They might lead to permanent functional and structural changes in the brain which are associated with permanent deviations or dysfunctions of reproduction.

II. GONADAL STEROIDS AND BRAIN SEXUAL DIFFERENTIATION

The male apparently lacks the mechanisms to regulate the cyclic release of gonadotropins, including positive feedback in his gonadotropin response to estrogen. However, this is not determined by the neuronal genome, because early neonatal castration produces "feminine male rats" which are capable of secreting gonadotropins in a cyclic pattern. Conversely, aromatizable androgen or estrogen injected into neonatal female rats masculinizes or defeminizes a number of neuroendocrine parameters.[3,4] Perinatal secretion of sex steroids is also important in determining whether an adult animal behaves as a male or a female. Female rat pups injected with aromatizable androgen or estrogen around the time of birth can be permanently masculinized as well as defeminized. These female rats mount more frequently than normal females (masculinization), but do not or only rarely show lordosis (defeminization), whereas male rats castrated at birth display female levels of lordosis (feminization). Together, these findings suggest that the sex difference in neuroendocrine and behavioral responses is determined by perinatal hormonal environment. This may raise the question as to whether the main organizational action of gonadal steroids is to enhance the development of neural pathways involved in male sexual behavior and suppress those of the female.

In behavioral defeminization of male rats, for example, it has been proposed that testosterone inhibits the neonatal development of a lordosis facilitation center located in the ventromedial hypothalamus. Testosterone implants in this region have been reported to be most effective in eliminating lordosis behavior in adulthood.[8] However, the surgical transection of descending fibers (presumably of septal origin) just above the anterior commissure has been found to potentiate lordosis in normal males.[9,10] As in normal females, lesions in the pontine central gray abolish lordosis in those males receiving the surgical transection of the dorsal extrahypothalamic descending fibers.[11] These findings indicate that the neural mechanisms for lordosis expression are still left intact or only slightly affected by the organizational action of neonatal testicular androgen in male rats. This is hard to reconcile with a view that neonatal testicular androgen destroys the neural mechanism for facilitation of lordosis. Instead, behavioral defeminization in normal male rats seems to be associated with the development of a strong forebrain lordosis-inhibiting system. In normal female rats, the inhibitory system is easily disinhibited by circulating ovarian hormones. However, the inhibition of this system in males is much stronger than in females, being insensitive to any hormonal priming to induce lordosis in males.[12] In fact, lordosis cannot be seen in adult male rats unless the inhibitory system is removed surgically by cutting the descending fibers. Therefore, the organizational action of androgen seems to build in a strong inhibitory neural circuit in the lordosis-regulating system.

III. GONADAL STEROIDS AND STRUCTURAL SEX DIFFERENCE IN THE NEUROENDOCRINE BRAIN

The medial preoptic area (POA) of the rat is one of the sexually dimorphic regions. Gorski and his co-workers have described an intensely stained neuron group with a striking sex difference which they have called the sexually dimorphic nucleus of the POA (SDN-POA).[13] The SDN-POA occupies more neurons in males than in females. These characteristics are dependent on perinatal steroid conditions,[14] suggesting the presence of a critical period during which the SDN-POA is most sensitive to androgen because continuous perinatal exposure to androgen is not essential for stimulation of the development of the SDN-POA (Figure 1).[15] Analogues of the rat SDN-POA have also been identified in gerbils,[16] ferrets,[17] and guinea pigs.[18] Recently, the presence of an analogue of the SDN-POA in the human brain has been demonstrated.[19] Morphometric analysis showed that the SDN is 2.5 times as large in men as in women. The functional significance of the SDN-POA is not clear at present, but it is presumably concerned with masculine behavior.

Other than the SDN-POA, another cell group comprising the sexually dimorphic nuclear complex has been identified in the rostroventral periventricular gray of the POA and its adjacent area, being larger and more densely cellular in female rats than in males (Figure 2).[20,21] This structure corresponds to the medial preoptic nucleus (MPN)[20,21] and is thought to play a critical role in regulating the cyclic release of gonadotropins in female rats, because small lesions confined to this part block spontaneous ovulation, inducing anovulatory persistent estrous syndrome.[22]

Recently, the reduction of the nuclear volume of the MPN has been found to depend on perinatal androgen.[23] As shown in Figure 3, the MPN is significantly smaller in the females exposed to testosterone propionate (TP) than in normal females ($P < 0.01$), but the degree of the volume reductions is not as marked as that in the females injected with TP neonatally. In the male similarly treated with TP, the MPN seems to be affected by TP. Its volume in the males receiving TP at day 21 of pregnancy decreased significantly, compared to normal males ($P < 0.01$). However, the volume of the MPN of the males could not be altered by neonatal castration. This failure of neonatal castration of males to reverse the nuclear volume to a female size may be due to the organizational action of endogenous androgen exerted before neonatal castration.

Sex difference in nuclear volume has also been found in the ventromedial nucleus (VMN),[24] the bed nucleus of the stria terminalis,[18] and the medial amygdaloid nucleus (MAN).[25] In the VMN, for instance, the volume is significantly greater in males than in females. Castration of neonatal male rats on the day of birth decreases the volume to a level comparable to that of normal females, whereas the volume is no longer influenced by castration at 7 days of age. This suggests that the volume of this nucleus is not determined entirely by the genetic factor, but rather the postnatal growth of the VMN is modified by the organizational action of testicular hormones. The critical period for this neonatal imprinting by testicular androgen in the developing VMN seems to be before day 7.

IV. SYNAPTIC PARAMETERS OF THE SEXUALLY DIMORPHIC BRAIN

At an electron microscopic level, sex difference in synaptic organization has been found in several regions. According to the site of the synaptic contact, three types of synapses are

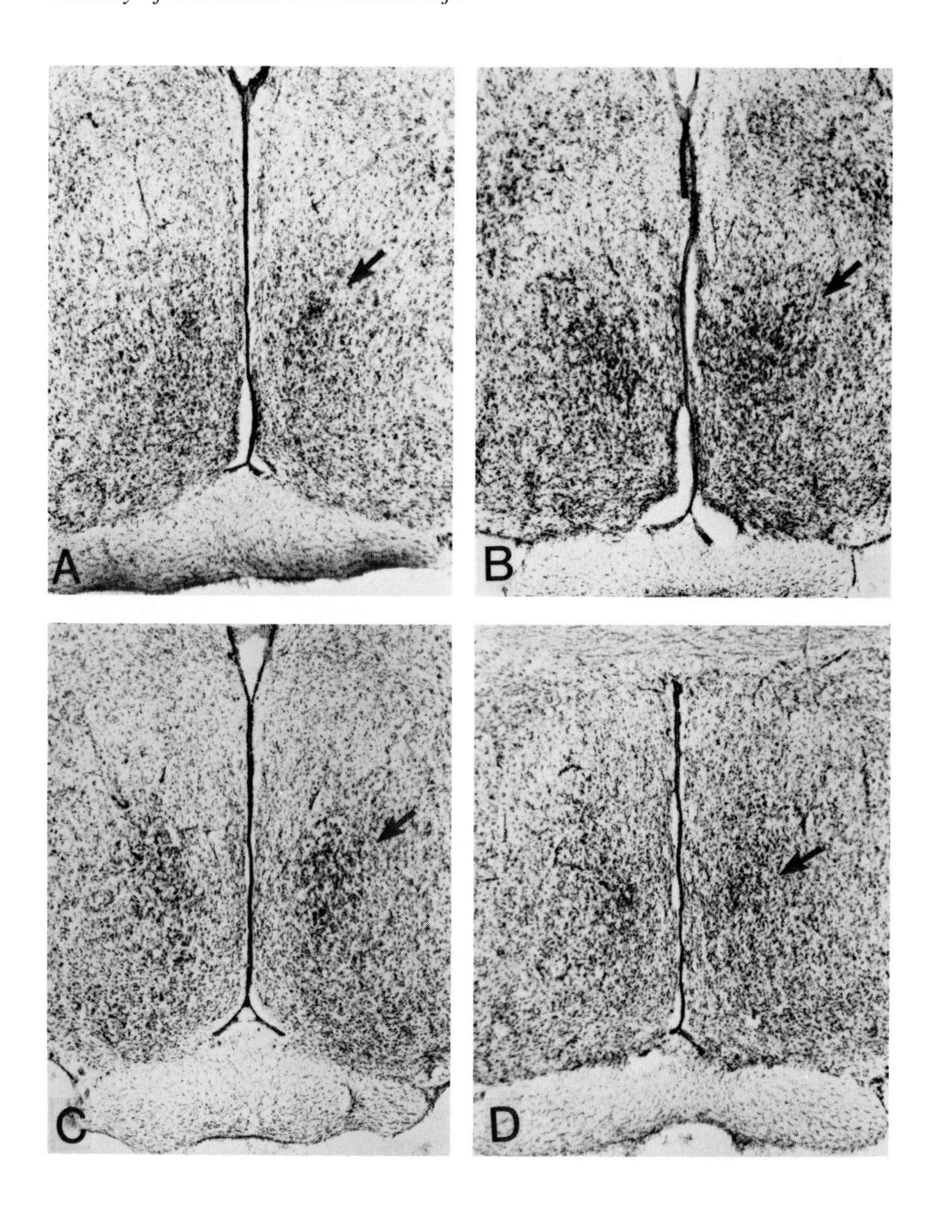

FIGURE 1. Histology of the SDN-POA which is indicated by an arrow. (A) POA of an oil-treated control rat. (B) POA of a female rat exposed to TP from day 15 to 22 of pregnancy. Note the enlargement of the nuclear size of the SDN-POA. (C) POA of a female rat treated with TP on day 17 of pregnancy. Note the enlarged SDN-POA. (D) POA of a female rat treated with TP on day 21 of pregnancy. The size of the SDN-POA is almost identical to that of the control.

roughly classified: shaft synapses made on the dendritic shaft (Figure 4), spine synapses on dendritic spine (Figure 4), and somatic synapses on the cell body. In the arcuate nucleus (ARCN), the number of spine synapses is approximately twice as many in females as in males, and the number of somatic synapses in females is twice that in males.[26] Interestingly, neonatal castration of males increases the number of spine synapses to the female level. On the other hand, the female synaptic pattern can be reversed by neonatal injection of TP. In the dorsomedial POA, the situation is similar to that in the ARCN. No difference in the number of shaft synapses is found between the two sexes, whereas the number of spine

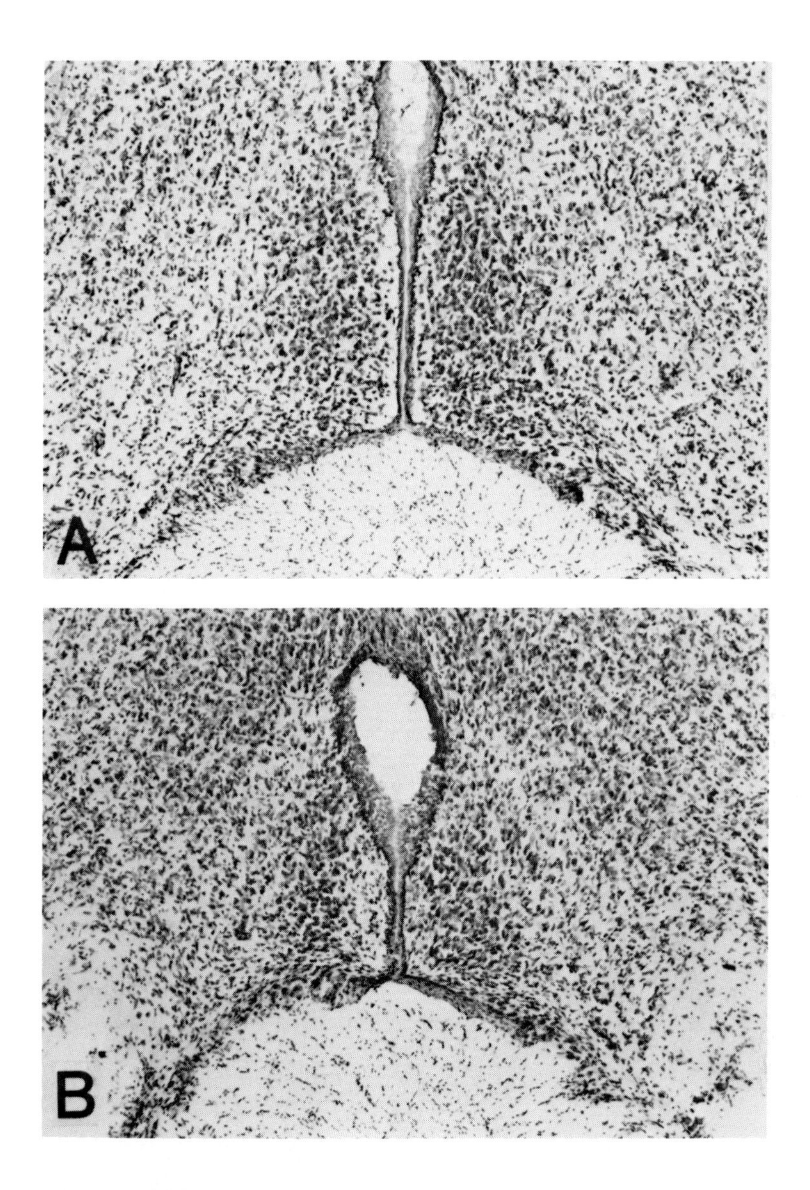

FIGURE 2. Histology of the MPN in the rostroventral periventricular POA. (A) MPN of a normal female rat. (B) MPN of a normal male. Note decrease in size. (C) MPN of a female treated with TP for the first 7 days. The size is almost comparable to that of the male.

synapses of nonamygdaloid origin is greater in female rats than in males.[27] Neonatal castration of males also causes an increase in the number of spine synapses to almost the same level as in females. In the suprachiasmatic nucleus, however, the incidence of spine synapses is higher in males than in females.[28,29] A characteristic feature of the sexually dimorphic pattern of the MAN is dependent on a difference in the number of shaft synapses.[30] These findings suggest that synaptic organization may vary according to the genomic responses of the individual nuclei to organizational action of sex steroids.

In the VMN, there is a regional difference in distribution of sex steroid receptors.[31] This may be reflected in a regional difference in synaptic pattern of the male VMN. As shown in Figure 5, the number of shaft and spine synapses in the ventrolateral part of the VMN (VL-VMN) where sex steroid receptors are abundant is significantly greater than that in the

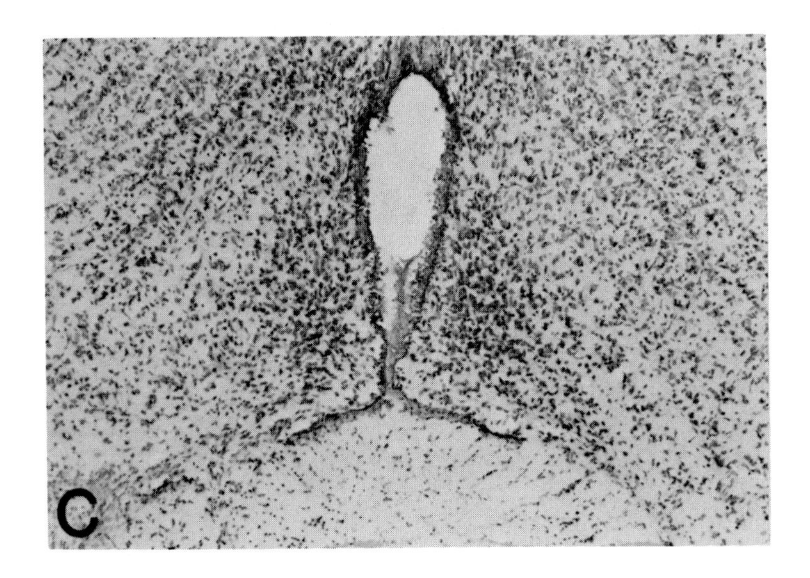

FIGURE 2C.

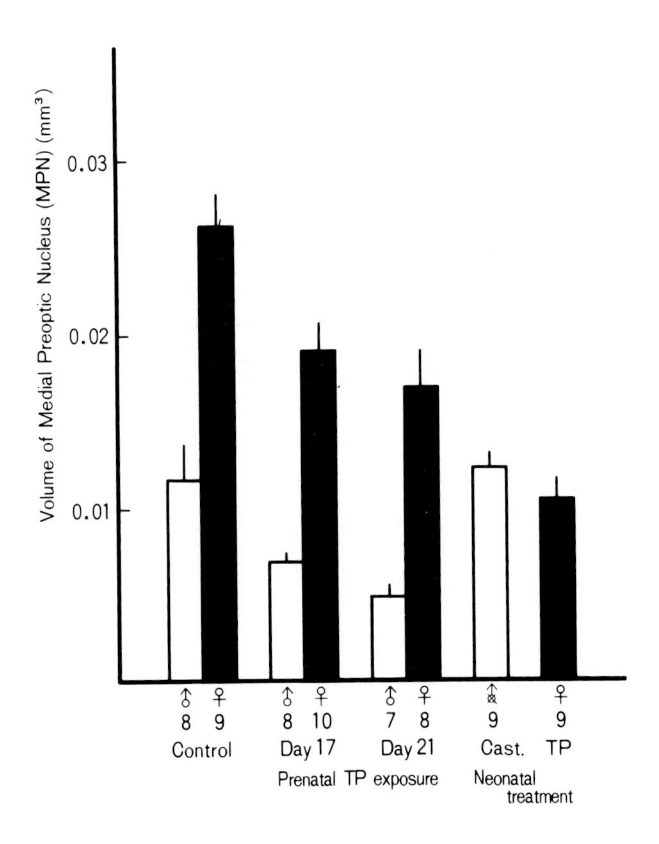

FIGURE 3. Effect of perinatal androgen exposure on the volume of the MPN. In the groups receiving prenatal treatment with TP, the animals were exposed to TP through the mothers at day 17 or 21 of pregnancy. In the group receiving neonatal treatment, males were castrated on the day of birth (Cast.) and females were injected with TP for the first 7 days (TP). Experimental and control animals were sacrificed at 90 days of age.

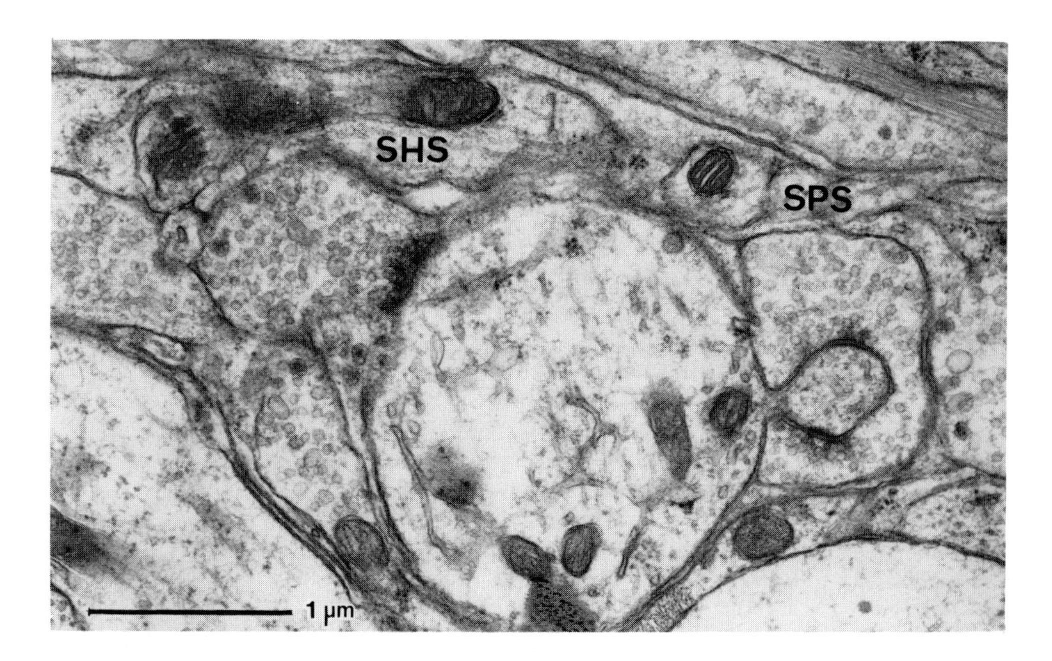

FIGURE 4. Shaft synapse contacting on a dendritic shaft (SHS) and spine synapse on a dendritic spine (SPS) in the neuropil of the VMN. (From *Endocrinology and Physiology of Reproduction*, Leung, P. C. K. et al., Eds., Plenum Press, New York, 1987, 14. With permission.)

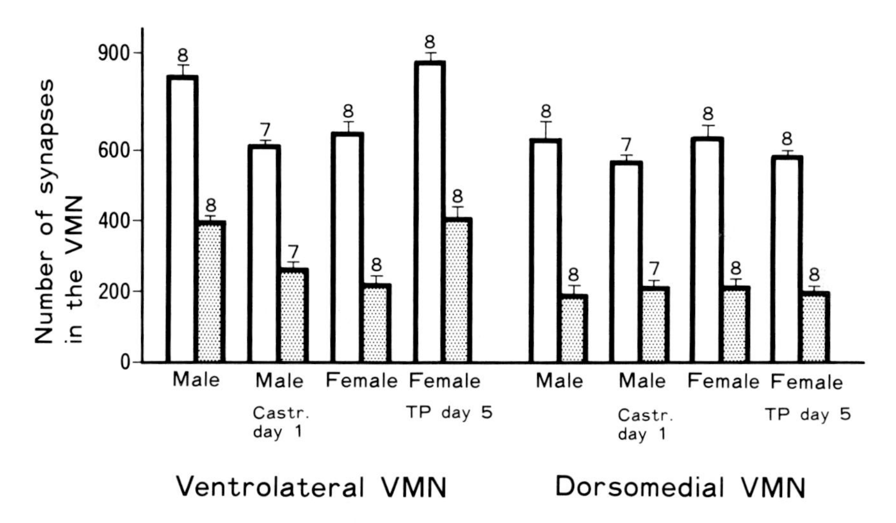

FIGURE 5. The number of shaft (open column) and spine (dotted column) synapses in the ventrolateral and dorsomedial parts of the VMN per 10,000 μm in normal males and females, males castrated on the day of birth, and females treated with TP. (From *Endocrinology and Physiology of Reproduction*, Leung, P. C. K. et al., Eds., Plenum Press, New York, 1987, 15. With permission.)

dorsomedial part of the VMN (DM-VMN) where sex steroid receptors are almost absent. The synaptic organization of the male DM-VMN is almost identical to that of female VL-VMN. This implies that the synaptic number of the VL-VMN is greater in males than in females, suggesting the presence of sexually dimorphic synaptic organization in the VL-VMN.[32] Neonatal treatment of females with TP significantly increases the number of shaft and spine synapses in the VL-VMN to a level comparable to that of nontreated females, but not in the DM-VMN. Neonatal castration of males significantly reduces the number of shaft and spine synapses in the VL-VMN to levels comparable to those of normal females. These findings confirm the importance of the neonatal sex steroid environment for the development of the sexually dimorphic pattern of the VMN. Since the neuropil of the female VL-VMN is not exposed to organizational action of androgen during neonatal days, it develops almost the same synaptic pattern as the DM-VMN, which lacks sex steroid receptors. A high correlation between the sexually dimorphic synaptic pattern and the presence of neurons containing sex steroid receptors indicates that the occurrence of synaptic sexual differentiation is rather specific to the sex steroid-sensitive neuronal system. Other regions which show a sexually dimorphic synaptic pattern are all (except the suprachiasmatic nucleus) known to be abundant in sex steroid receptors.

V. GONADAL STEROIDS AND SYNAPTOGENESIS IN THE DEVELOPING NEUROENDOCRINE BRAIN

The synaptic population in the hypothalamic and limbic structures is quite small at neonatal age and then increases progressively during the course of postnatal development. In the ARCN, synaptic density remains less than 50% of the adult level at 21 days of age and reaches the plateau around the onset of puberty.[33,34] The synaptic number of the VMN increases considerably within the first 20 days. At 20 days of age, the numerical density of shaft and spine synapses in the DM-VMN and VL-VMN is approximately 70 to 80% of that at 45 days of age when it reaches the maximal level. No further increase in the number of synapses occurs at 100 days of age.[35] This tendency has also been detected in the POA.[36,37] In the MAN, synaptogenesis is almost completed before 21 days of age.[38] This indicates that the maturation rate of neural circuit formation is different among these brain regions.

The neuropil matrix of these areas in the hypothalamic and limbic brain is in an immature state at neonatal age, and major neural circuit networks for operating postpubertal neuroendocrine and behavioral regulations are not yet established at this period. The sexually undifferentiated neuropil of the neural substrates which contain abundant steroid receptors could be subjected to organizational action of sex steroids. As a matter of fact, estrogen markedly stimulates axonal and dendritic differentiation and synapse formation during postnatal development.[1,6] In the MAN, for example, estrogen specifically promotes the formation of shaft synapses.[30,38] This provides clear evidence for a mechanism underlying synaptic sexual differentiation of this nucleus, because the sex difference can be attributed to a significant increase in the number of shaft synapses in response to organizational action of androgen in males. In addition, shaft synaptogenesis in newborn medial amygdaloid tissue grafted into the anterior eye chamber of adult ovariectomized rats is also markedly facilitated by estrogen given via the host.[39] This may suggest that estrogen directly affects graft tissues, presumably the postsynaptic elements, to form shaft synapses.

VI. CONSEQUENCES OF PERINATAL EXPOSURE TO NORMAL OR ABNORMAL LEVELS OF GONADAL STEROIDS IN THE NEUROENDOCRINE BRAIN

According to Toran-Allerand,[6] gonadal steroids exert organizational effects on the neuroendocrine brain in a cascading manner. Stimulation of axonal and dendritic growth by steroids could lead to sex differences in neuron size, neuron number, and synaptic pattern. Compared to steroid-insensitive neurons, those stimulated by sex steroids might have more effective synaptic contacts which can promote their subsequent survival during a period of cell death. Furthermore, the perinatal sex steroid environment might cause initiation of cell division of the neurons, increased migration into the certain neuronal groups, and/or prevention of neuronal death. It is puzzling therefore that a significant reduction of the nuclear volume in the MPN occurs following perinatal treatment with androgen. Although the reason for a discrepancy between the SDN-POA and the MPN is not known, these two groups have been suggested to participate in different sexually dimorphic functions — the SDN, for example, regulating male sexual behavior, and the MPN serving as a key station to trigger neural signals for the cyclic release of gonadotropins. A possible difference in manner of genomic response to organizational action of androgen between these two neural substrates could be assumed.

The process of brain masculinization or defeminization is not always a physiological phenomenon because the functional and morphological alterations in the neuroendocrine mechanisms of gonadotropin secretion and behavioral functions vary in degrees according to the dosage, the time, and the duration of hormonal injection.[40] Treatment of newborn female rats with androgen or estrogen from the day of birth to day 30 has been reported to cause a permanent severe suppression of secretion and synthesis of not only luteinizing hormone (LH) but also follicle-stimulating hormone (FSH), resulting in persistent anovulatory diestrous syndrome.[41,42] In these cases, it is apparent that the degree of hypothalamic masculinization or defeminization goes far beyond the normal level. In the MAN, the synaptic number of females treated with large doses of estrogen from days 1 to 30 is significantly greater than that of normal males. A similar excessive stimulation of the synaptogenesis by a long-term estrogen treatment has been reported in the developing ARCN, resulting in disturbances of the regulatory mechanisms of gonadotropin secretion.[7,43] Since the neuronal environment of the neuroendocrine brain is quite easily accessible to available sex steroids during perinatal days, the excess or insufficient steroid levels during these days may cause excessive neuronal survival or death and may cause overproduction or underproduction of synapses. Consequently, these may bring about a less standard or abnormal neural circuit pattern in the neuroendocrine brain. A possible outcome would be some defective morphological and/or functional development of the neuroendocrine brain. Thus, abnormal concentration of gonadal steroids during perinatal development of the neuroendocrine brain can be considered as one of the important etiogenetic factors of the permanent developmental disorders in the neuroendocrine regulatory system.

Prenatal exposure to diethylstilbestrol (DES) in women has been reported to be associated with a highly increased risk for a rare form of cancer, clear-cell adenocarcinoma of the vagina or cervix. Also, DES-exposed women have a poorer prognosis for pregnancy outcome and may be at higher risk for menstrual irregularities and hirsutism with associated neuroendocrine abnormalities.[44] Since DES as a nonsteroidal estrogen is known to be free from fetoplacental inactivation and to reach the intracellular estrogen receptors, pre- or perinatal DES exposure should have profound effects on the sexual differentiation of the brain.[5,45] Recently, it has been reported that prenatal DES exposure causes long-lasting changes in gonadotropin secretory patterns of infant rhesus monkeys,[46] suggesting the possibility of direct effects of prenatal DES on sexually dimorphic areas of the primate brain. In a behavioral analysis, it has been found that DES-exposed women show a statistically significant shift towards bi- or homosexuality.[47]

Dörner has postulated that sexual preferences, in particular of homo- and heterotype, may be determined by the structural organization of the developing brain and that homosexuality could reflect abnormal sexual differentiation of the brain.[48] According to a retrospective study of Dörner et al., out of about 800 homosexual males, highly significantly more homosexuals were born in East Germany during the stressful war and early postwar period of World War II than in the years before or after this critical period.[49] In addition, 100 bi- or homosexual men as well as heterosexual men of similar age were asked about stressful events that may have occurred during their prenatal life. A highly significantly increased incidence of prenatal stress was found in bisexual and, in particular, in homosexual men as compared to heterosexual men.[51] They have speculated that stressful pre- or perinatal events may be an etiogenetic factor for homosexuality in human males, because prenatal exposure to stressful conditions through the mother may affect fetal androgen secretion.

There are reports indicating that positive estrogen feedback effect on LH secretion, which is typically seen in normal females but absent in normal males, is observed in a group of homosexual males,[51,52] suggesting that homosexual men may possess, at least in part, a more female-type differentiated brain based on a possible androgen deficiency during brain differentiation. This type of neuroendocrine response could be an effective biological marker for sexual orientation. Goy and others have found that female rhesus monkeys, androgenized either prenatally or immediately postnatally, show anatomical and behavioral masculinization. However, these androgenized females eventually menstruate and ovulate.[3] Human females androgenized *in utero* as a result of the adrenogenital syndrome do show tomboyish behavior during childhood but they also become fertile subsequently, though with a somewhat delayed menarche.[53] These findings suggest the possibility that the positive feedback characteristic of the neuroendocrine brain of female primates and women could be immune to fetal androgens, although sexually dimorphic behaviors may be affected by androgen. Therefore, further investigations including larger samples are needed to confirm that a prenatal endocrine state (elevated or depressed levels of androgen) may modify the fetal brain differentiation so as to produce less standard neuroendocrine responses. In this connection, it is potentially of significance to note a report describing that a phenotypic girl patient with XY gonadal agenesis was capable of showing a positive feedback LH response to estrogen.[54]

ACKNOWLEDGMENTS

The work of the authors' laboratory was supported by research grants from the Ministry of Education, Science and Culture and the Ministry of Health and Welfare of Japan.

REFERENCES

1. **Arai, Y.,** Synaptic correlates of sexual differentiation, *Trends Neurosci.,* 4, 291, 1981.
2. **Arnold, A. P. and Gorski, R. A.,** Gonadal steroid induction of structural sex differences in the central nervous system, *Annu. Rev. Neurosci.,* 7, 423, 1984.
3. **Goy, R. W. and McEwen, B. S.,** *Sexual Differentiation of the Brain,* MIT Press, Cambridge, Mass., 1980.
4. **Maclusky, N. J. and Naftolin, F.,** Sexual differentiation of the central nervous system, *Science,* 211, 1294, 1981.
5. **Arai, Y., Mori, T., Suzuki, Y., and Bern, H. A.,** Long-term effects of perinatal exposure to sex steroids and diethylstilbestrol on the reproductive system of male mammals, *Int. Rev. Cytol.,* 84, 235, 1983.
6. **Toran-Allerand, C. D.,** On the genesis of sexual differentiation of the central nervous system: morphogenic consequences of steroid exposure and possible role of α-feto-protein, in *Sex Differences in the Brain: The Relation Between Structure and Function,* De Vries, G. J., De Bruin, J. P. C., Uyling, H. B. M., and Corner, M. A., Eds., Elsevier, Amsterdam, 1984; *Prog. Brain Res.,* 61, 63, 1984.
7. **Arai, Y. and Matsumoto, A.,** Synapse formation of the hypothalamic arcuate nucleus during post-natal development in the female rat and its modification by neonatal estrogen treatment, *Psychoneuroendocrinology,* 3, 31, 1978.
8. **Nadler, R. J.,** Intrahypothalamic exploration of androgen-sensitive brain loci in neonatal female rats, *Trans. N.Y. Acad. Sci. Ser. II,* 34, 572, 1972.
9. **Yamanouchi, K. and Arai, Y.,** Female lordosis pattern in the male rat induced by estrogen and progesterone: effect of interruption of the dorsal inputs to the preoptic area and hypothalamus, *Endocrinol. Jpn.,* 22, 243, 1975.
10. **Yamanouchi, K. and Arai, Y.,** Lordosis behaviour in male rats: effect of differentiation in the preoptic area and hypothalamus, *J. Endocrinol.,* 76, 381, 1978.
11. **Yamanouchi, K. and Arai, Y.,** Presence of a neural mechanism for the expression of female sexual behaviours in the male rat brain, *Neuroendocrinology,* 40, 393, 1985.

12. **Yamanouchi, K. and Arai, Y.,** Heterotypical sexual behavior in male rats: individual difference in lordosis response, *Endocrinol. Jpn.*, 23, 179, 1976.
13. **Gorski, R. A., Gordon, J. H., Shryne, J. E., and Southam, A. M.,** Evidence for a morphological sex difference within the medial preoptic area of the rat brain, *Brain Res.*, 148, 333, 1978.
14. **Döhler, K. D., Coquelin, A., Davis, F., Hines, M., Shryne, J. E., and Gorski, R. A.,** Differentiation of the sexually dimorphic nucleus in the preoptic area of the rat brain is determined by the perinatal hormone environment, *Neurosci. Lett.*, 33, 295, 1982.
15. **Ito, S., Murakami, S., Yamanouchi, K., and Arai, Y.,** Prenatal androgen exposure, preoptic area and reproductive functions in the female rat, *Brain Dev.*, 8, 463, 1986.
16. **Commins, D. and Yahr, P.,** Adult testosterone levels influence the morphology of a sexually dimorphic area in the mongolian gerbil brain, *J. Comp. Neurol.*, 224, 132, 1984.
17. **Tobet, S. A., Gallagher, C. A., Zahniser, D. J., Cohen, M. H., and Baum, M. J.,** Sexual dimorphism in the preoptic/anterior hypothalamic area of adult ferrets, *Endocrinology*, 112 (Suppl.), 240, 1983.
18. **Hines, M., Davies, F. C., Coquelin, A., Goy, R. W., and Gorski, R. A.,** Sexually dimorphic regions in the medial preoptic area and the bed nucleus of the stria terminals of the guinea pig brain: a description and an investigation of their relationship to gonadal steroids in adulthood, *J. Comp. Neurol.*, 144, 193, 1985.
19. **Swaab, D. F. and Fliers, E.,** A sexually dimorphic nucleus in the human brain, *Science*, 228, 1112, 1985.
20. **Bleier, R., Byne, W., and Siggelkow, I.,** Cytoarchitectonic sexual dimorphism of the medial preoptic and anterior hypothalamic areas in guinea pig, rat, hamster, and mouse, *J. Comp. Neurol.*, 212, 118, 1982.
21. **Bleier, R. and Byne, W.,** Septum and hypothalamus, in *The Rat Nervous System*, Vol. 1, Paxinos, G., Ed., Academic Press, Sydney, 1985, 87.
22. **Terasawa, E., Wiegand, S. J., and Bridson, W. E.,** A role for the medial preoptic nucleus on afternoon of proestrus in female rats, *Am. J. Physiol.*, 238, E533, 1980.
23. **Ito, S., Murakami, S., Yamanouchi, K., and Arai, Y.,** Perinatal androgen exposure decreases the size of the sexually dimorphic medial preoptic nucleus in the rat, *Proc. Jpn. Acad.* Ser. B, 62, 408, 1986.
24. **Matsumoto, A. and Arai, Y.,** Sex difference in volume of the ventromedial nucleus of the hypothalamus in the rat, *Endocrinol. Jpn.*, 30, 277, 1983.
25. **Mizukami, S., Nishizuka, M., and Arai, Y.,** Sexual difference in nuclear volume and its ontogeny in the rat amygdala, *Exp. Neurol.*, 79, 569, 1983.
26. **Matsumoto, A. and Arai, Y.,** Sexual dimorphism in "wiring pattern" in the hypothalamic arcuate nucleus and its modification by neonatal hormonal environment, *Brain Res.*, 190, 238, 1980.
27. **Raisman, G. and Field, P. M.,** Sexual dimorphism in the neuropil of the preoptic area of the rat and its dependence on neonatal androgen, *Brain Res.*, 54, 1, 1973.
28. **Güldner, F. H.,** Sex dimorphism of axo-spine synapses and postsynaptic density material in the suprachiasmatic nucleus of the rat, *Neurosci. Lett.*, 28, 145, 1982.
29. **Le Blond, C. B., Morris, S., Karakiulakis, G., Powell, R., and Thomas, P. J.,** Development of sexual dimorphism in the suprachiasmatic nucleus of the rat, *J. Endocrinol.*, 95, 137, 1982.
30. **Nishizuka, M. and Arai, Y.,** Sexual dimorphism in synaptic organization in the amygdala and its dependence on neonatal hormone environment, *Brain Res.*, 212, 31, 1981.
31. **Pfaff, D. W. and Keiner, M.,** Atlas of estradiol-concentrating cells in the central nervous system of the female rats, *J. Comp. Neurol.*, 151, 121, 1973.
32. **Matsumoto, A. and Arai, Y.,** Male-female difference in synaptic organization of the ventromedial nucleus of the hypothalamus in the rat, *Neuroendocrinology*, 42, 232, 1986.
33. **Matsumoto, A. and Arai, Y.,** Developmental changes in synaptic formation in the hypothalamic arcuate nucleus of female rats, *Cell Tissue Res.*, 169, 143, 1976.
34. **Matsumoto, A. and Arai, Y.,** Effect of androgen on sexual differentiation of synaptic organization in the hypothalamic arcuate nucleus: an ontogenetic study, *Neuroendocrinology*, 33, 166, 1981.
35. **Matsumoto, A. and Arai, Y.,** Development of sexual dimorphism in synaptic organization in the ventromedial nucleus of the hypothalamus in rats, *Neurosci. Lett.*, 68, 165, 1986.
36. **Reier, P. J., Cullen, M. J., Froelich, J. S., and Rothchild, I.,** The ultrastructure of the developing medial preoptic nucleus in the postnatal rat, *Brain Res.*, 122, 415, 1977.
37. **Lawrence, J. M. and Raisman, G.,** Ontogeny of synapses in a sexually dimorphic part of the preoptic area in the rat, *Brain Res.*, 183, 466, 1986.
38. **Nishizuka, M. and Arai, Y.,** Organizational action of estrogen on synaptic pattern in the amygdala: implications for sexual differentiation of the brain, *Brain Res.*, 213, 422, 1981.
39. **Nishizuka, M. and Arai, Y.,** Synapse formation in response to estrogen in the medial amygdala developing in the eye, *Proc. Natl. Acad. Sci. U.S.A.*, 79, 7024, 1982.
40. **Arai, Y.,** Some aspects of the mechanism involved in steroid-induced sterility, in *Steroid Hormones and Brain Function,* Sawyer, C. H. and Gkorski, R. A., Eds., University of California Press, Berkeley, 1971, 185.

41. **Takawaki, K.,** Reproductive organs and anterior hypophysis of neonatally androgenized female rat, *Sci. Rep. Tokyo Woman's Christian Coll.*, 1, 31, 1968.
42. **Matsumoto, A., Asai, T., and Wakabayashi, K.,** Effects of X-ray irradiation on the subsequent gonadotropin secretion in normal and neonatally estrogenized female rats, *Endocrinol. Jpn.*, 22, 233, 1975.
43. **Matsumoto, A. and Arai, Y.,** Effect of estrogen on early postnatal development of synaptic formation in the hypothalamic arcuate nucleus of female rats, *Neurosci. Lett.*, 2, 79, 1976.
44. **Herbst, A. L. and Bern, H. A., Eds.,** *Developmental Effects of Diethylstilbestrol in Pregnancy*, Thieme-Stratton, New York, 1981.
45. **Slaughter, M., Wilen, R., Ryan, K. J., and Naftolin, F.,** The effect of low dose diethylstilbestrol administration in neonatal female rats, *J. Steroid Biochem.*, 8, 321, 1977.
46. **Fuller, G. B., Yates, D. E., and Helton, W. C.,** Diethylstilbestrol reversal of gonadotropin patterns in infant rhesus monkeys, *J. Steroid Biochem.*, 15, 497, 1981.
47. **Meyer-Bahlburg, H. F. L. and Ehrhardt, A. A.,** Prenatal diethylstilbestrol exposure; behavioral consequences in humans, *Monogr. Neural Sci.*, 12, 90, 1986
48. **Dörner, G.,** Sexual differentiation of the brain, *Vitam. Horm. (N.Y.)*, 38, 325, 1980.
49. **Dörner, G., Geier, T., Ahreus, L., Krell, L., Munx, G., Sieler, H., Kittfner, E., and Muller, H.,** Prenatal stress as possible aetiogetic factor of homosexuality in human males, *Endokrinologie*, 75, 365, 1980.
50. **Dörner, G., Schenk, B., Schmeidel, B., and Ahrens, L.,** Stressful events in prenatal life of bi- and homosexual men, *Exp. Clin. Endocrinol.*, 81, 83, 1983.
51. **Dörner, G., Rohde, W., Stahl, F., Krell, L., and Masius, W. G.,** A neuroendocrine predisposition for homosexuality in men, *Arch. Sex Behav.*, 4, 1, 1975.
52. **Gladue, B. A., Green, R., and Hellman, R.,** Neuroendocrine response to estrogen and sexual orientation, *Science*, 225, 1496, 1984.
53. **Ehrhaldt, A. A. and Meyer-Bahlburg, H. F. L.,** Effects of prenatal sex hormones in gender related behavior, *Science*, 211, 1312, 1981.
54. **Rios, E. P., Herrera, J., Bermudez, J. A., Rocha, G., Lisker, R., Morato, T., and Perez-Palacios, G.,** Endocrine and metabolic studies in a XY patient with gonadal agenesis, *J. Clin. Endocrinol. Metab.*, 39, 540, 1974.

Chapter 3

MESENCHYMAL-EPITHELIAL INTERACTIONS AS A MECHANISM FOR REGULATING HORMONALLY INDUCED EPITHELIAL DIFFERENTIATION AND GROWTH

G. R. Cunha, R. M. Bigsby, P. S. Cooke, and A. A. Donjacour

TABLE OF CONTENTS

I. Introduction 22

II. Mesenchymal-Epithelial Interactions in Development of the Male Genital Tract 22

III. Role of Mesenchyme as a Mediator of Androgenic Effects upon Epithelium 24

IV. Role of Stroma in Morphogenesis and Growth of Estrogen-Target Epithelia 25

Acknowledgments 32

References 34

I. INTRODUCTION

Mesenchymal-epithelial interactions are known to be of utmost importance during development. In many developing organs mesenchyme regulates epithelial growth, induces specific patterns of ductal branching morphogenesis, specifies epithelial morphology and spatial organization, and elicits specific patterns of epithelial cytodifferentiation and functional activity.[1-5] In the genital tract, mesenchymal-epithelial interactions are also known to regulate expression of specific hormone receptors.[6,7] Although it was previously thought that hormonal effects on epithelial cells are elicited directly by intraepithelial receptors, analysis of androgenic and estrogenic responses in male and female genital tracts, respectively, suggests that a variety of hormonal effects elicited by sex steroids in epithelial cells are mediated via indirect processes which are regulated by mesenchyme.

While all organogenetic processes are initiated prenatally, organs of the reproductive system are rudimentary or immature at birth, and most of the morphogenetic processes as well as the onset of functional activity occur over extended periods postnatally. In rats and mice, organogenesis of the reproductive tract occurs in the period from the latter third of gestation to about 2 months postpartum, during which time new tissue architecture is laid down. It is likely that postnatal morphogenesis proceeds by the same fundamental mechanisms that are operative during earlier fetal periods. However, the question arises as to the exact role of stromal-epithelial interactions in adulthood after morphological and functional maturity is attained.

For all adult reproductive organs, epithelial morphology and functional activity must be maintained as epithelial cells continually senesce, die, and are replaced. As epithelial cells are replaced, cellular proliferation must be regulated to maintain normal morphology and function. Morphogenetic events in adult reproductive organs and the mammary gland closely resemble the primary developmental events that occur in the perinatal period. This is particularly true of seasonal breeders in which morphogenetic processes occur cyclically in both males and females, but also applies to females during estrous and menstrual cycles, pregnancy, and lactation. During all types of reproductive cycles, parenchymal and stromal elements may be degraded and later regenerated from rudimentary precursors. Morphogenetic and growth periods are usually followed by expression of functional activity.

In this review, the role of mesenchymal-epithelial interactions during fetal and postnatal periods, including adulthood, will be examined. For the purposes of this review, the term stroma will be used to designate loose fibrous connective tissue of the adult which contains fibroblasts and smooth muscle cells as the predominant cell types. Mesenchyme is defined as undifferentiated loose embryonic connective tissue and is the precursor of stroma.

II. MESENCHYMAL-EPITHELIAL INTERACTIONS IN DEVELOPMENT OF THE MALE GENITAL TRACT

The male genital tract develops from two embryonic anlagen: the Wolffian ducts and urogenital sinus (UGS). The Wolffian duct, whose epithelium is mesodermal in origin, gives rise to the epididymis, ductus deferens, seminal vesicle, and ejaculatory ducts. The urogenital sinus, whose epithelium is derived from endoderm, gives rise to the prostate, bulbourethral glands, urethra, and periurethral glands, and contributes substantially to the urinary bladder.[8,9] Within the prostatic complex each lobe (ventral, dorsolateral, and coagulating gland) has a unique pattern of ductal branching, while the seminal vesicle has its own distinctively folded mucosa.[10,11]

Morphogenesis of the prostate and seminal vesicle is dependent upon an interaction between epithelium and mesenchyme, as well as the presence of androgens.[12-14] Glandular morphogenesis fails to occur when androgens are absent,[13,14] when urogenital epithelium is

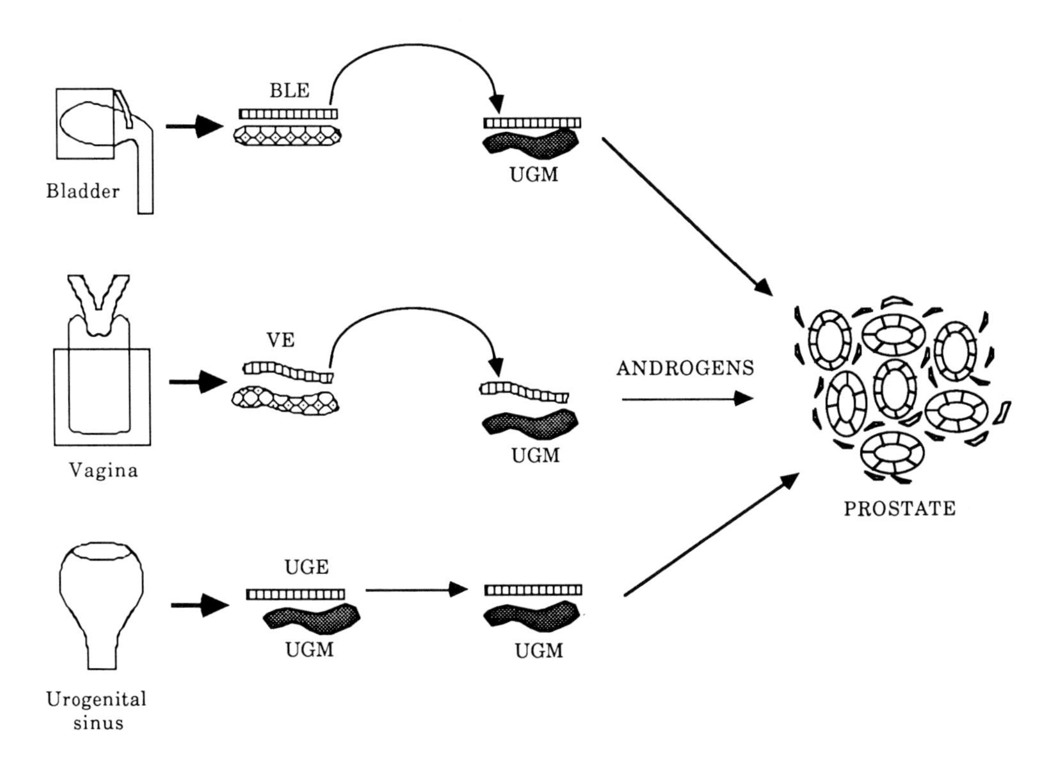

FIGURE 1. Diagrammatic summary of tissue recombination experiments involving instructive and permissive prostatic inductions. Urogenital sinus, bladder, and vagina were separated into epithelial and mesenchymal components following trypsinization or treatment with EDTA. Epithelia isolated from urogenital sinus (UGE), bladder (BLE), and vagina (VE) were recombined with UGM and grown in intact male hosts. In response to endogenous androgens of host origin, each epithelium is induced to form prostatic glandular structures.[15-19]

grown in the absence of mesenchyme, or when prospective glandular epithelium is grown in association with mesenchyme from a nonurogenital organ.[12] In tissue recombination studies (Figure 1), prostatic ductal morphogenesis can be induced by urogenital sinus mesenchyme (UGM) in epithelia of the urogenital sinus (UGE), fetal or adult urinary bladder (BLE), neonatal or immature (≤20 days) vagina (VE), or the adult prostate.[1,15-20] Seminal vesicle morphogenesis can be elicited by seminal vesicle mesenchyme in epithelial cells of the caudal portion of the Wolffian duct (seminal vesicle rudiment) or the cranial portion of the Wolffian duct which normally would give rise to the epididymis and ductus deferens.[15,21] In both cases glandular induction occurs only when two conditions are fulfilled: (1) when intact male hosts (which provide adequate levels of androgens) are employed and (2) when UGM is utilized.[12] These early observations suggested that androgenic effects on epithelial cells are mediated by the mesenchyme.[12,15,22,23]

In tissue recombinants composed of UGM + BLE, the mesenchyme elicits the development of solid prostatic epithelial cords that exhibit a high rate of cellular proliferation.[24] These solid cords subsequently canalize, and the epithelial cells differentiate into a tall simple columnar secretory epithelium. Secretory cytodifferentiation occurs only if the epithelium is associated with UGM.[1,25] The secretory phenotype induced in BLE is distinctly prostatic by several criteria (Table 1).[1,6,7,18]

Prostatic epithelium induced in UGM + BLE recombinants expresses androgen receptors, prostate-specific antigens, and androgen dependency for DNA synthesis. Also, mapping of proteins synthesized by UGM + BLE recombinants by two-dimensional polyacrylamide gel electrophoresis (PAGE) demonstrates a distinctly prostatic pattern. From this body of evidence we have concluded that UGM plays an important role in inducing and specifying

Table 1
EPITHELIAL CHARACTERISTICS IN PROSTATE, URINARY BLADDER, AND HETEROTYPIC TISSUE RECOMBINANTS

Type of analysis or feature	Specimen: Bladder	Prostate	UGM + BLE
Histology	Transitional	Glandular	Glandular
Electron microscopy	Nonsecretory asymmetric membrane	Secretory symmetric membrane	Secretory symmetric membrane
Histochemistry			
Alkaline phosphatase	+	−	−
Alcian blue	−	+	+
Nonspecific esterase	±	+	+
Prostate antigens	−	+	+
Androgen receptors	−	+	+
Androgen-dependent DNA synthesis	−	+	+
Protein synthesis (2-D gels)	Bladder	Prostate	Prostate-like

From Cunha, G. R., Fujii, H., Neubauer, B. L., Shannon, J. M., Şawyer, L. M., and Reese, B. A., *J. Cell Biol.,* 96, 1662, 1983; and Neubauer, B. L., Chung, L. W. K., McCormick, K., Taguchi, O., Thompson, T. C., and Cunha, G. R., *J. Cell Biol.,* 96, 1671, 1983.

certain aspects of prostatic epithelial morphogenesis, growth, cytodifferentiation, and gene expression.

III. ROLE OF MESENCHYME AS A MEDIATOR OF ANDROGENIC EFFECTS UPON EPITHELIUM

For years it has been tacitly assumed that effects of testosterone or dihydrotestosterone on androgen target cells result from binding of the hormone to specific androgen receptors within the responding cells. This idea is derived from the following observations: (1) androgen receptors are demonstrable in prostatic extracts as well as in isolated prostatic epithelial and stromal cells[26-34] and (2) androgens elicit growth of the prostate and specifically stimulate epithelial proliferation as judged by analysis of epithelial labeling index with ^{3}H-thymidine and by analysis of mitotic index.[35-37] Therefore, stimulation of prostatic epithelial DNA synthesis has been thought to result from the receptor-mediated action of androgens within prostatic epithelial cells themselves. However, this correlation does not by any means establish a causal relationship between epithelial androgen receptors and androgen-induced epithelial proliferation. An alternate model for the mechanism by which androgens induce epithelial proliferation in target organs has been proposed. This model has been derived in part from experiments with the androgen-insensitive testicular feminization (Tfm) mouse. Male Tfm mice have testes which produce adequate levels of testosterone but fail to masculinize due to a defective androgen receptor system.[38] Externally such mice are feminized, and the UGS forms a vagina instead of forming prostate. In addition, these mice lack both Müllerian and Wolffian duct derivatives. Recent studies utilizing tissue recombinants prepared with wild-type and Tfm tissues demonstrate that for both the developing prostate and the mammary gland, mesenchyme is the actual target and mediator of androgenic effects upon the epithelium.[17,39-42] Despite the insensitivity of Tfm tissues to androgens,[38] epithelial cells of the embryonic UGS or the bladder of Tfm fetuses can be induced to form prostatic tissue when associated with UGM derived from androgen receptor-positive wild-type fetuses, provided the tissue recombinants are grown in the presence of androgens.[17,39,42] By contrast, when Tfm mesenchyme is utilized, prostatic development is never observed, whether the epithelium is derived from either wild-type or Tfm mice (Figure 2). These observations are in complete agreement with the earlier reports of normal androgenic response in tissue recombinants composed of wild-type mesenchyme and Tfm epithelium of the developing mammary gland.[40,41] All studies of this type demonstrate that the mesenchyme is the actual target and mediator of androgenic effects upon the epithelium, a concept proposed earlier from studies on developing male accessory sexual glands.[15,22,23] The importance of mesenchyme as a primary androgen target tissue is corroborated by the finding that androgen receptors in wild-type mice are initially present in mesenchymal cells, but appear to be absent during perinatal periods in developing prostatic epithelial cells.[25,34,43] In fact, when the autoradiographic localization of dihydrotestosterone (^{3}H-DHT) is compared in fetal androgen target organs of wild-type mice and their Tfm counterparts, the only apparent difference is the presence of androgen receptors in wild-type mesenchyme and their absence in Tfm mesenchyme.[44]

While it appears that mesenchymal cells play an important role during prostatic morphogenesis, it is possible that this concept may be applicable only to the developing gland. To examine this possibility, tissue recombinants were prepared by combining wild-type UGM with either Tfm or wild-type BLE (UGM + wild-type BLE or UGM + Tfm BLE). Mature prostatic ducts develop following 1 month of growth of these tissue recombinants in intact male hosts. In these tissue recombinants wild-type epithelial cells express androgen receptors, while Tfm epithelial cells remain devoid of androgen receptors.[17,18,24,42,45,46] Analysis of these tissue recombinants has shown that stroma is also important in regulating epithelial proliferation in the mature prostate. For example, the prostatic ductal mucosa in UGM + wild-type BLE and UGM + Tfm BLE recombinants undergoes atrophic changes when male hosts bearing both types of recombinants are castrated.[45,46] Biochemical analysis demonstrates that DNA synthesis is stimulated in UGM + Tfm BLE recombinants when testosterone is administered to castrated hosts. The kinetics of this induction are similar to those reported for wild-type prostate.[45] More recently, these findings have been refined and confirmed through parallel autoradiographic analysis utilizing labeling index with ^{3}H-thymidine to measure epithelial proliferation.[46] Prostatic epithelial labeling index is virtually identical in mature prostatic ducts in both UGM + wild-type BLE and UGM + Tfm BLE recombinants following androgenic stimulation (Figure 3). Since Tfm epithelial cells of UGM + Tfm BLE recombinants lack androgen receptors,[24,46] these data demonstrate that androgen-induced prostatic epithelial proliferation does not require the presence of epithelial androgen receptors, but instead suggest an indirect mechanism involving trophic factors or regulators produced by androgen receptor-positive stromal cells.

IV. ROLE OF STROMA IN MORPHOGENESIS AND GROWTH OF ESTROGEN-TARGET EPITHELIA

As described above, many lines of evidence support an indirect, stromally mediated mechanism of androgenic action on androgen-target epithelium. Several parallel observations suggest that similar mechanisms may account for estrogenic effects on target epithelia. As in males, development of the epithelial parenchyma of estrogen-target organs of the female genital tract is induced and specified by the mesenchyme.[47,48] For example, uterine stroma (UtS) induces neonatal VE to undergo uterine differentiation, while vaginal stroma (VS) induces neonatal uterine epithelium (UtE) to undergo vaginal differentiation.[47-49] Grafts that express vaginal differentiation (VS + UtE) synthesize proteins that, when analyzed by two-

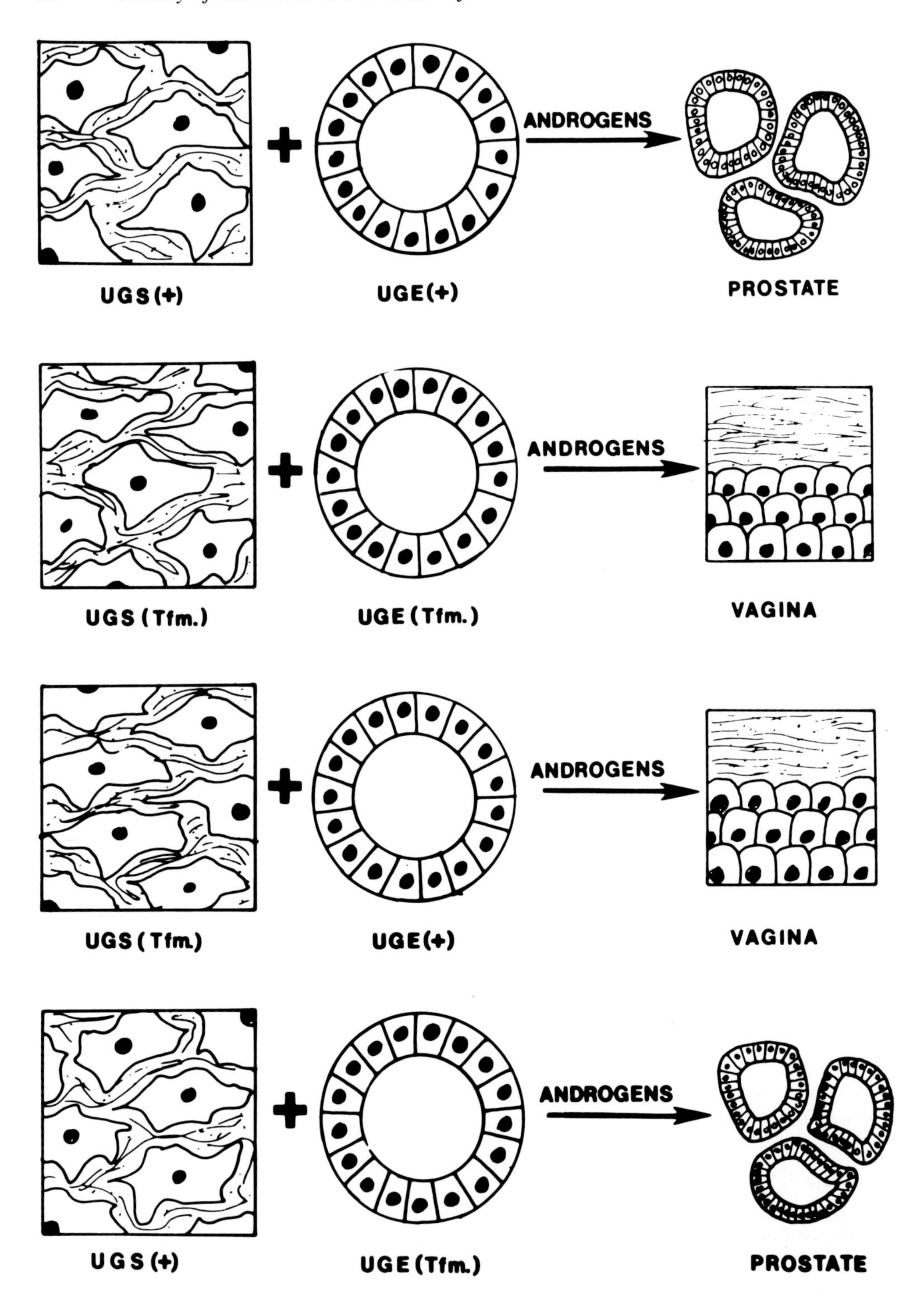

FIGURE 2. Recombination experiments using epithelium and mesenchyme of urogenital sinuses from Tfm and wild-type mouse embryos. Wild-type mesenchyme (UGS +) induced prostatic development in the epithelium, regardless of its source (Tfm or wild type). Conversely, epithelium does not form prostate when associated with Tfm mesenchyme (UGS Tfm), but rather exhibits vaginal differentiation. (From Cunha, G. R., Chung, L. W. K., Shannon, J. M., and Reese, B. A., *Biol. Reprod.*, 22, 19, 1980. With permission.)

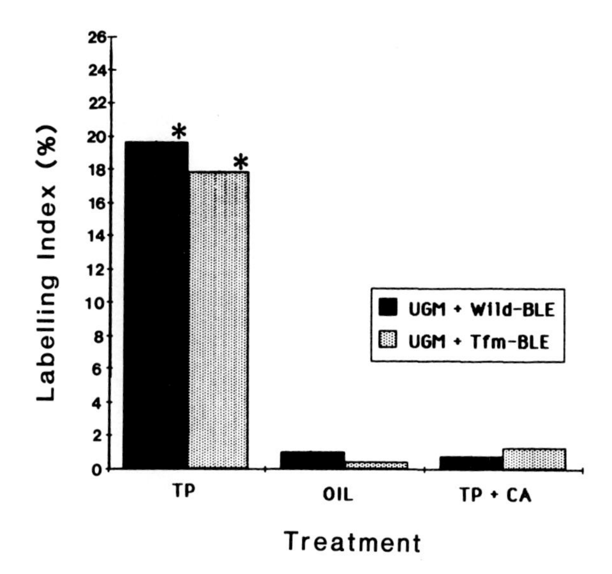

FIGURE 3. ^{3}H-thymidine labeling index of tissue recombinants treated with testosterone propionate (TP), oil, or TP + cyproterone acetate (CA), an antiandrogen. Data represent mean values based on analysis of random measurements of 17 to 39 prostatic ductal cross sections per treatment group representing 1000 to 2000 epithelial cells. Both UGM + wild-type BLE and UGM + Tfm BLE recombinants showed significantly higher epithelial labeling indices than oil- or TP + CA-treated tissue recombinants ($P < 0.0001$). (From Sugimura, Y. and Cunha, G. R., *Prostate,* 9, 217, 1986. With permission.)

dimensional PAGE, give a protein map that is very similar to that of the vagina. Likewise, UtS + VE recombinants (which express uterine differentiation) produce a uterine profile of proteins.[50] Moreover, the induced VE of VS + UtE recombinants grown in intact cycling female hosts produces alternating cornified and mucified layers through the estrous cycle in concert with the VE of the host.[47,49,50] This cyclical change in epithelial phenotype is one of the unique functional characteristics of VE.[49] Thus, in developing estrogen-target organs, the mesenchyme induces specific patterns of epithelial morphogenesis, cytodifferentiation, and functional activity.

Although estrogens do not appear to be required for development of the female genital tract, the fetal and neonatal female genital tracts of many species are sensitive to exogenous estrogens. The extensive literature documents an array of teratogenic effects and numerous epithelial abnormalities elicited by estrogenic compounds in the developing female genital tract of humans as well as laboratory animals (see Chapters 5 and 10). At the cellular level, exogenous estrogens have been shown to elicit substantial increases in uterine and vaginal epithelial proliferation in neonatal mice,[51-54] vaginal epithelial cornification,[55-57] and uterine epithelial hypertrophy.[58] These effects have been elicited by estradiol, diethylstilbestrol, and triphenylethylene compounds such as tamoxifen, nafoxidine, Clomid® (clomiphene citrate), as well as the phytoestrogen coumestrol.[55,57-62] Thus, the estrogenic sensitivity of epithelial cells of the developing female genital tract is well established.

Parallels with androgen-dependent development of the male genital tract suggest the possibility that these estrogenic effects upon UtE and VE of neonatal mice are mediated via the mesenchymal cells. Autoradiographic studies demonstrate the presence of nuclear estrogen-binding sites in mesenchymal cells of the developing murine uterus and vagina.[54,63,64] By contrast, analysis of estrogen uptake after exposure to ^{3}H-estradiol or ^{3}H-moxestrol in

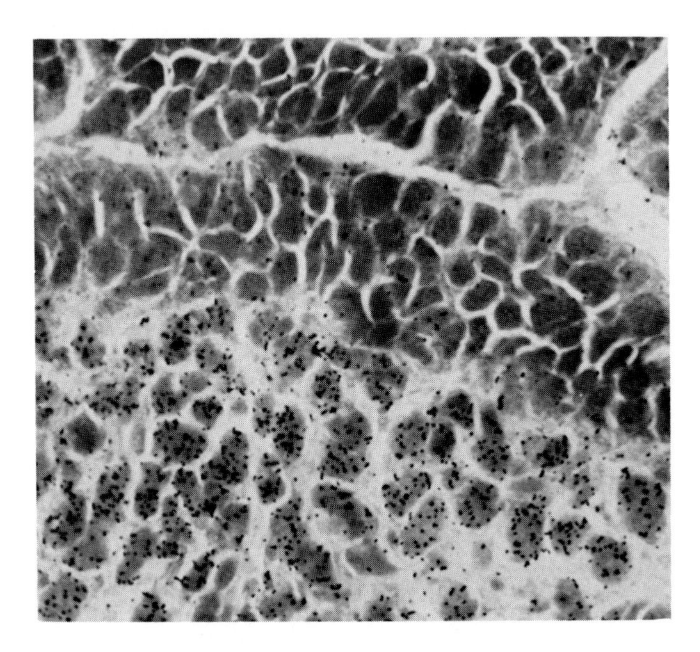

FIGURE 4. Autoradiogram showing localization of ^{3}H-estradiol in the uterus of a 4-day-old mouse. The uterus was cut into pieces which were incubated with ^{3}H-estradiol for 1 hr at 37°C. The tissues were then washed and frozen in liquid propane. Four-micron frozen sections were thaw-mounted onto emulsion-coated slides. After 12 weeks of exposure, the slides were developed and stained with hematoxylin and eosin. The mesenchymal cells show intense nuclear localization of ^{3}H-estradiol and therefore contain estrogen receptors. The lack of any concentration of silver grains over the nuclei of the epithelial cells indicates that these cells lack estrogen receptors. (From Bigsby, R. M. and Cunha, G. R., *Endocrinology,* 119, 390, 1986. With permission.)

vitro or in vivo (Figure 4) indicates that VE and UtE of newborn mice do not have nuclear estrogen-binding sites (estrogen receptors).[63] These autoradiographic findings have been confirmed recently by biochemical studies.[54] When epithelial cells of 4- and 5-day-old Balb/c mice were incubated with ^{125}I-estradiol over a concentration range of 0.15 to 10 n*M,* binding was neither saturable nor specific (Figure 5). However, parallel analysis of UtE cells derived from 20-day-old mice demonstrated saturable, specific uptake of radiolabeled estradiol. In agreement with the autoradiographic studies, UtS cells of both 4- and 20-day-old mice exhibited estrogen receptor activity.[54]

Even though the UtE of 4-day-old mice is devoid of estrogen receptors, these cells respond to exogenous estrogen in vivo by increasing their rate of proliferation (Figure 6). The lag between estrogen administration and increased DNA synthesis is identical to that reported in ovariectomized adult mice.[65] In addition, neonatal VE and UtE cells that have expressed an estrogenic response continue to lack estrogen receptors.[54,66] These results suggest that estrogen-induced proliferation of UtE is mediated by mesenchymal estrogen receptors in the neonatal mouse. Since the time course of estrogen response is similar to that of the adult, it is possible that the proliferative response of adult UtE is also mediated by a similar indirect mechanism, even though adult UtE has estrogen receptors.

Cell culture studies cast further doubt on the idea that estrogens act as direct mitogens on normal estrogen target epithelial cells. There are numerous reports of primary cultures of normal VE and UtE cells in which estrogen does not have mitogenic effects (Table 2). Some neoplastic epithelial cells from estrogen target organs are mitogenically responsive to estrogen in culture,[83-85] but estrogen is not mitogenic for other tumor cells in vitro despite

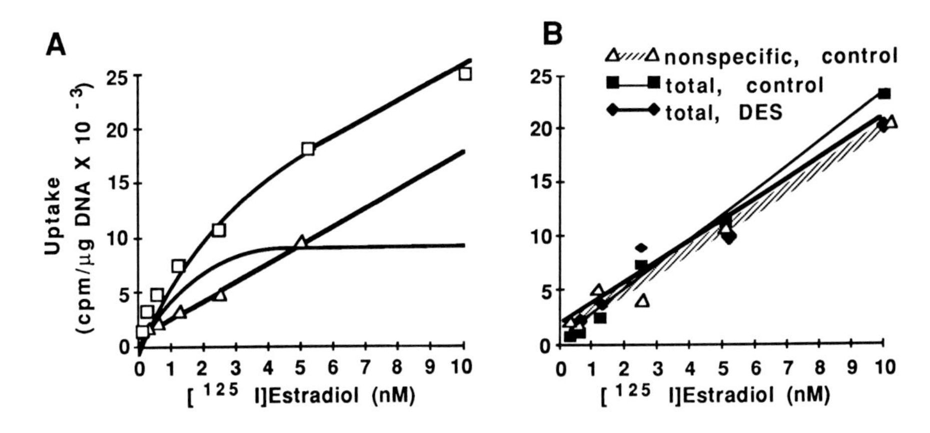

FIGURE 5. Whole-cell uptake of [^{125}I]estradiol. Epithelial cells and mesenchymal cells of uteri from 4-day-old Balb/c mice were isolated following enzymatic digestion with trypsin and collagenase. Suspensions of mesenchymal cells (A) or epithelial cells (B) were incubated with different concentrations of [^{125}I]estradiol in the presence or absence of a 200-fold excess of moxestrol. Cells were collected on glass fiber filters and washed with a buffered salt solution. In A, the solid line without symbols represents the specific uptake by mesenchymal cells, i.e., the difference between total uptake (□, without moxestrol) and nonspecific uptake (△, with moxestrol). In B, epithelial cells taken from control animals or diethylstilbestrol-treated animals were also incubated in [^{125}I]estradiol, with or without moxestrol; there was no difference between total uptake of either control or diethylstilbestrol-treated epithelial cells in relation to nonspecific uptake of control epithelial cells, indicating a lack of specific binding. (From Bigsby, R. M. and Cunha, G. R., *Endocrinology,* 119, 390, 1986. With permission.)

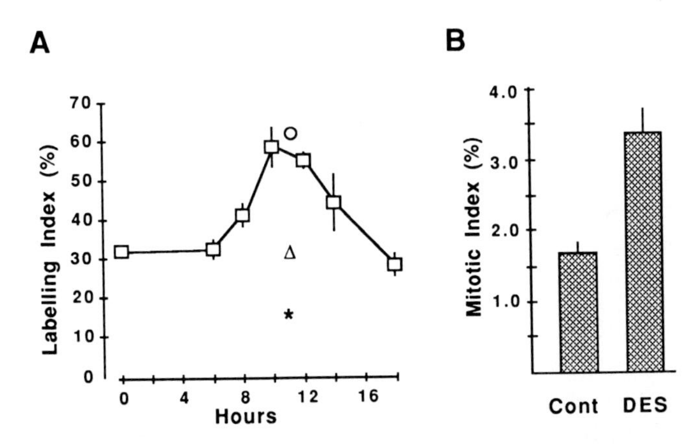

FIGURE 6. Thymidine labeling index (A) and mitotic index (B) of uterine epithelial cells of 5-day-old mice following diethylstilbestrol treatment. (A) ^{3}H-thymidine was injected into each animal at the times indicated following a single 5-µg injection of diethylstilbestrol (□), E_2 (○), dihydroxytestosterone (△), or dexamethasone (*). One hour later the animals were killed and the uterine horns processed for autoradiography. Labeling index of the epithelial cells is expressed at percent. (B) Animals were treated with colchicine 16 to 18 hr after diethylstilbestrol treatment for 2 hr and then their uteri were processed for histological examination. The number of cells arrested in metaphase was counted and expressed as a percent. The mean index ± SEM of three to five animals at each time is presented. (From Bigsby, R. M. and Cunha, G. R., *Endocrinology,* 119, 390, 1986. With permission.)

the fact that the tumors are estrogen-dependent in vivo. This capacity to respond mitogenically to estradiol may be a key feature of the neoplastic character of these particular tumor cell lines. The single report by Ishiwata et al.,[70] in which estradiol stimulated the proliferation of cultured normal endometrial epithelial cells (as well as cells derived from endometrial carcinoma), suffers from the lack of any demonstration of steroid specificity and the requirement of nearly micromolar concentrations of estradiol for effectiveness. Thus, epithelial cells, which are normally responsive to estrogens in vivo, usually do not respond mitogenically to estrogens in vitro when grown as pure epithelial cultures. However, there are a few reports demonstrating that estrogen stimulates DNA synthesis and cellular proliferation when the epithelia are grown as mixed stromal and epithelial cell cultures or as tissue explants in organ culture (Table 2). Although these reports in the literature have been sporadic and seldom repeated, they do indicate that an estrogenic effect is more likely to occur if the stroma is present in the culture system. A recent report showed that the pH indicator phenol red, which is in most culture media, interferes with the estrogenic response of MCF-7 cells, a tumorigenic breast cell line.[86] It remains to be seen if an estrogen effect can be obtained in primary cultures of pure epithelial cells grown in medium free of phenol red.

Several hypotheses have been proposed to explain the inability of estradiol to stimulate normal epithelial cell proliferation in vitro. Culture conditions may selectively favor the proliferation of estrogen-independent cells, or the epithelium may become transformed in culture so that it no longer requires estradiol for growth. Alternatively, estrogens and other sex steroids may cause epithelial growth by acting indirectly via stroma-derived growth regulators not present in pure epithelial cultures. To test some of these possibilities, we have examined the ability of cultured UtE and VE to reexpress normal morphology and hormone responsiveness when recombined with homologous stroma.

UtE and VE from ovariectomized 40-day-old mice were grown in collagen gels for 1 to 2 weeks using serum-free medium.[72,74] Total DNA content of the VE and UtE cultures increased four- and four- to eightfold, respectively, in the absence of estradiol. Proliferation of these epithelia was not stimulated by estradiol.[72,74,87] The VE in serum-free culture does not keratinize or mucify normally. Nonetheless, these epithelial cells do possess functional estrogen receptors insofar as exposure of these epithelia to estradiol in vitro decreased cytosolic estrogen receptor content, increased nuclear estrogen receptor levels, and increased cytosolic progesterone receptor content.[87] Therefore, the lack of proliferation in response to estradiol in vitro is not due to an inoperative estrogen receptor system.[87]

In an attempt to reestablish differentiation and growth responses absent in vitro, cultured VE was recombined with fresh VS and the resultant tissue recombinants were grown under the renal capsule of intact female hosts for 4 weeks. Under these conditions the cultured VE reexpressed normal histotypic features (stratified squamous morphology and alternating keratinized and mucified layers) characteristic of the changes which occur during the estrous cycle (Figure 7). The epithelium in recombinants composed of UtS and cultured UtE was also morphologically normal.[49] In these tissue recombinants, estrogen-dependent proliferation of both UtE and VE was also reestablished (Figure 8). While these observations do not prove that estrogens elicit their effects upon target epithelial cells via stromal mediators, it is evident that loss of differentiated features in UtE and VE in vitro and loss of growth responsiveness to estrogens in vitro can be regained following reassociation with homologous stroma and transplantation of the tissues in vivo. These observations emphasize the importance of stroma, since direct grafting of cultured UtE and VE in vivo is incompatible with continued epithelial survival.[49]

For decades the estrogenicity of a compound has been assessed in terms of its ability to cause uterine growth or VE proliferation and cornification.[88,89] VE is atrophic and only two to three layers thick in ovariectomized mice, but thickens and cornifies following injection of estradiol or other estrogenic compounds. However, vaginal hyperplasia and cornification

Table 2
PROLIFERATION OF UTERINE AND VAGINAL EPITHELIA IN CULTURE: A SURVEY OF THE LITERATURE IN WHICH ESTROGENS WERE TESTED FOR MITOGENIC EFFECTS ON REPRODUCTIVE TRACT EPITHELIA IN CELL OR ORGAN CULTURE SYSTEMS

Citation	Organ	Species	Estrogenic effect[a]
Isolated Target Epithelia in Primary Culture[b]			
Flaxman et al.[67]	Vagina	Mouse	None
Liszczak et al.[68]	Uterus	Human	None
Kirk et al.[69]	Uterus	Human	None
Ishiwata et al.[70]	Cervix	Human	Increased mitotic index
Casimiri et al.[71]	Uterus	Rat	None
Iguchi et al.[72]	Vagina	Mouse	None
Tomooka et al.[73]	Uterus	Mouse	None
Iguchi et al.[74]	Uterus	Mouse	None
Bigsby (unpublished)	Uterus	Rabbit	None
Cooke, Fujii, Cunha (unpublished)	Vagina	Mouse	None
Organ Culture and Mixed Endometrial Cell Culture[c]			
Maurer et al.[75]	Uterus (organ)	Cow	Increased mitosis
Demers et al.[76]	Uterus (organ)	Human	Cell crowding
Nordqvist[77]	Uterus (organ)	Human	None
Stadnicka et al.[78]	Uterus (organ)	Mouse	None (necrotic)
	Vagina (organ)	Mouse	Increased stratification
Flaxman et al.[79]	Vagina (organ)	Mouse	None
Gerschenson et al.[80]	Uterus (mixed)	Rabbit	Increased labeling index
Kaufman et al.[81]	Uterus (organ)	Human	Increased thymidine incorporation
Wilbanks et al.[82]	Cervix (mixed)	Human	None

[a] In pure epithelial cell cultures and mixed stromal-epithelial cell cultures the effect of estrogens on proliferation was assessed by determining the numbers of cells in the culture dish, assessing the incorporation of radiolabeled thymidine (labeling index or cpm/g DNA), or determining the mitotic index following colchicine block. In the organ cultures the histological appearance of the tissue was sometimes interpreted to indicate a mitogenic effect of the hormone, e.g., epithelial crowding or increased squamous stratification, but no definitive measure of DNA synthetic activity or mitotic activity was made.

[b] Epithelial cells were reported to be isolated and grown as pure cell cultures, i.e., without any stromal cell contamination.

[c] Epithelial cells mixed with stromal fibroblasts were plated and grown together (mixed) or pieces of the organ were cultured as intact explants (organ).

can be elicited as well by daily mechanical stimulation of the vagina.[90] A permanent irreversible state of vaginal hyperplasia and cornification can also be elicited by treating newborn mice briefly with estradiol or diethylstilbestrol during the perinatal period; this hyperplasia persists in adult ovariectomized mice.[55,59,60,91] The resulting condition has been called ovary-independent persistent vaginal cornification (or hyperplasia).[92] It is of interest that during the induction of this condition by estrogen treatment during the early neonatal period, the VE cells are devoid of nuclear estrogen-binding sites.[63,66] Since the vaginal mesenchymal cells possess estrogen receptors at this time, the induction of ovary-independent persistent vaginal cornification appears to be a mesenchyme-mediated process.[63,66] Continued expres-

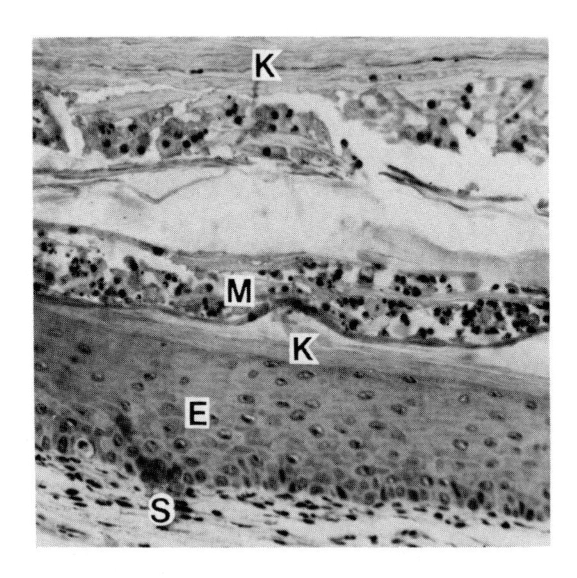

FIGURE 7. A tissue recombinant composed of fresh VS (S) + cultured VE (E) grown in an intact female host for 4 weeks. Note that the epithelium of this recombinant is stratified, and alternating sloughed layers of mucified (M) and keratinized (K) cells lie above the epithelium, indicating that normal cycling of epithelial differentiation has occurred in response to ovarian hormones. (Magnification × 320.) (From Cooke, P. S., Uchima, F.-D. A., Fujii, D. K., Bern, H. A., and Cunha, G. R., *Proc. Natl. Acad. Sci. U.S.A.*, 83, 2109, 1986. With permission.)

sion of this condition in adulthood is dependent upon stromal-epithelial interactions in which the mesenchyme also plays an important role. For instance, if epithelium from a normal adult vagina is recombined with VS from an adult mouse that was neonatally estrogenized, the epithelium exhibits ovary-independent hyperplasia; the reciprocal recombination (normal VS + neonatally estrogenized VE) also results in ovary-independent hyperplasia.[93] It appears, therefore, that neonatal exposure of the developing vagina to exogenous estrogens results in permanent alterations in both the epithelium and stroma, which ultimately cause a form of abnormal VE differentiation known as ovary-independent persistent vaginal cornification.

In summary, in both the male and female genital tracts stromal cells appear to play key roles in regulating a variety of androgenic and estrogenic effects in epithelial cells. Clearly, mesenchymal cells induce and specify patterns of epithelial morphology and cytodifferentiation, as well as function. Mitogenic effects of androgens and estrogens appear to also have stringent stromal requirements. Mesenchymal cells play an important role in both normal and abnormal epithelial differentiation in the female genital tract. Studies utilizing serum-free in vitro approaches are currently underway in several laboratories to define the mechanism of these interactions between epithelium and mesenchyme.

ACKNOWLEDGMENTS

The authors would like to thank Ed Cleary for typing the manuscript and Simona Ikeda for preparation of the figures. This study was supported in part by the following grants: March of Dimes #1-837 and NIH Grants AM 32157, HD 17491, CA 05388, HD 21919, IF32 HD 06520, and SF32 HD 06580.

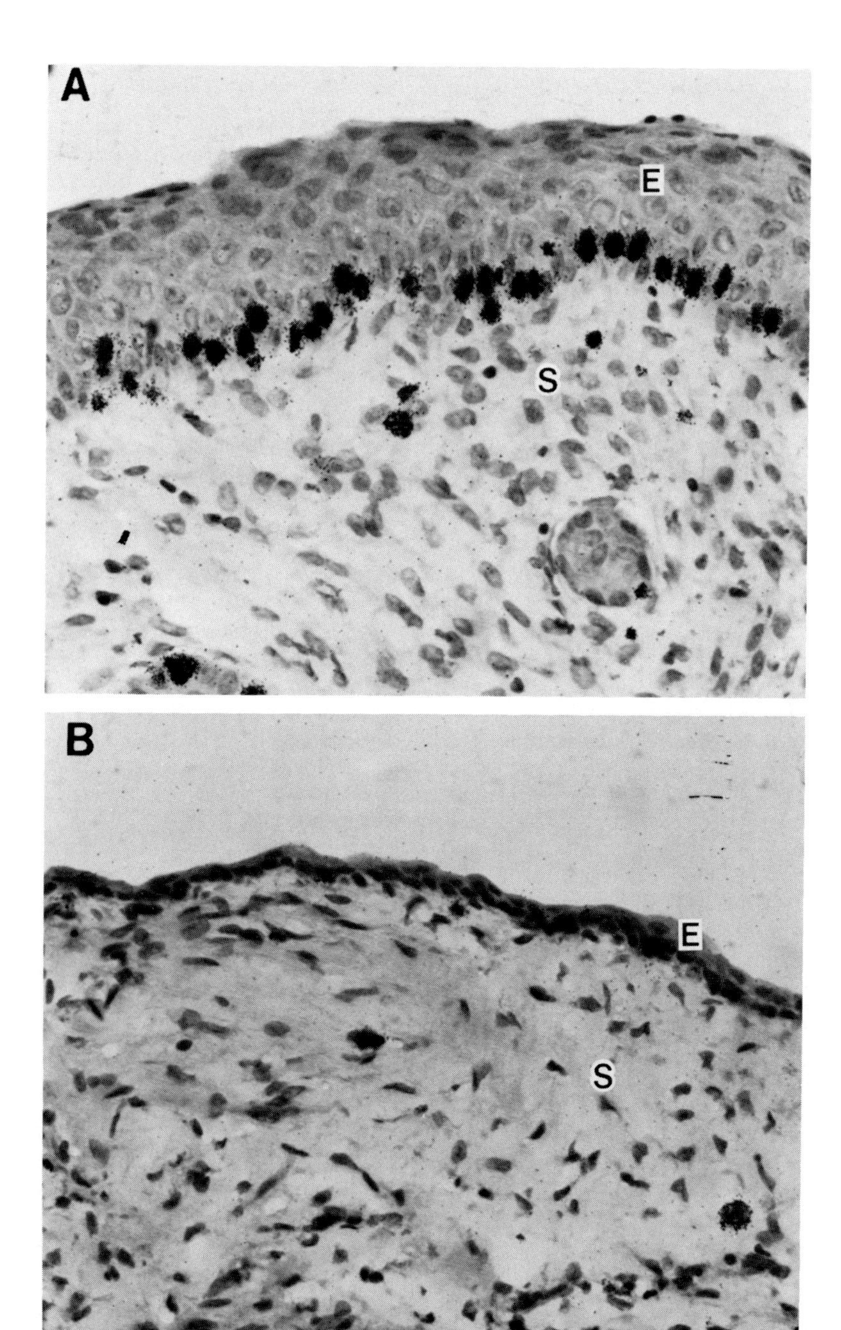

FIGURE 8. Tritiated thymidine labeling of fresh VS (S) + cultured VE (E) recombinants in ovariectomized hosts. The hosts had been injected with either 100 ng of estradiol in oil (A) or oil alone (B) 18 hr previously. The epithelium of recombinants in E_2-injected hosts (A) was stratified and heavily labeled, while that in oil-injected hosts was atrophic, consisting of only two to three cell layers which were sparsely labeled. (Magnification × 400.) (From Cooke, P. S., Uchima, F.-D. A., Fujii, D. K., Bern, H. A., and Cunha, G. R., *Proc. Natl. Acad. Sci. U.S.A.*, 83, 2109, 1986. With permission.)

REFERENCES

1. **Cunha, G. R., Fujii, H., Neubauer, B. L., Shannon, J. M., Sawyer, L. M., and Reese, B. A.,** Epithelial-mesenchymal interactions in prostatic development. I. Morphological observations of prostatic induction by urogenital sinus mesenchyme in epithelium of the adult rodent urinary bladder, *J. Cell Biol.*, 96, 1662, 1983.
2. **Sengel, P.,** *Morphogenesis of Skin*, Cambridge University Press, New York, 1976, 1.
3. **Ekblom, P.,** Basement membrane proteins and growth factors in kidney differentiation, in *The Role of Extracellular Matrix in Development*, Trelstad, R. L., Ed., Alan R. Liss, New York, 1984, 173.
4. **Bernfield, M. R., Cohn, R. H., and Banerjee, S. D.,** Glycosaminoglycans and epithelial organ formation, *Am. Zool.*, 13, 1067, 1973.
5. **Bernfield, M. R., Banerjee, S. D., Koda, J. E., and Rapraeger, A. C.,** Remodeling of the basement membrane as a mechanism of morphogenetic tissue interaction, in *The Role of Extracellular Matrix in Development*, Trelstad, R. L., Ed., Alan R. Liss, New York, 1984, 545.
6. **Cunha, G. R., Reese, B. A., and Sekkinstad, M.,** Induction of nuclear androgen-binding sites in epithelium of the embryonic urinary bladder by mesenchyme of the urogenital sinus of embryonic mice, *Endocrinology*, 107, 1767, 1980.
7. **Neubauer, B. L., Chung, L. W. K., McCormick, K., Taguchi, O., Thompson, T. C., and Cunha, G. R.,** Epithelial-mesenchymal interactions in prostatic development. II. Biochemical observations of prostatic induction by urogenital sinus mesenchyme in epithelium of the adult rodent urinary bladder, *J. Cell Biol.*, 96, 1671, 1983.
8. **Felix, W.,** The development of the urogenital organs, in *Manual of Human Embryology*, Keibel, R. and Mall, F. P., Eds., J. B. Lippincott, Philadelphia, 1912, 869.
9. **Cunha, G. R.,** Development of the male urogenital tract, in *Urologic Endocrinology*, Rajfer, J., Ed., W. B. Saunders, Philadelphia, 1986, 6.
10. **Sugimura, Y., Cunha, G. R., and Donjacour, A. A.,** Morphogenesis of ductal networks in the mouse prostate, *Biol. Reprod.*, 34, 961, 1986.
11. **Lung, B. and Cunha, G. R.,** Development of seminal vesicles and coagulating glands in neonatal mice. I. The morphogenetic effects of various hormonal conditions, *Anat. Rec.*, 199, 73, 1981.
12. **Cunha, G. R.,** Epithelial-stromal interactions in development of the urogenital tract, *Int. Rev. Cytol.*, 47, 137, 1976.
13. **Jost, A.,** Gonadal hormones in the sex differentiation of the mammalian fetus, in *Organogenesis*, DeHaan, R. L. and Urpsrung, H., Eds., Holt, Rinehart & Winston, New York, 1965, 611.
14. **Price, D. and Ortiz, E.,** The role of fetal androgens in sex differentiation in mammals, in *Organogenesis*, DeHaan, R. L. and Ursprung, H., Eds., Holt, Rinehart & Winston, New York, 1965, 629.
15. **Cunha, G. R.,** Epithelio-mesenchymal interactions in primordial gland structures which become responsive to androgenic stimulation, *Anat. Rec.*, 172, 179, 1972.
16. **Cunha, G. R.,** Age-dependent loss of sensitivity of female urogenital sinus to androgenic conditions as a function of the epithelial-stromal interaction in mice, *Endocrinology*, 95, 665, 1975.
17. **Cunha, G. R. and Lung, B.,** The possible influences of temporal factors in androgenic responsiveness of urogenital tissue recombinants from wild-type and androgen-insensitive (Tfm) mice, *J. Exp. Zool.*, 205, 343, 1978.
18. **Cunha, G. R., Lung, B., and Reese, B.,** Glandular epithelial induction by embryonic mesenchyme in adult bladder epithelium of Balb/c mice, *Invest. Urol.*, 17, 302, 1980.
19. **Cunha, G. R. and Vanderslice, K. D.,** Identification of species origin of mammalian cells in histological sections, *Stain Technol.*, 59, 7, 1984.
20. **Norman, J. T., Cunha, G. R., and Sugimura, Y.,** The induction of new ductal growth in adult prostatic epithelium in response to an embryonic prostatic inductor, *Prostate*, 8, 209, 1986.
21. **Higgins, S. J. and Cunha, G. R.,** unpublished observations.
22. **Cunha, G. R.,** Tissue interactions between epithelium and mesenchyme of urogenital and integumental origin, *Anat. Rec.*, 172, 529, 1972.
23. **Cunha, G. R.,** Support of normal salivary gland morphogenesis by mesenchyme derived from accessory sexual glands of embryonic mice, *Anat. Rec.*, 173, 205, 1972.
24. **Shannon, J. M. and Cunha, G. R.,** Characterization of androgen binding and deoxyribonucleic acid synthesis in prostate-like structures induced in testicular feminized (Tfm/Y) mice, *Biol. Reprod.*, 31, 175, 1984.
25. **Cunha, G. R., Chung, L. W. K., Shannon, J. M., Taguchi, O., and Fujii, H.,** Hormone-induced morphogenesis and growth: role of mesenchymal-epithelial interactions, *Recent Prog. Horm. Res.*, 39, 559, 1983.
26. **Liao, S.,** Molecular actions of androgen, in *Biochemical Actions of Hormones*, Litwak, G., Ed., Academic Press, New York, 1977, 351.

27. **Stumpf, W. E. and Sar, M.,** Autoradiographic localization of estrogen, androgen, progestin, and glucocorticosteroid in "target tissues" and "non-target tissues", in *Receptors and Mechanism of Action of Steroid Hormones,* Pasqualini, J. R., Ed., Marcel Dekker, New York, 1976, 41.
28. **Krieg, M., Schlenker, A., and Voigt, K.-D.,** Inhibition of androgen metabolism in stroma and epithelium of the human benign prostatic hyperplasia by progesterone, estrone, and estradiol, *Prostate,* 6, 233, 1985.
29. **Krieg, M., Klotzl, G., Kaufmann, J., and Voigt, K. D.,** Stroma of human benign prostatic hyperplasia: preferential tissue for androgen metabolism and oestrogen binding, *Acta Endocrinol.,* 96, 422, 1981.
30. **Tilley, W. D., Horsfall, D. J., McGee, M. A., Henderson, D. W., and Marshall, V. R.,** Distribution of oestrogen and androgen receptors between the stroma and epithelium of the guinea-pig prostate, *J. Steroid Biochem.,* 22, 713, 1985.
31. **Lahtonen, R., Bolton, N. J., Konturri, M., and Vihko, R.,** Nuclear androgen receptors in the epithelium and stroma of human benign prostatic hypertrophic glands, *Prostate,* 4, 129, 1983.
32. **Jung-Testas, I., Groyer, M.-T., Bruner-Lorand, J., Hechter, O., Baulieu, E.-E., and Robel, P.,** Androgen and estrogen receptors in rat ventral prostate epithelium and stroma, *Endocrinology,* 109, 1287, 1981.
33. **Sirett, D. A. N., Cowan, S. K., Janeczko, A. E., Grant, J. K., and Glen, E. S.,** Prostatic tissue distribution of 17β-hydroxy-5α-androstan-3-one and of androgen receptors in benign hyperplasia, *J. Steroid Biochem.,* 13, 723, 1979.
34. **Shannon, J. M. and Cunha, G. R.,** Autoradiographic localization of androgen binding in the developing mouse prostate, *Prostate,* 4, 367, 1983.
35. **Tuohimaa, P. and Niemi, M.,** Cell renewal and mitogenic activity of testosterone in male sex accessory glands, in *Male Accessory Sex Organs: Structure and Function in Mammals,* Brandes, D., Ed., Academic Press, New York, 1974, 329.
36. **Bruchovsky, N., Lesser, B., van Doorn, E. V., and Craven, S.,** Hormonal effects on cell proliferation in rat prostate, *Vitam. Horm. (N.Y.),* 33, 61, 1975.
37. **Coffey, D. S.,** The effects of androgens on DNA and RNA synthesis in sex accessory tissue, in *Male Accessory Sex Organs: Structure and Function* Brandes, D., Ed., Academic Press, New York, 1974, 303.
38. **Ohno, S.,** *Major Sex-Determining Genes,* Springer-Verlag, New York, 1979, 1.
39. **Lasnitzki, I. and Mizuno, T.,** Prostatic induction and interaction of epithelium and mesenchyme from normal wild-type and androgen-insensitive mice with testicular feminization, *J. Endocrinol.,* 85, 423, 1980.
40. **Drews, U. and Drews, U.,** Regression of mouse mammary gland anlagen in recombinants of Tfm and wild-type tissues: testosterone acts via the mesenchyme, *Cell,* 10, 401, 1977.
41. **Kratochwil, K. and Schwartz, P.,** Tissue interaction in androgen response of the embryonic mammary rudiment of mouse: identification of target tissue of testosterone, *Proc. Natl. Acad. Sci. U.S.A.,* 73, 4041, 1976.
42. **Cunha, G. R., Chung, L. W. K., Shannon, J. M., and Reese, B. A.,** Stromal-epithelial interactions in sex differentiation, *Biol. Reprod.,* 22, 19, 1980.
43. **Takeda, H., Mizuno, T., and Lasnitzki, I.,** Autoradiographic studies of androgen-binding sites in the rat urogenital sinus and postnatal prostate, *J. Endocrinol.,* 104, 87, 1985.
44. **Cunha, G. R.,** Androgenic effects upon prostatic epithelium are mediated via trophic influences from stroma, in *New Approaches to the Study of Benign Prostatic Hyperplasia,* Kimball, F. A., Buhl, A. E., and Carter, D. B., Eds., Alan R. Liss, New York, 1984, 81.
45. **Cunha, G. R. and Chung, L. W. K.,** Stromal-epithelial interactions: induction of prostatic phenotype in urothelium of testicular feminized (Tfm/y) mice, *J. Steroid Biochem.,* 14, 1317, 1981.
46. **Sugimura, Y. and Cunha, G. R.,** Androgenic induction of deoxyribonucleic acid synthesis in prostate-like glands induced in the urothelium of testicular feminized (Tfm/y) mice, *Prostate,* 9, 217, 1986.
47. **Cunha, G. R.,** Stromal induction and specification of morphogenesis and cytodifferentiation of the epithelia of the Mullerian ducts and urogenital sinus during development of the uterus and vagina in mice, *J. Exp. Zool.,* 196, 361, 1976.
48. **Cunha, G. R. and Fujii, H.,** Stromal-parenchymal interactions in normal and abnormal development of the genital tract, in *Developmental Effects of Diethylstilbestrol (DES) in Pregnancy,* Herbst, A. and Bern, H. A., Eds., Thieme-Stratton, New York, 1981, 179.
49. **Cooke, P. S., Uchima, F.-D. A., Fujii, D. K., Bern, H. A., and Cunha, G. R.,** Restoration of normal morphology and estrogen responsiveness in cultured vaginal and uterine epithelia transplanted with stroma, *Proc. Natl. Acad. Sci. U.S.A.,* 83, 2109, 1986.
50. **Cunha, G. R., Shannon, J. M., Taguchi, O., Fujii, H., and Meloy, B. A.,** Epithelial-mesenchymal interactions in hormone-induced development, in *Epithelial-Mesenchymal Interactions in Development,* Sawyer, R. H. and Fallon, J. F., Ed., Praeger Scientific Press, New York, 1983, 51.
51. **Forsberg, J.-G.,** An estradiol mitotic rate inhibiting effect in the Mullerian epithelium in neonatal mice, *J. Exp. Zool.,* 175, 369, 1970.
52. **Kimura, T., Kawashima, S., and Nishizuka, Y.,** Effects of prenatal treatment with estrogen on mitotic activity of vaginal anlage cells in mice, *Endocrinology,* 27, 739, 1980.

53. **Eide, A.,** The effect of oestradiol on DNA synthesis in the neonatal mouse uterus and cervix, *Cell Tissue Res.,* 156, 551, 1975.
54. **Bigsby, R. M. and Cunha, G. R.,** Estrogen stimulation of DNA synthesis in epithelium lacking estrogen receptors, *Endocrinology,* 119, 390, 1986.
55. **Takasugi, N.,** Cytological basis for permanent vaginal changes in mice treated neonatally with steroid hormones, *Int. J. Cytol.,* 44, 193, 1976.
56. **Cunha, G. R., Lee, A. K., and Lung, B.,** Electron microscopic observations of vaginal development in untreated and neonatally estrogenized Balb/c Crgl mice, *Am. J. Anat.,* 152, 343, 1978.
57. **Burroughs, C. D., Bern, H. A., and Stoksdtad, E. L. R.,** Prolonged vaginal cornification and other changes in mice treated neonatally with coumestrol, a plant estrogen, *J. Toxicol. Environ. Health,* 15, 51, 1985.
58. **McCormack, S. and Clark, J. H.,** Clomid administration to pregnant rats causes abnormalities of the reproductive tract in offspring and mothers, *Science,* 204, 629, 1979.
59. **Forsberg, J.-G. and Kalland, T.,** Neonatal estrogen treatment and epithelial abnormalities in the cervicovaginal epithelium of adult mice, *Cancer Res.,* 41, 721, 1981.
60. **McLachlan, J. A., Newbold, R. R., and Bullock, B. C.,** Long term effects on the female mouse genital tract associated with prenatal exposure to diethylstilbestrol, *Cancer Res.,* 40, 3988, 1980.
61. **Taguchi, O. and Nishizuka, Y.,** Reproductive tract abnormalities in female mice treated neonatally with tamoxifen, *Am. J. Obstet. Gynecol.,* 151, 675, 1985.
62. **Burroughs, C. D., Williams, B. A., Mills, K. T., and Bern, H. A.,** Genital tract abnormalities in female C57BL/Crgl mice exposed neonatally to phytoestrogens (cuomestrol and zearalenone), *Cancer Res.,* 27, 220, 1986.
63. **Cunha, G. R., Shannon, J. M., Vanderslice, K. D., Sekkingstad, M., and Robboy, S. J.,** Autoradiographic analysis of nuclear estrogen binding sites during postnatal development of the genital tract of female mice, *J. Steroid Biochem.,* 17, 281, 1982.
64. **Stumpf, W. E., Narbaitz, R., and Sar, M.,** Estrogen receptors in the fetal mouse, *J. Steroid Biochem.,* 12, 55, 1980.
65. **Martin, L., Finn, C. A., and Trinder, G.,** Hypertrophy and hyperplasia in the mouse uterus after oestrogen treatment: an autoradiographic study, *J. Endocrinol.,* 56, 133, 1973.
66. **Cunha, G. R., Shannon, J. M., Taguchi, O., Fujii, H., and Chung, L. W. K.,** Mesenchymal-epithelial interactions in hormone-induced development, *J. Anim. Sci.,* 55 (Suppl.), 14, 1982.
67. **Flaxman, B. A., Chopra, D. P., and Newman, D.,** Growth of mouse vaginal epithelial cells in vitro, *In Vitro,* 9, 194, 1973.
68. **Liszczak, T. M., Richardson, G. S., MacLaughlin, D. T., and Kornblith, P. L.,** Ultrastructure of human endometrial epithelium in monolayer culture with and without steroid hormones, *In Vitro,* 13, 344, 1977.
69. **Kirk, D., King, R. J. B., Heyes, J., Peachey, L., Hirsch, P. J., and Taylor, W. T.,** Normal human endometrium in cell culture, *In Vitro,* 14, 651, 1978.
70. **Ishiwata, I., Okumura, H., Nozawa, S., Kurihara, S., and Yamada, K. I.,** Effects of estradiol-17β on growth and differentiation of benign and malignant human uterine cervical squamous cells in vitro, *Acta Cytol.,* 22, 555, 1978.
71. **Casimiri, V., Rath, N. C., Parvez, H., and Psychoyos, A.,** Effect of sex steroids on rat endometrial epithelium and stroma cultured separately, *J. Steroid Biochem.,* 12, 293, 1980.
72. **Iguchi, T., Uchima, F. D. A., Ostrander, P. L., and Bern, H. A.,** Growth of normal mouse vaginal epithelial cells in and on collagen gels, *Proc. Natl. Acad. Sci. U.S.A.,* 80, 3743, 1983.
73. **Tomooka, Y., DiAugustine, R. P., and McLachlan, J. A.,** Proliferation of mouse uterine epithelial cells in vitro, *Endocrinology,* 118, 1011, 1986.
74. **Iguchi, T., Uchima, F. D. A., Ostrander, P. L., Hamamoto, S. T., and Bern, H. A.,** Proliferation of normal mouse uterine luminal epithelial cells in serum-free collagen gel culture, *Proc. Jpn. Acad.,* 61, 292, 1985.
75. **Maurer, H. R., Rounds, D. E., and Raiborn, C. W.,** Effects of oestradiol on calf endometrial tissue in vitro, *Nature (London),* 213, 182, 1967.
76. **Demers, L. M., Csermely, T., and Hughes, E. C.,** Culture of human endometrium. II. Effects of estradiol, *Obstet. Gynecol.,* 36, 269, 1970.
77. **Nordqvist, S.,** The synthesis of DNA and RNA in normal human endometrium in short-term incubation in vitro and its response to oestradiol and progesterone, *J. Endocrinol.,* 48, 17, 1970.
78. **Stadnicka, A., Szoltys, M., and Witkowska, W.,** Effect of estradiol on the mouse female reproductive system in organ culture, *Acta Histochem.,* 46, 267, 1973.
79. **Flaxman, B. A., Chopra, D. P., and Harper, R. A.,** Autoradiographic analysis of hormone-independent development of the mouse vaginal epithelium in organ culture, *In Vitro,* 10, 42, 1974.
80. **Gerschenson, L. E., Berliner, J., and Yang, J. J.,** Diethylstilbestrol and progesterone regulation of cultured rabbit endometrial cell growth, *Cancer Res.,* 34, 2873, 1974.

81. **Kaufman, D. G., Adamec, T. A., Walton, L. A., Carney, C. N., Melin, S. A., Genta, V. M., Mass, M. J., Dorman, B. H., Rodgers, N. Y., Photopulos, G. J., Powell, J., and Grisham, J. W.,** Studies of human endometrium in organ culture, *Methods Cell Biol.*, 21B, 1, 1980.
82. **Wilbanks, G. D., Leipus, E., and Tsurumoto, D.,** Tissue culture of the human cervix, *Methods Cell Biol.*, 21B, 29, 1980.
83. **Lippman, M. E., Huff, K. K., Jakesz, R., Hecht, T., Kasid, A., Bates, S., and Dickson, R. B.,** Estrogens regulate production of specific growth factors in hormone-dependent human breast cancer, *Ann. N.Y. Acad. Sci.*, 464, 11, 1986.
84. **Soto, A. M. and Sonnenschein, D.,** Mechanism of estrogen action on cellular proliferation: evidence for indirect and negative control on cloned breast tumor cells, *Biochem. Biophys. Res. Commun.*, 122, 1097, 1984.
85. **Holinka, C. F., Hata, H., Kuramoto, H., and Gurpide, E.,** Responses to estradiol in a human endometrial adenocarcinoma cell line (Ishikawa), *J. Steroid Biochem.*, 24, 85, 1986.
86. **Berthois, Y., Katzenellenbogen, J. A., and Katzenellenbogen, B. S.,** Phenol red in tissue culture media is a weak estrogen: implication concerning the study of estrogen-responsive cells in culture, *Proc. Natl. Acad. Sci. U.S.A.*, 83, 2496, 1986.
87. **Uchima, F. D. A., Edery, M., Uguchi, T., and Bern, H. A.,** Estrogen induces progesterone receptor but not proliferation of mouse vaginal epithelium in vitro, *Cancer Res.*, 25, 206, 1984.
88. **Allen, E.,** The oestrous cycle in the mouse, *Am. J. Anat.*, 30, 297, 1922.
89. **Clark, J. H. and Peck, E. J.,** *Female Sex Steroids, Receptors and Function*, Springer-Verlag, New York, 1979, 1.
90. **Clark, J. H. and Guthrie, S. C.,** The estrogenic effects of clomiphene during the neonatal period in the rat, *J. Steroid Biochem.*, 18, 513, 1983.
91. **Bern, H. A. and Talamantes, F. J.,** Neonatal mouse models and their relation to disease in the human female, in *Developmental Effects of Diethylstilbestrol (DES) in Pregnancy*, Herbst, A. L. and Bern, H. A., Eds., Thieme-Stratton, New York, 1981, 129.
92. **Bern, H. A., Jones, L. A., Mori, T., and Young, P. N.,** Exposure of neonatal mice to steroids: long term effects on the mammary gland and other reproductive structures, *J. Steroid Biochem.*, 6, 673, 1975.
93. **Cunha, G. R., Lung, B., and Kato, K.,** Role of the epithelial-stromal interaction during the development and expression of ovary-independent vaginal hyperplasia, *Dev. Biol.*, 56, 52, 1977.

Chapter 4

HISTOGENESIS OF IRREVERSIBLE CHANGES IN THE FEMALE GENITAL TRACT AFTER PERINATAL EXPOSURE TO HORMONES AND RELATED SUBSTANCES

John-Gunnar Forsberg

TABLE OF CONTENTS

I. Introduction ... 40

II. Normal Development of the Female Genital Tract ... 40

III. Induction Mechanisms of Permanent Vaginal Cornification ... 42
- A. General Aspects on Induction of Permanent Vaginal Cornification ... 42
- B. Cellular Heterogeneity in Vaginal Epithelium with Ovary-Independent Cornification ... 42
- C. The Concept of A and B Cells ... 43
- D. Mitotic Activity in the Neonatal Vaginal Epithelium under Estrogen Influence ... 44
- E. Different Opinions on Developmental Mechanisms of Ovary-Independent Cornified Vaginal Epithelium ... 44

IV. Neoplastic Transformation of the Squamous Vaginal Epithelium after Perinatal Estrogen, Androgen, or Progesterone Treatment ... 45

V. Estrogen Effects in the Columnar Epithelium of the Cervicovaginal Region after Prenatal or Neonatal Exposure ... 46
- A. Vaginal and Cervical Adenosis ... 46
- B. Immediate Effects by Estradiol and DES in the Neonatal Period of Mice ... 46
- C. The Cervicovaginal Epithelium from Immature and Adult Mice after Neonatal Estrogen Treatment ... 48
- D. Adenosis-Associated Malignancy in Mice ... 49
- E. Effects in the Cervicovaginal Epithelium after Treatment with Tamoxifen, Clomiphene, and Nafoxidine in Neonatal Life (Mouse) ... 50
- F. Effects in the Nonhuman Primate Genital Tract after Fetal Exposure to DES ... 50
- G. Effect of DES on the Human Fetal Vaginal Epithelium ... 50
- H. Vaginal Adenosis in Women Exposed to DES during Fetal Life ... 51

VI. Uterine Effects after Neonatal Estrogen Treatment ... 51
- A. Uterine Estrogen Response and Squamous Metaplasia ... 51
- B. Uterine Hyperplasia and Neoplasia ... 52

VII. Oviductal Morphology in Perinatally Estrogen-Treated Females ... 53

VIII. Are the Permanent Effects Induced by Estrogen Perinatally Receptor Mediated or Nonreceptor Associated? ... 53

A. Effects Secondary to a Disturbed Ovarian Function 53
B. Genetic or Epigenetic Effects 54

IX. Conclusions 55

References 56

I. INTRODUCTION

Morphogenetic effects are just one example of the different effects produced by steroids. The insect steroid hormone ecdysone is of importance for metamorphosis. During normal sexual differentiation in mammals, exposure of the genital primordia to testicular androgen allows expression of the male genotype into a male phenotype. Estrogens do not seem to have a similar morphogenetic effect on the female genotype. It is well documented that by exposing immature males to exogenous estrogens or females to androgens, the normal sexual phenotype can be influenced.

Estrogen treatment of females during the period of genital organ differentiation has diverse and adverse permanent effects, sometimes resulting in malignancy; some androgens and progesterone may have similar effects. For many years, these long-term effects were considered to be of laboratory interest only and not related to a clinical situation. When an association was described between treatment of women with synthetic estrogens and genital malignancy in the female offspring at puberty, researchers' attitudes changed abruptly.[1] Animal experiments on permanent estrogen effects were now considered model systems for the human situation. A great deal has been learned from animal studies about possible mechanisms of action of estrogen, but the role of estrogens in "transplacental carcinogenesis" in man is not clear.

In the following, different types of permanent developmental effects induced by estrogens and some other agents will be discussed, with particular attention to what is known about induction mechanisms.

II. NORMAL DEVELOPMENT OF THE FEMALE GENITAL TRACT

To understand histogenesis of estrogen-induced changes in the genital tract it is necessary to have some knowledge of normal development. Only a broad review will be given here. A more comprehensive review was published by Forsberg in 1973.[2]

In the mouse, the common canal formed through the fusion of the lower parts of the Müllerian ducts (the paramesonephric ducts) is the anlage of the upper vagina (upper three fifths — the Müllerian vagina) and the common cervical canal (CCC).[3] The lower part of the vaginal anlage (two fifths — the sinus vagina) is the dorsal part of the urogenital sinus, connected to the ventral urethral part through an epithelial bridge. Later the urogenital sinus undergoes a complete division through deepening of folds on the lateral walls. At birth, the Müllerian vagina has a high pseudostratified columnar epithelium and the border between this epithelial type and that in the sinus vagina is distinct. The fornices define the border between the vagina and the CCC.

A main controversy in vaginal morphogenesis is whether the adult vaginal epithelium is derived (more or less) from Müllerian epithelium or whether the latter is replaced by epithelium from the urogenital sinus.[3] Because vaginal morphogenesis is different in different species, no overall description is possible.[2]

Based on results from several different types of studies, researchers concluded that the pseudostratified columnar epithelium in the mouse Müllerian vagina undergoes a transformation into a squamous epithelium.[2] Parallel with this transformation is a high mitotic activity. Cells arising in the columnar epithelium migrate basally, forming a basal epithelial zone of small cuboidal cells which are the origin of the adult squamous epithelium. Superficially in the epithelium, there is a zone of remaining columnar cells. Under estrogen stimulation, the cells in the basal zone proliferate and cornify with shedding of the superficial columnar cells. The transformation progresses into the CCC until it reaches the level of the squamo-columnar junction, situated in the uppermost part of the CCC of the NMRI mouse strain used in our laboratory. Vaginal morphogenesis is similar in the mouse and the rat but initial stages occurring in the mouse after birth appear in the rat shortly before birth.

Support for a dual origin of the mouse vagina was obtained from studies of mice carrying the gene testicular feminization.[4] Experiments with stromal-epithelial recombinations have shown the capacity of typical Müllerian epithelium from the uterine region to differentiate into a vaginal epithelial type under the influence of vaginal stroma.[5] A dual origin of the rat vagina was also described by Del Vecchio.[6]

Quite another opinion on mouse vaginal development has recently been presented.[7,8] The vagina was described as developing by downgrowth of the lower fused part of the Müllerian and Wolffian ducts along the sinus ridges on the dorsal wall of the urogenital sinus. The Müllerian ducts were found to successively engulf and replace the sinus ridges as well as the Wolffian epithelium. Thus the mouse vagina would be entirely derived from Müllerian ducts and the term "sinus vagina" would be a misnomer. Even though the origin of the lower vaginal part is thus still controversial, the term "sinus vagina" will be used in the following for the lower two fifths of the mouse vagina.

Human vaginal development is more complicated than that in laboratory animals.[3] The primary vaginal anlage in the human is formed from the fused Müllerian ducts with columnar epithelium. The next stage in morphogenesis is reminiscent of that in the mouse: beginning adjacent to the urogenital sinus, the columnar epithelium undergoes a transformation into a low squamous epithelium. While this transformation process progresses upwards, a solid plate of cells similar to those in the urogenital sinus appears between the dorsal wall of the urogenital sinus and the transformed Müllerian epithelium in the vaginal anlage. This cell plate is called the vaginal plate. At the same time as the columnar Müllerian epithelium changes its character into a squamous type, this is resorbed by the upwards-growing vaginal plate. Finally, the Müllerian epithelium is replaced by the vaginal plate in the whole vaginal region. Again starting in the vicinity of the urogenital sinus, the vaginal plate proliferates and a cornified epithelium and lumen appear in the center.

The nature of the vaginal plate cells is obscure. Several authors have ascribed to them an origin of the urogenital sinus, but a Wolffian origin has also been suggested.[3] The present author's conclusion was that the vaginal plate may be of sinus origin but a Wolffian contribution could not be excluded.[3] A Wolffian origin was supported in later studies.[9] The opinion on the growth direction of the vaginal plate differs — most authors describe an upwards-growing plate, while others favor the view of a downward growth.[9,10] Several problems in human vaginal morphogenesis are thus unsolved, but today there seems to be agreement on one point: the definite vaginal epithelium is not of Müllerian origin.

The theory of basic steps in human vaginal morphogenesis (the primary Müllerian vaginal anlage and the replacement of the Müllerian epithelium with sinus epithelium) has gained support in recent years.[11] The capacity of human columnar Müllerian epithelium to transform into an immature squamous epithelium was also corroborated in experiments in which the upper uterine cervix and lower uterine corpus were grafted onto nude mice.[12]

Thus, while in rodents the vagina has a dual origin, in humans the Müllerian epithelium is expelled from the vaginal region, but this process is preceded by a transformation of the columnar Müllerian epithelium.

III. INDUCTION MECHANISMS OF PERMANENT VAGINAL CORNIFICATION

A. General Aspects on Induction of Permanent Vaginal Cornification

Early papers by Gardner[13] and Takasugi et al.[14] demonstrated that treatment of neonatal female mice with testosterone or estradiol could result in a permanent cornified vaginal epithelium. Later it was shown that treatment must start within the first 3 days after birth to obtain this effect.[15-17]

The permanent cornification may be of an ovary-dependent or ovary-independent type. The ovary-dependent type is induced by low daily doses of estradiol or testosterone (0.1 to 0.5 μg/day in BALB/c mice) or progesterone (100 μg/day) and is considered a result of disturbed hypothalamic pituitary control, resulting in no luteinizing hormone (LH) peaks and anovulatory ovaries.[18-21]

Ovary-independent vaginal cornification is induced by higher daily doses of estradiol or testosterone (5 to 25 μg/day for 5 days in BALB/c mice but only 0.1 μg estradiol or 5 μg testosterone per day in mice belonging to the RIII strain).[15,16,18] Even daily doses of 100 μg dihydrotestosterone in the neonatal period are effective, and 17α-hydroxyprogesterone caproate was reported to result in both ovary-dependent and ovary-independent cornification.[22-24] Cholesterol, progesterone, or cortisone are ineffective.[15] The vaginal epithelium retains its cornified type after ovariectomy alone or in combination with adrenalectomy and/or hypophysectomy, or after transplantation into ovariectomized hosts.[14-16,25,26] Injections of large amounts of progesterone, cortisol and deoxycorticosterone, or testosterone prevents vaginal cornification in only 20 to 40% of ovariectomized mice with persistent vaginal cornification.[25] Grafting experiments indicate the presence of estrogens in females with ovary-independent cornification.[18] The epithelium is not completely hormone-independent (as indicated by a reduced height after ovariectomy), but it has a markedly reduced sensitivity to estrogens.[16] Vaginal homogenate has a reduced binding capacity for ^{3}H-estradiol, and nuclear estrogen receptors are not detectable, probably because of the estrogen being rapidly lost from nuclear binding proteins.[27,28]

Ovary-independent cornification can be induced by culturing pieces of the neonatal vagina in vitro in the presence of estradiol for 2 to 3 days before grafting into a 4-day-old host, which indicates a direct estrogen effect on the vaginal cells.[29] Because only estradiol and not testosterone was effective in vitro, different mechanisms of action seem to come into play.[15,23] Aromatization of testosterone into estrogen is not probable because of the high doses of estradiol necessary and because a nonaromatizable androgen such as dihydrotestosterone is active.[22]

Studies on mitotic activity in different vaginal regions suggest that only the upper and Müllerian-derived vaginal part may be involved in ovary-independent cornification.[22,30] With a single dose of 50 μg estradiol on day 1 or 3 after birth, the incidence of irreversible vaginal cornification was lower than when females were exposed to the same dose *in utero* on day 17 of fetal life. It was concluded that the vaginal epithelium is more susceptible to estradiol in fetuses than in neonates.[31] Even though the mouse placenta retards passage of diethylstilbestrol (DES) into the fetal compartment, there is a threefold accumulation of radioactivity — relative to fetal plasma — in the reproductive tract.[32,33] Depending on the strain used, neonatal treatment with progesterone (20 μg/day) or deoxycorticosterone acetate (20 μg/day) may result in ovary-independent proliferation and mucification.[23]

B. Cellular Heterogeneity in Vaginal Epithelium with Ovary-Independent Cornification

Vaginal epithelium from adult females with ovary-independent permanent vaginal cornification shows alterations in the cell cycle and in the response to estrogen.[34] The vaginal epithelium in these females seems to be a heterogeneous cell population, some cells not being affected by neonatal estrogen treatment. Support for this view was obtained from, for example, studies involving repeated treatment periods with similar or different types of inhibitory steroids (progesterone, testosterone, cortisol, and deoxycorticosterone acetate). In some experiments using croton oil and progesterone, the results argue for a changed relative proportion of affected and nonaffected cells.[16]

Because high doses of vitamin A inhibit vaginal cornification in normal females, experiments were undertaken to study this situation in females with ovary-independent vaginal cornification.[35] Intravaginal instillation of vitamin A acetate into these ovariectomized females resulted in an increased mitotic rate. It was postulated that vitamin A selected for basal cells with short mitotic cycles.[15] Basal cells with pycnotic nuclei were postulated to have inherently long cycles. Ovary-independent cornification was also prevented by vitamin A administered simultaneously with estradiol, but more than 2 months had to pass after the combined treatment before the suppression was evident.[36-38] A new cell type, "light cells", was described after estradiol-vitamin A treatment, and these cells replaced the dense basal cells.[37] Since a similar suppression was observed after transplantation of the vagina into estradiol-vitamin A-treated hosts (and also in the vagina of a female parabiotically joined to an estradiol-vitamin A-treated partner), the occurrence of an estradiol-vitamin A-induced circulating suppressive factor was indicated.[36] Vitamin A alone does not induce the suppressive factor. The vaginal epithelia of estradiol- and vitamin A-treated females responded normally to postpubertal estradiol treatment.[38] Again it was hypothesized that vaginal epithelium with ovary-independent cornification is a heterogeneous cell population, with estrogen responder and nonresponder cells. Vitamin A seems to result in permanently altered cells being gradually replaced by a population of normal cells.

These results, which argue for cellular heterogeneity in the vaginal epithelium with ovary-independent cornification, are essential to keep in mind because they could mean a heterogeneous origin of the cells or a common stem cell and daughter cells selecting different differentiation pathways.

C. The Concept of A and B Cells

During differentiation of the epithelium in the Müllerian vagina (C57B1/Tw mice), clusters of so-called "A" cells were found among the columnar cells.[16,23,39,40] Under estrogen treatment large polygonal "B" cells (diameter 12 to 15 μm) with numerous cytoplasmic processes appeared in and around the clusters of smaller A cells (5 to 6 μm). Nodules of B cells were supposed to fuse to form a sheet of squamous epithelium capable of proliferation and cornification in the absence of estrogen. The A cells apparently transformed into B cells, and intermediate cell types were described.[40] The A cells were seen at the junction of the sinus and the Müllerian vagina and also in the fornicocervical region.[16,23]

The A cells were of two types: one light A cell with few free ribosomes and one dark A cell.[16,23] The light A cells had degraded or incomplete desmosomal structures and were interpreted as undifferentiated cells destined to be eliminated during normal differentiation. During estrogen treatment, there was an initial increase in the number of dark A cells, but after repeated injections the number decreased when large B cells appeared and aggregated in nodules attached to the basal membrane.

A cells have a high activity of acid phosphatases and β-glucoronidase related to the content of lysosomes.[16] When neonatal mice were primed with lysosomal membrane stabilizers and later treated with estradiol beyond the normal critical period for induction of ovary-independent vaginal cornification, the latter type of response was obtained.[16,23] Thus A cells, after lysosomal stabilization, were supposed to survive longer than normally and were possibly transformed into B cells. However, no cluster of A cells was seen after treatment with lysosomal stabilizers.[16]

In a later study, A cells could not be identified clearly and the nature of the undifferentiated cells was considered obscure.[41] Nodules of B cells formed isolated islands under the columnar epithelium in the upper part of the vagina, but these islands had no connection with proliferating, similarly situated cells in the lower part of the vagina, thereby excluding sinus epithelium origin.[41] In experiments using ^{3}H-thymidine, the labeling pattern was different in the columnar cells, in the B-cell nodules, and in the cuboidal sinus cells.[41] It was suggested that B cells arise directly or indirectly from undifferentiated cells, the nature of which is unclear.

In our experiments using NMRI mice, we have not been able to distinguish between A and B cells. This cellular heterogeneity was not observed either in BALB/c Crgl mice.[42] It is noteworthy that A and B cells were described only from the junction between the sinus vagina and Müllerian vagina and from the fornicocervical region. Actually, they should be evident along the whole Müllerian vagina. The occurrence of degenerative cells is in accordance with the degeneration in the epithelium associated with proliferation.[43,44] Quantitative studies of degeneration and proliferation showed these processes to be closely linked phenomena. The presence of mitotic and chromosomal aberrations suggested that at a pronounced proliferation some mitoses could go wrong.[45,46] A similar association between cell proliferation, cell degeneration, and occurrence of mitotic disturbances has been described from other highly proliferating tissues, such as the neural epithelium.[47,48]

D. Mitotic Activity in the Neonatal Vaginal Epithelium under Estrogen Influence

In NMRI mice, neonatal estrogen treatment does not lead to increased proliferation, but to reduced mitotic activity in the columnar epithelium of the Müllerian vagina.[49,50] The effect was seen as early as 18 hr after the first injection with estradiol. Opposite results were described from C57B1/Tw mice.[30] Both 1 and 30 μg estradiol resulted in 4 to 7 times higher mitotic rate than in controls, 24 hr after the first injection. With an increasing number of injections the mitotic rate dropped to control level with 1 μg estradiol and below control level with 30 μg. Secondary peaks were seen with both doses after cessation of treatment. The mitotic rate of the B cells was significantly higher than that in the epithelium lining the proximal part of the vagina. In the study by Forsberg,[49] the mitotic rate was analyzed in the morphologically uniform columnar epithelium of the Müllerian vagina only, before any signs of a basal zone appeared. As soon as this occurs the cells of the basal zone divide vigorously under estrogen influence. The illustrations in the paper by Iguchi et al.[30] could indicate that the basal zone appears earlier in the C57B1/Tw strain than in NMRI mice. When fetuses from this strain were exposed to estradiol from day 17 of fetal life, the mitotic inhibiting effect in the columnar epithelium of the Müllerian vagina and cervix was evident in late fetal life and on day 3 after birth.[51] The cells of the sinus vagina were stimulated to proliferate.

Because neonatal DES treatment was found to influence the ratio of cGMP to cAMP in the vagina of 40-day-old females, the basic mechanism in induction of ovary-independent vaginal cornification was suggested to be alteration of cell division control.[52]

E. Different Opinions on Developmental Mechanisms of Ovary-Independent Cornified Vaginal Epithelium

The postulated cell heterogeneity in ovary-independent cornified vaginal epithelium could be explained in different ways. The suggestion of estrogen-induced survival of cells normally destined to die (A cells) is an attractive concept for which further support is needed. In some studies it was difficult to define A cells. It is questionable whether a proliferative increase of short duration, as described by Iguchi et al.,[30] could have such a dramatic effect as to change the destiny of the cells. In other studies, in which survival of normally dying cells was also proposed to have developmental effects, growth of the cells was accelerated to proliferate to such a level and for such a time that they were withdrawn from normal differentiation.[47,48] The general significance of the hypothesis of surviving cells is inconsistent with the finding of an estrogen-induced proliferative inhibition in the Müllerian vagina.[49,51] The ability to postpone induction of permanent cornification by using lysosomal stabilizers does not necessarily mean that cell death is causally related to permanent cornification.[16,23] The membrane stabilizers could have an effect on the estrogen mechanism of action itself.[53,54]

Another alternative is that the undifferentiated cells are the columnar cells in the Müllerian vagina. The cell heterogeneity in the adult vagina is then a result of an estrogen effect at the nascency of cells destined for the basal zone. The interaction between estrogen and responding cells could involve a pattern of changes, depending on type of interaction: receptor interaction with native estrogen or metabolites or formation of different types of DNA adducts (further discussed in Section VIII.B). Cellular selection mechanisms could later result in only some cell types being allowed to survive and expand.

Under estrogen treatment, a continuous hyperplastic sheet of basal cells appears in the lower part of the Müllerian vagina; in the upper part isolated nodules of proliferating cells are seen (B cells, in the terminology of Takasugi). This seems to reflect the normal progression of the differentiation process. In the early phase of neonatal estrogen treatment, cells destined for the basal zone already occur and are numerous in the low Müllerian vagina, but occur with decreasing incidence at higher levels. Estrogen, through its mitotic inhibiting effect, could prevent formation of new basal cells, resulting in scattered nodules.

Neonatal estrogen treatment of cyproterone acetate-induced feminized male mice resulted in cornification in the middle part of the vaginal anlage, derived from the urogenital sinus; in the upper third, presumably of Müllerian origin, the epithelium was hyperplastic but the incidence of cornification low.[55] A possible participation of sinus cells in the development of estrogen-independent persistent vaginal cornification was suggested. According to the present author, the important finding was the squamous epithelium, and the failure to cornify could be a reflection of a cornification gradient.[56]

IV. NEOPLASTIC TRANSFORMATION OF THE SQUAMOUS VAGINAL EPITHELIUM AFTER PERINATAL ESTROGEN, ANDROGEN, OR PROGESTERONE TREATMENT

Precancerous and cancerous changes are seen in the vaginal epithelium at an advanced age in mice treated with estrogen or testosterone neonatally or prenatally.[23,57-59] A disruption of the basal lamina and reduction of desmosomes was observed at 3 months and was frequent at 10 months. At the latter age protrusions and penetrations of cytoplasmic processes from the basal cells occurred through gaps in the basal lamina.[60] A special cell type (C cells) was identified at the tips of hyperplastic downgrowths of basal cells.[23] Cancer cells were postulated to arise from C cells, following a sequence of A cells to B cells to C cells. After neonatal estrogen treatment, the basal cell layer of the vaginal epithelium had a mixed population of light and dark basophilic cells (large number of free ribosomes in electron microscopy) on days 5 and 8 after birth.[42] The basophilic cells had about four times more hemidesmosomes per basal cell cross section than the light cells. The authors suggest a hormone-induced selection of vaginal cells whose expression of ovary-independent cornification is associated with a selective number of hemidesmosomes, with these cells having an increased potential for abnormal differentiation, including neoplasia.[42]

Progesterone also results in hyperplastic lesions and downgrowths, but to a lesser degree than estradiol.[19] Tumors arising in females treated with estrogen neonatally are of a squamous type, but those in progesterone-treated females are mixed with both squamous and gland-like components.[20,61]

Treatment of BALB/c mice with 0.1 μg estradiol for 5 days after birth resulted in 16% of the females having hyperplastic lesions, including epidermoid carcinomas, at an age of about 17 months.[15] Ovariectomy completely inhibited the incidence of these lesions, indicating that endogenous hormones were of importance. When the daily neonatal dose of estradiol was increased to 25 μg/day, the incidence of hyperplastic lesions increased to 81% but was reduced to 38% in similarly treated females after ovariectomy.[15] Thus, the vaginal epithelium which is so altered to produce ovary-independent vaginal cornification may produce hyperplastic and carcinoma-like lesions in the absence of ovaries.[15] Vaginal and cervical carcinomas were observed in 20- to 26-month-old mice treated with 2 mg DES on the day of birth, and vaginal tumors developed in rats transplacentally exposed to DES.[62,63]

The tumors that occur are generally restricted and metastases do not occur in the animal of origin.[64] The tumors grow in syngeneic hosts and finally kill the animals; they are transplantable for several generations, and the transplants invade surrounding tissues.[52,61,65] A short-term secondary challenge to estrogen may result in increased mitotic rate and more lesions, while a continuous adult exposure may have no appreciable effect or even temporarily reduce the severity of the histologic lesions or reduce the incidence.[15,64,66]

Vaginal concretions are often seen in the vaginal lumen from neonatally estrogen-treated females.[67] Urine leakage from the common urethral-vaginal opening in these females into the vagina has been suggested as a possible factor in the origin of the concretions.[16] Other reports failed to confirm a correlation between concretions and hypospadia.[68] Because no concretions and no vaginal tumors develop in C3H mice, a possible promoting action has been suggested for the concretion.[16]

V. ESTROGEN EFFECTS IN THE COLUMNAR EPITHELIUM OF THE CERVICOVAGINAL REGION AFTER PRENATAL OR NEONATAL EXPOSURE

A. Vaginal and Cervical Adenosis

The overwhelming majority of papers published on the effects of estrogen treatment in the perinatal life of rats or mice have focused on the response of the squamous epithelium. In 1969, we reported on what was then called an atypical epithelium in the uterine cervix and vaginal fornix of outbred NMRI mice, treated with 5 μg estradiol for the first 5 days after birth.[69] When the females were killed at the age of 1 month there were large regions in the vaginal fornix and uterine cervix where the normal squamous epithelium was replaced with a columnar epithelium. DES, injected according to the same schedule as estradiol, resulted in the same type of changes.[70] The heterotopic columnar epithelium (HCE) was interpreted as a remaining epithelium which had not undergone the normal transformation into a squamous epithelium. Later in life, the HCE formed gland-like downgrowths into the stroma, which was defined as adenosis. While HCE and adenosis were originally described in NMRI mice, the results have been verified in other strains.[71,72]

In our terminology only a wavy HCE, but without actual downgrowths, does not allow the designation adenosis. Sometimes the term adenosis has been used in too wide a sense and included HCE.[73,74] The term "cervical adenosis" has been used for nonneoplastic changes observed after continuous feeding of C3H/HeN-MTV mice with DES, starting at weaning.[75] However, "cervical adenosis" was not defined, nor was the normal position of the cervical squamo-columnar junction.

B. Immediate Effects by Estradiol and DES in the Neonatal Period of Mice

Treatment of neonatal female mice with daily doses of 5 μg estradiol-17β had a profound effect on the proliferative activity in the columnar epithelium of the Müllerian vagina, in the uterine cervix, and in the uterine horns. However, the effects were different in the different regions.[49] Eighteen hours after the first injection (on the day of birth), the mitotic activity in the uterine epithelium reached a peak which was about sixfold as high as the mitotic rate level in the same region in controls. The peak was of short duration because

24 hr later the mitotic activity was similar in control and experimental females, only later to decrease to lower levels in estradiol-treated females. In the uterine cervix and in the morphologically homogeneous columnar epithelium of the Müllerian vagina, no indications for an increased mitotic activity were found. Instead, 18 hr after the first injection, the mitotic activity was lower in the latter regions of estradiol-treated females than in controls. In the Müllerian vagina it was possible to follow the mitotic activity after two injections only because the epithelium undergoes transformation into a basal and superficial epithelial zone. In the uterine cervix, the mitotic activity was low 18 hr after the second injection and remained at a very low level under continuous estradiol treatment. The mitogen response was thus different in the uterine epithelium, in the untransformed columnar epithelium in the upper part of the vagina, and in the uterine cervix.

Later on, these experiments were repeated under DES treatment. A depressed mitotic activity in 6-day-old females was seen after treatment with a daily dose 10^{-5} μg DES and an almost complete inhibition was seen after 10^{-1} μg DES or higher doses.[50] In autoradiographic studies, no labeling was seen in vaginal columnar cells in 5-day-old mice treated with daily doses of 20 μg estradiol for 4 days and given ^{3}H-thymidine 24 hr before death.[41]

In autoradiograms after ^{3}H-thymidine labeling, the labeling index was decreased in the cervical epithelium but increased in the uterine epithelium, 5 hr after an estradiol injection on day 1 of life.[76] Estradiol prolonged the G1 phase of the cervical cell cycle and to a lesser extent also the S phase, but shortened the cell cycle in the uterine epithelium.[77]

In plasma clot cultures of uterine and cervical epithelium, estradiol inhibited the growth of cervical epithelium but had no effect on the uterine epithelium, which points to a direct effect on the cervical cells.[78]

The estradiol- or DES-induced mitotic inhibition later subsides. The mitotic activity in the HCE and in the basal cells of the vaginal epithelium was similar in 15-day-old DES-treated and control mice, and after puberty there is a pronounced mitotic activity in HCE regions.[79,80]

It is postulated that estrogen-induced mitotic inhibition is essential for the untransformed Müllerian columnar epithelium to remain and the failure of the normal transformation to occur. Because the response of neonatal fornicovaginal epithelium to DES resembled that of neonatal and adult uterine epithelium more than it resembled the response of adult fornicovaginal epithelium, it was suggested that cells stimulated to a uterine-like response might be the origin of HCE and adenosis.[81] However, in adult life adenosis cells and uterine cells differ ultrastructurally.[82]

Both mitotic rate studies and the experiments involving ^{3}H-thymidine indicate that the estrogen-proliferative inhibiting effect ensues rapidly after the first injection. A possible association exists between estrogen-induced mitotic inhibition and estrogen effects on adenylate cyclase activity in the neonatal period. Estradiol increased the biochemical activity of adenylate cyclase at 3 days after birth but was without effect at 14 days.[54] Cytochemical studies demonstrated the same time-dependent increase in enzyme activity and localized the activity to the cells in the superficial epithelial zone.[83] Studies on the activity of cAMP-dependent protein kinase isoenzymes in the neonatal uterine cervix failed to reveal any difference between control and estrogen-treated females.[84] Thus, the role of the increased adenylate cyclase activity in the neonatal uterine cervix is still unknown. An aqueous extract of uterine epithelium has been shown to have a rapid, transient, and organ-specific mitotic inhibiting effect by delaying cells from entering G2.[85]

Because estrogens can be metabolized into highly reactive metabolites, the study of a possible genotoxic effect of DES in the neonatal uterine cervix was of interest.[86-88] In organ cultures, the mouse fetal genital tract can metabolize DES via reactive intermediates to Z,Z-dienestrol.[89]

Cervical stromal and epithelial cells from neonatal mice were cultured separately in the presence of various concentrations of DES or estradiol-17β. As controls for organ specificity, kidney cells were cultured in the presence of DES. A concentration of 10^{-7} to 10^{-5} *M* DES resulted in a significantly increased number of sister chromatid exchanges (SCEs) in diploid stromal cells; estradiol-17β had no effect, nor did DES on kidney cells under the same experimental conditions.[90] Pretreatment of the females with phenobarbital (inducer of P-450 microsomal monooxygenases) resulted in a significant increase of SCEs in the presence of DES in the culture medium while α-naphthoflavone and indomethacin reduced the number of SCEs to an intermediate level. When both α-naphthoflavone and indomethacin were present in the medium, no DES effect was seen, which points to involvement of both microsomal monooxygenases and prostaglandin synthetase in induction of SCE.[90]

In cultures of cervical epithelial cells, 10^{-6} and 10^{-5} *M* concentrations of DES increased the number of SCEs per diploid cell.[91] In contrast to the situation for stromal cells, 10^{-5} *M* estradiol-17β had a significant effect in epithelial cells, but quantitatively less than DES.

DES had no effect on the incidence of cells in the tetraploid range in the stroma, nor in the epithelium. However, in the latter the incidence of cells in the tetraploid range was high (20 to 30%), both in the absence and presence of DES in the medium.

Three results emerge from those studies: DES is more potent in inducing SCEs than estradiol, DES does not induce gross chromosomal aberrations in the neonatal uterine cervix, and an organ specificity has been demonstrated, since kidney cells from the same females as the cervical cells were not influenced. The more profound effect in the stromal cells than in the epithelium is noteworthy and could reflect an important stroma-DES interaction. Vaginal stroma from female mice with estrogen-induced ovary-independent vaginal cornification can elicit this response in normal vaginal epithelium.[92]

C. The Cervicovaginal Epithelium from Immature and Adult Mice after Neonatal Estrogen Treatment

As mentioned earlier, neonatal estrogen treatment results in more or less widespread regions in the upper part of the vagina, the vaginal fornices, and the CCC, having a columnar epithelial lining (HCE). The incidence of HCE is low in the upper part of the vagina (<10%) while in the vaginal fornices and CCC, both the incidence and extension of HCE are striking (80 to 100%).[50] The occurrence of HCE is specific for estrogens; it is not seen after neonatal treatment with testosterone, 5α-dihydrotestosterone, progesterone, or *trans*-stilbene. HCE also occurs after neonatal treatment with dienestrol or estradiol-17α. DES was about 100-fold as potent as estradiol-17β and about 500-fold as potent as estradiol-17α.[50] A single dose of 5 μg DES on the day of birth also induces HCE. When the 5-day treatment period was postponed from the day of birth until day 4 or 6 of life, no HCE was seen in the vagina and vaginal fornices and the incidence in the CCC was reduced. This is in line with the age-related and low-to-higher-level developmental gradient earlier described.

In outbred NMRI mice, daily doses of 10^{-2} μg DES per day, but not lower doses, resulted in a low (20%) incidence of HCE, increasing to 90% at 5 μg/day. C57B1/6 mice were about as sensitive as NMRI mice, while BALB/c mice were more resistant.[50] The critical dose for the ICR/JC1 mice was 10^{-1} μg/day.[79] In the latter strain, a high incidence of HCE was seen after treatment with 200 or 2000 μg DES per day, starting on day 15 of pregnancy.[79]

After puberty, the HCE gets a more wavy character, finally forming gland-like downgrowths into the stroma; therefore the term adenosis can now be used. This progression is ovary-dependent and does not occur in ovariectomized females.[50,69] The adenosis can be circumscript in the cervix and penetrating only shortly into the stroma or it can be of a more general nature, engaging the whole cervical canal and penetrating deeply into the stroma. At 9 months, 50% of the females treated with 5 μg DES per day neonatally had adenosis.[50] In adult females, a process of metaplasia occurs, both in the HCE and in the adenosis elements. The squamous epithelium seems to undermine the columnar epithelium which is sloughed.[50]

The importance of estrogen, both for induction and growth of adenosis, is also evident from grafting experiments.[73] Different regions from the uterovaginal anlage from 1-day-old BALB/c or C57/B1 females were grafted under the kidney capsule of adult hosts. The term adenosis in that study includes both HCE and adenosis in the present author's terminology. Adenosis was never seen in grafts from the posterior part of the vaginal anlage, originating from the urogenital sinus. Adenosis was not seen in grafts grown in ovariectomized hosts, but it did occur in grafts from intact and estradiol-treated ovariectomized females 1 month after transplantation. Grafts were allowed to grow for a further month, but now in the absence of estradiol or in progesterone-treated hosts had no adenosis. The adenosis from 1 month must thus have disappeared during the second month in the absence of estradiol or in the presence of progesterone.

Ultrastructurally, the cells of HCE and adenosis appear devoid of major cytoskeletal features such as the terminal web and lose polarity of cytoplasmic organization.[82] The adenosis cells were different from uterine gland cells.

The occurrence of HCE and adenosis in the cervicovaginal region from neonatally estrogen-treated female mice, as described by Forsberg[69], was for several years only incidentally referred to in other papers. No HCE or adenosis was found in a study of female mice exposed to DES on days 9 through 16 of gestation.[68] Only 1 out of 31 10- to 14-week-old mice exposed to ethinyl estradiol *in utero* had gland-like structures in the vaginal fornices.[93] Vaginal or cervical adenosis was seen in CD-1 mice or hamsters treated with DES prenatally and in rats exposed to DES transplacentally.[33,59,63,94] Both HCE and gland-like downgrowths were described from the vaginal fornices in four different mouse strains treated perinatally with estradiol benzoate or DES.[71] Treatment with 2 μg estradiol benzoate or 2 μg DES neonatally only resulted in an incidence of these lesions of 68 and 100%, respectively, in 30- to 36-day-old females. In the BALB/cCrgl strain, a combined pre- and postnatal treatment with estradiol benzoate resulted in a tendency for glands to be present in the lower part of the CCC.[71] The authors finally presented evidence for a reduced distribution of HCE regions from day 10 to older age stages studied, which was interpreted as a developmental arrest being rectified with increasing age. In our experience a similar reduction of HCE extension is most evident in the upper part of the vagina during the prepubertal period.

D. Adenosis-Associated Malignancy in Mice

Among 23 preparations from 17-month-old, neonatally DES-treated females, 8 had well-differentiated local malignancy in the adenosis regions. In 4 cases cervical adenocarcinoma was seen, in 1 case an epidermoid malignancy, and in 3 cases the changes were of a mixed type. No similar changes were seen in 36- to 52-week-old females treated with 5 μg estradiol-17β per day in the neonatal period.[50] Epidermoid malignancy has also been described from 13-month-old DES mice.[95]

Neonatal exposure to DES resulted in 75% of the females having vaginal adenosis at 35 days. Prenatal treatment resulted in a significantly lower incidence: 15% of females had adenosis at 1 month and 10% at 18 months.[96] Age-dependent metaplastic changes have been described from regions with vaginal adenosis.[59] The common finding in 12- to 18-month-old females was excess vaginal cornification. Vaginal epidermoid tumors were found in 5 out of 20 females and 1 out of 35 exposed females had a well-differentiated vaginal adenocarcinoma. Squamous metaplasia of the cervical epithelium was seen in 25% of treated females and in 1% of controls. Cervical stromal stimulation was evident in the form of occasional leiomyoma, stromal cell sarcoma, and leiomyosarcoma. In a later study 2 cases of vaginal adenocarcinomas were reported from 91 prenatally DES-treated females.[96]

Based on earlier discussed results on DES-induced genotoxic effects, HCE and adenosis may contain dormant malignant cells, which because of some factor acting later in life may be stimulated to malignant degeneration. Among such possible factors are ovarian hormones, a bypass of deviating cells from immune control because of DES-induced depressed cytotoxic and immune functions in the same females, or a viral factor.[97] The latter has gained interest in human DES-exposed offpsring.[98] Female mice treated with 5 μg DES for 5 days after first have a reduced activity of natural killer lymphocytes and an increased incidence of 3-methylcholanthrene-induced sarcomas induced by a single low dose of the carcinogen.[99] Prenatal DES exposure in hamsters produced a significant increase in carcinogenic response after exposure to 7,12-dimethylbenz(*a*)anthracene in postnatal life.[100] In fact, neonatal treatment with steroids and related substances may increase the risk for, rather than induce, tumors *de novo*.[20]

E. Effects in the Cervicovaginal Epithelium after Treatment with Tamoxifen, Clomiphene, and Nafoxidine in Neonatal Life (Mouse)

Triphenylethylene derivates (tamoxifen, clomiphene) and nafoxidine have a varying estrogen agonist-antagonist spectrum in different species and tissues. Clomiphene, tamoxifen, and nafoxidine are estrogen agonists in the mouse uterus and vagina.[101,102] When neonatal female mice were treated with daily doses of 5 μg of these drugs for the first 5 days after birth, HCE occurred in the cervicovaginal region at 8 weeks after birth. Tamoxifen, ICI 47.699 (*cis*-isomer of tamoxifen), and clomiphene were more effective than nafoxidine in inducing HCE, and even more so than estradiol and DES.[103] "More effective" means larger areas covered with HCE. With clomiphene and tamoxifen, large areas with HCE were not only seen in the vaginal fornices and CCC, but also in a large region in the upper part of the vagina, which is only rarely involved after treatment with estradiol and DES. Gland-like downgrowths also appeared earlier than with estradiol or DES. These results are similar to those reported by Taguchi and Nishizuka.[104]

Tamoxifen has a long half-life which could be related to its pronounced effect.[105] When neonatal females were treated with a long-acting estradiol preparation (estradiol-benzoate), this had no effect different from the free steroid.[103]

While the estradiol- or DES-induced HCE progressed into adenosis and, in the case of DES, into carcinoma-like changes, the HCE in females treated with clomiphene and tamoxifen underwent a pronounced metaplasia. The large regions with HCE in the upper vagina at 8 weeks were reduced to small scattered islands at 6 months, and the metaplasia also engaged the cervical region.[103] The reduction in the incidence of HCE in the cervicovaginal regions is thus age-dependent.[106]

Female rats treated with clomiphene on day 5 or 12 of pregnancy and killed 15 weeks later had epithelial changes in the vagina and uterine cervix reminiscent of adenosis.[107]

While corpora lutea are never seen in ovaries from DES- or estradiol-treated NMRI mice (standard dose 5 μg/day for 5 days neonatally), they do occur in clomiphene-exposed ovaries and also occasionally in tamoxifen and nafoxidine ovaries.[103] This could point to some difference in ovarian function being responsible for the pronounced HCE metaplasia.

F. Effects in the Nonhuman Primate Genital Tract after Fetal Exposure to DES

In newborn *Cebus apella* monkeys exposed to DES during fetal life, the upper half of the vagina was patent and contained numerous crypts and glands originating from the columnar surface epithelium.[108] In unexposed females, the vagina was filled with a solid cord of squamous cells, except in the vicinity of the vaginal fornices and vaginal portio, where the lumen was lined with a squamous epithelium. Vaginal and cervical anomalies and vaginal adenosis were also observed in adult *Macaca mulatta* females after fetal DES exposure.[109]

G. Effect of DES on the Human Fetal Vaginal Epithelium

Results from grafting human fetal vaginal anlage into DES-treated nude mice indicated that DES could inhibit transformation of the columnar Müllerian epithelium into a squamous type.[12,110]

This is the mechanism proposed above for the mouse model. However, in man a further step occurs after formation of squamous Müllerian epithelium, namely its replacement by an upwards-growing epithelium.[2] Probably, the latter is not able to replace untransformed columnar epithelium. Based on pathological findings of different types of adenosis (mucinous type, tuboendometrial type), it was postulated that DES-exposed human vaginal stroma impedes upward growth of sinus epithelium, leaving Müllerian epithelium to be influenced to differentiate into tuboendometrial type of adenosis.[111]

H. Vaginal Adenosis in Women Exposed to DES during Fetal Life

In women exposed to DES in fetal life after the 18th week of gestation, anomalies are rarely seen in the uterine cervix and vagina.[112-114] Clinically manifest vaginal adenosis has been reported with an incidence of 35 to 90% in different series after early exposure.[115] The incidence is higher and the adenosis more widespread in DES women than in unexposed women.[115] Vaginal adenosis is a dynamic condition which is progressively replaced with squamous epithelium through metaplasia.[116,117] The incidence and extension of DES-induced adenosis is closely associated with both the timing of fetal exposure to DES and age at clinical examination.[116]

The association between adenosis and clear-cell adenocarcinoma has long been a matter of discussion: is adenosis a precancerous stage or are adenosis and clear-cell adenocarcinoma two independent conditions? Both can be coexistent, but the problem is to convincingly demonstrate transitional stages between the two. In a study of 20 patients, 16 had foci of atypical adenosis or ectropion and were immediately adjacent to the cancer in 14 cases.[118] This proximity of tuboendometrial type of atypical adenosis and infrequency of atypical adenosis at a distance from the tumor are highly suggestive evidences of an association.[119] The mucinous type of adenosis is unlikely to be involved in malignancy.[118] These findings form a link to the mouse model where morphologic malignancy occurs within adenosis regions.[50]

Another problem is whether the metaplastic changes taking place in the adenosis regions involve an increased risk to vaginal dysplasia and squamous neoplasia.[120] Sometimes metaplasia has been considered as a healing process for adenosis.[114,121] For several years the opinions were controversial, but after summarizing the data, no conclusive evidence indicated that intraepithelial neoplasia was more common in the vaginas of DES-exposed women than in appropriate controls.[122] In a later report from the National Collaborative Diethylstilbestrol Adenosis Project, a twofold to fourfold increase in incidence rates of dysplasia and carcinoma *in situ* was described after prenatal exposure to DES, which was attributed to the greater extent of squamous metaplasia.[123] If the metaplastic epithelium is more sensitive than normal epithelium to some factor, e.g., herpes- or papillomavirus, this could explain the different incidence in controls and DES-exposed women.[98]

VI. UTERINE EFFECTS AFTER NEONATAL ESTROGEN TREATMENT

A. Uterine Estrogen Response and Squamous Metaplasia

Uteri from adult but neonatally estradiol-treated mice have usually been described as thread-like or medium sized.[25,57] Uterine hypoplasia also occurred in Syrian hamsters exposed to DES prenatally and in daughters of women treated with DES during pregnancy.[124,125] DES has been reported to accelerate development of the immature mouse uterus.[126]

The mean diameter of uteri from BALB/cCrgl mice treated with daily doses of 0.1 μg DES for 5 days after birth was similar to that of controls at 1 month after birth but reduced at 2 to 11 months.[127] Continuous estradiol treatment of ovariectomized control or DES females resulted in uterine growth, but DES uteri were smaller than control uteri.[127]

Treatment of neonatal female NMRI mice with 10^{-1}, 1, or 5 μg DES per day for 5 days after birth resulted in a significantly reduced uterine wet weight at 28 days; a daily dose of 10^{-2} μg had no effect.[128] With the highest dose (5 μg/day) almost no growth took place from day 12, while with lower doses an accelerated growth occurred after day 21. The uterine weight in 28-day-old females treated with 5 μg DES neonatally was slightly lower than in females ovariectomized on day 1 of life. When control or neonatally DES-treated (5 μg/day) females were challenged with 1 μg estradiol on days 25 to 27 and killed on day 28, the weight increase was 4.9-fold in controls and 4.2-fold in uteri from DES females; in females ovariectomized on day 1 of life and challenged with estradiol the weight increase was 7.7-fold.[128] The low initial uterine weight of DES females was thus not per se responsible for the less dramatic weight increase in DES females compared with those ovariectomized on day 1. The results from the study by Forsberg et al.[128] using immature females challenged with a short-term treatment with estradiol and those by Ostrander et al.[127] using adult female mice and continuous estradiol exposure from polyethylene capsules are in general agreement: uteri from neonatally DES-treated females are smaller than those from controls, both respond to a later challenge with estradiol, the response is slightly lower in DES uteri than in control uteri, and the former starting from a lower level do not reach the same size as control uteri. A reduced response of neonatally estrogen-exposed uteri to a short-term challenge with estrogen later in life has earlier been described for uterine wet weight, mitotic rate, and cell number.[34,129-131] In rats, the reduced uterotropic response to a challenge with estradiol in immature life was no longer seen at puberty or maturity.[132]

The reduced uterine response to estradiol in neonatally estrogen-treated animals could be related to a defective estrogen-receptor mechanism. A reduced level of cytosolic receptor and reduced ability to synthesize receptors under estrogen influence has been reported; the ability of estradiol to induce prolactin receptors after ovariectomy was not influenced.[127,131] Uterine estrogen-receptor characteristics were similar in controls and neonatally DES-treated hamsters, but DES females had enhanced estrogen metabolism, resulting in low systemic levels of estradiol.[133]

Squamous metaplasia of the uterine epithelium is seen after prenatal or neonatal estrogen or testosterone treatment in rats and mice, even as early as 6 weeks after birth.[57,62,68,134-138] Isolated nodules of metaplastic epithelium occur which indicate that an upward growth of squamous epithelium from the vaginal region is not obligate.[68,138] In females older than 3 months, subcolumnar basal cells similar to those seen in the vaginal region appeared in the uterine epithelium.[138] Similar cells were seen by Ostrander et al.;[127] they were never observed in control uteri. These cells would be able to proliferate and become the origin of the metaplastic epithelium.

The importance of endogenous or exogenous estrogens for metaplastic development in perinatally estrogen-treated females has been stressed. Possibly, estrogen sensitizes the uterine epithelium to metaplastic effects of later estrogen exposure.[137] Uterine squamous metaplasia was only occasionally seen in ovariectomized control females after continuous estradiol treatment, but occurred in 67 to 94% of neonatally DES-treated and estradiol-challenged mice.[127] A combined progesterone and estradiol challenge reduced the incidence of metaplasia.

Another effect of neonatal estrogen treatment is disruption of the inner circular muscular layer and thinning of the outer longitudinal layer.[127]

B. Uterine Hyperplasia and Neoplasia

Cystic endometrial glandular hyperplasia has been described in female offspring older than 5 months after estrogen exposure in fetal life.[59,139]

A continuous challenge with progesterone resulted in adenomyosis (aberrant growth of uterine glands and stroma into the muscularis) with a similar incidence in control and neonatally DES-treated females; estradiol induced cystic glandular hyperplasia.[127] Adenomyosis in the offspring was seen after continuous DES feeding of the mothers from day 7 of gestation through the day of birth.[140] Grafting of an ectopic pituitary gland indicated that hyperprolactinemia was a major endocrine factor in adenomyosis genesis.[140]

Prenatal DES exposure with adequate doses results in uterine tumors at an advanced age in both mice and rats.[59,141] Treatment of pregnant CD-1 mice with an intraperitoneal injection of DES at 16 days and 16 hr of gestation or later resulted in 14 out of 143 female offspring having uterine adenocarcinomas at an age of 19 to 20 months.[135] Leiomyomas were observed in 3 out of 64 controls. When females exposed to DES in fetal life were mated to normal control males, this second generation offspring had an increased incidence of uterine tumors, while the incidence of commonly occurring tumor types in this strain was similar in controls and in females with past generation DES exposure.[142] The mechanism for this second generation effect is unknown. Among 18 female rats exposed to DES on the 19th day after conception 3 endometrial polyps and 1 fibroma were found.[141] No uterine tumors were found among 34 controls.

Prenatal exposure to DES causes a significant increase in the carcinogenic response of the hamster uterus to later DES treatment.[124]

VII. OVIDUCTAL MORPHOLOGY IN PERINATALLY ESTROGEN-TREATED FEMALES

As early as 1940, Greene et al.[143] described abnormal oviductal morphology in adult rats exposed to estradiol dipropionate from day 12 to 13 of gestation until day 18 to 21. There was a dose-related inhibition of the normal convoluted oviduct. With the highest estrogen doses used, convolution was absent and the oviducts were straight structures, lateral to the ovaries. Oviductal malformations were also described from adult rats and mice exposed to DES or estradiol prenatally.[139,144-146]

The oviduct structure was similar to that found in normal fetuses: oviduct uncoiled, short, adherent to and wrapped around the ovary. The authors interpreted the findings as a developmental arrest of the oviduct.[144,145] Microscopically, the incidence of salpingitis was increased; the epithelium was of a secretory type and sometimes hyperplastic with mucosal folds extending through the muscularis. The mucosa of the isthmus region was lined by numerous abnormal nonciliated cells, while the ciliated cells were normal. Large cysts adjacent to the ovary and oviduct could be related to both Müllerian and Wolffian duct alterations.[144] A progressive epithelial hyperplasia and pseudogland formation also occurred after neonatal DES treatment.[147]

VIII. ARE THE PERMANENT EFFECTS INDUCED BY ESTROGEN PERINATALLY RECEPTOR MEDIATED OR NONRECEPTOR ASSOCIATED?

A. Effects Secondary to a Disturbed Ovarian Function

Perinatal treatment of female mice and rats with estrogen or testosterone results in disturbances in the hypothalamic-pituitary regulation, evident as acyclic gonadotropin release in adult life with no LH surges.[148] These central effects are reflected peripherally in disturbed ovarian morphology. The typical ovarian picture is the absence of corpora lutea and hypertrophy of the interstitial tissue.[57] We have never observed corpora lutea in ovaries from NMRI mice treated with 10^{-4} μg or higher daily doses of DES for the first 5 days after birth.[149] On the other hand, female offspring from mothers treated during gestation had examples of ovaries with corpora lutea, even with the highest dose used (100 μg DES per kilogram maternal body weight), but the interstitial tissue was dominant.[144,150] Recently, neonatal treatment with estrogens and aromatizable androgens has been shown to induce polyovular follicles, but the mechanism behind this has not yet been demonstrated.[128,151]

In ovaries from neonatally DES-treated mice we have observed a successive reduction in the number of follicles with increasing age.[80] After fetal exposure to DES, the number of follicles was similar to that in controls up to 2 months; after this age there was an increase in the number of large follicles in DES females, and by 6 months the number of follicles was reduced.[144]

B. Genetic or Epigenetic Effects

Fetal exposure to DES results in qualitative and quantitative changes in protein synthesis in the genital tract. A protein apparent on day 14 of gestation (with a molecular weight of approximately 70,000 and pI 5.8) was greatly diminished by DES; this was true also 17 days after birth.[152] The protein was not diminished after adult exposure to estradiol-17β. The fact that estrogens can irreversibly change the protein pattern means that they interact profoundly in developmental processes.

In such examples as oviductal developmental arrest and proliferative inhibition in the cervicovaginal columnar epithelium, estrogens might interact with responder cells, making them unresponsive to a growth-stimulating factor, or estrogens may interact with the factor itself or the cell synthesizing it. Such an effect could be thought of as a transitory mechanism and the result of an estrogen-receptor interaction. Premature hormonal exposure may not only amplify but also depress the response of receptors to hormonal challenge later in life, indicating that the maturing receptor may become damaged by early ligand exposure.[153] In immature female rats, estrogen regulates acutely and in a specific way the level of uterine epidermal growth factor.[154]

Autoradiographic analysis after exposure to ^{3}H-estrogens failed to demonstrate nuclear estrogen receptors in vaginal and uterine epithelia of neonatal mice, while they occurred in the stroma at all age stages, including fetal life.[155] Estrogen binding was neither saturable nor specific in epithelial cells from 4- to 5-day-old females, but it was both in 20-day-old mice.[156] Estrogen-induced effects on the proliferation rate in vaginal and uterine epithelia in the neonatal period might thus be mediated by growth factors from the stroma. However, there is also an increased specific synthesis after repeated estrogen injections in the columnar cells of the vaginal anlage which makes the absence of receptors questionable.[157]

The estrogen response in the neonatal period has previously been primarily studied in rats. A full estrogen response can be obtained when estradiol is given in adequate repeated doses.[158,159] Thus, autoradiographic and biochemical studies of the receptor in neonatal uterine or vaginal tissue from acutely low-dose estrogen-treated females may not reflect an in vivo situation under continuous treatment with higher doses. Nuclear receptors could be induced in uterine and vaginal epithelia under prolonged treatment with proper doses, and the role of the receptor mechanism in the induction of permanent estrogen-induced changes should not be excluded.

Some of the effects of estradiol on HeLa cells are similar to those of colcemide, and high doses of estradiol prevent the cells from reaching mitosis.[160,161] DES has colchicine-like effects in some cell systems, and spindle-fiber formation is inhibited.[162-164] At high doses, polymerization of microtubule formation may occur.[165,166] Several reports describe DES induction of polyploidy and aneuploidy.[162,167,168] However, when neonatal cervical stromal or epithelial cells were studied in vitro in the presence of varying concentrations of DES (10^{-5} to 10^{-8} *M*), no signs of a tendency to polyploidy or aneuploidy, nor an increased incidence of structurally abnormal chromosomes, were found.[90,91] In the stromal cells, less than 5% of the cells were in the tetraploid range; in the epithelium this figure was 20 to 30% and not influenced by DES. Also, we have not been able to find any evidence for metaphase arrest (colchicine-like effects) in the cervicovaginal anlage after treating neonatal female mice with DES. Thus, while DES could have an effect on microtubules and ploidy in vitro, it is questionable whether this mechanism is working in vivo in genital target cells. The increased incidence of SCEs induced in vitro could indicate that DES could have more delicate genome effects.[90,91] In human leukocytes, DES causes extensive DNA strand breakage, and in Syrian hamster embryo cells it induces unscheduled DNA synthesis.[169,170]

Steroidal and *trans*-stilbene estrogens (DES) induce cell transformation, and both estrogens induce renal carcinoma in Syrian hamsters, which is probably related to a nonestrogenic but catechol estrogen formation mechanism.[86,87,168,171,172] DES penetrates the placental barrier, and oxidative metabolites of DES and tissue-specific covalent binding to macromolecules were identified in rodent fetuses.[33,173-175] The mouse fetal genital tract can metabolize DES into Z,Z-dienestrol via a reactive semiquinone or quinone.[89] DES metabolism differs between target and nontarget organs and can be modulated by inducers.[176] Covalently modified DNA nucleotides were found in the Syrian hamster kidney after chronic DES exposure.[177] DES-quinone forms unstable adduct intermediates with DNA, which, however, decompose with time.[178] The quinone was supposed to be a possible precursor of a DES radical or DES-semiquinone.

Thus there are indications of estrogen-induced damage to cellular macromolecules which could be thought of as influencing the ability of cells to respond to a normal proliferative or differentiation stimulus during a critical developmental period.

IX. CONCLUSIONS

Estrogens are potent molecules which can induce a spectrum of developmental aberrations in females. Exposure at a proper time affects not only the normal differentiation of genital epithelia but also the pituitary-hypothalamic control and cellular immune functions. In recent years, a great deal has been learned about the metabolism of estrogens and formation of reactive intermediates. A detailed analysis of the interaction between an estrogen molecule or a metabolite and the target cell could open new perspectives for the understanding of both normal estrogen-controlled growth and estrogen-associated malignancy. The study of perinatally estrogen-treated animals allows the opportunity to analyze different factors in the carcinogenic process, both direct epithelial effects and secondary influences of depressed cellular immune function, and disturbed endocrine pattern. A further advantage is the seemingly successive progression from ovary-independent cornified epithelium into epidermoid carcinomas, or from undifferentiated columnar cervicovaginal epithelium via adenosis and metaplasia into adenocarcinomas or epidermoid carcinoma. In the endometrium, different types of pathological conditions are represented: metaplasia, cystic hyperplasia, adenomyosis, and malignancy. In addition, estrogens have developmental effects which can be classified as teratogenic, e.g., the developmental arrest of the oviduct, hypoplasia of the uterus, and clitoral hypospadia. Some of the effects seen with estrogen are also observed with testosterone and progesterone, but much less is known about metabolic activation and action mechanisms for these steroids than is the case with estrogens. Some aspects of estrogen effects are similar in man and rodents, e.g., the high incidence of adenosis and even the occurrence of malignancy, but the tumor types differ.

REFERENCES

1. **Herbst, A. L., Ulfelder, H., and Poskanzer, D. C.,** Adenocarcinoma of the vagina: association of maternal stilbestrol therapy with tumor appearance in young women, *N. Engl. J. Med.,* 284, 878, 1971.
2. **Forsberg, J.-G.,** Cervicovaginal epithelium: its origin and development, *Am. J. Obstet. Gynecol.,* 115, 1025, 1973.
3. **Forsberg, J.-G.,** Derivation and Differentiation of the Vaginal Epithelium, M.D. thesis, University of Lund, Lund, 1963.
4. **Cunha, G. R.,** The dual origin of vaginal epithelium, Am. J. Anat., 143, 387, 1975.
5. **Cunha, G. R., Chung, L. W. K., Shannon, J. M., and Reese, B. A.,** Stromal-epithelial interactions in sex differentiation, *Biol. Reprod.,* 22, 19, 1980.
6. **Del Vecchio, F. R.,** Zur Entwicklung der kaudalen Abschnitte der Müllerschen Gänge bei der Ratte, *(Rattus norvegicus), Acta Anat.,* 113, 235, 1982.
7. **Bok, G. and Drews, U.,** The role of the Wolffian ducts in the formation of the sinus vagina. An organ culture study, *J. Embryol. Exp. Morphol.,* 73, 275, 1983.
8. **Mauch, L. B., Thiedemann, K. V., and Drews, U.,** The vagina is formed by downgrowth of wolffian and mullerian ducts, *Anat. Embryol.,* 172, 75, 1985.
9. **Witschi, E.,** Development and differentiation of the uterus, in *Proc. 3rd Annu. Symp. Physiology and Pathology of Human Reproduction,* Mach, H. C. and Wayne, Ch. I., Eds., Wayne State University Press, Detroit, 1970, chap. 1.
10. **O'Rahilly, R.,** The development of the vagina in the human, *Birth Defects Orig. Artic. Ser.,* 13, 123, 1977.
11. **Ulfelder, H. and Robboy, S. J.,** The embryologic development of the human vagina, *Am. J. Obstet. Gynecol.,* 126, 769, 1976.
12. **Robboy, S. J.,Taguchi, O., and Cunha, G. R.,** Normal development of the human female reproductive tract and alterations resulting from experimental exposure to diethylstilbestrol, *Hum. Pathol.,* 13, 190, 1982.
13. **Gardner, W. U.,** Sensitivity of the vagina to estrogen: genetic and transmitted differences, *Ann. N.Y. Acad. Sci.,* 83, 145, 1959.
14. **Takasugi, N., Bern, H. A., and DeOme, K. B.,** Persistent vaginal cornification in mice, *Science,* 138, 438, 1962.
15. **Takasugi, N., Kimura, T., and Mori, T.,** Irreversible changes in mouse vaginal epithelium induced by early post-natal treatment with steroid hormones, in *The Post-Natal Development of Phenotype,* Kazda, S. and Denenberg, V. H., Eds., Academia Prague, Prague, 1970, 229.
16. **Takasugi, N.,** Cytological basis for permanent vaginal changes in mice treated neonatally with steroid hormones, *Int. Rev. Cytol.,* 44, 193, 1976.
17. **Kimura, T., Nandi, S., and DeOme, K. B.,** Nature of induced persistent vaginal cornification in mice. II. Effect of estradiol and testosterone on vaginal epithelium of mice of different ages, *J. Exp. Zool.,* 165, 211, 1967.
18. **Kimura, T., Basu, S. L., and Nandi, S.,** Nature of induced persistent vaginal cornification in mice. I. Effect of neonatal treatment with various doses of steroids, *J. Exp. Zool.,* 165, 71, 1967.
19. **Jones, L. A. and Bern, H. A.,** Long-term effects of neonatal treatment with progesterone alone and in combination with estrogen on the mammary gland and reproductive tract of female BALB/cfC3H mice, *Cancer Res.,* 37, 67, 1977.
20. **Jones, L. A. and Bern, H. A.,** Cervicovaginal and mammary gland abnormalities in BALB/cCrgl mice treated neonatally with progesterone and estrogen, alone or in combination, *Cancer Res.,* 39, 2560, 1979.
21. **Plapinger, L. and McEwen, B. S.,** Gonadal steroid-brain interactions in sexual differentiation, in *Biological Determinants of Sexual Behaviour,* Hutchison, J. B., Ed., John Wiley & Sons, New York, 1978, 153.
22. **Ohta, Y. and Iguchi, T.,** Development of the vaginal epithelium showing estrogen-independent proliferation and cornification in neonatally androgenized mice, *Endocrinol. Jpn.,* 23, 333, 1976.
23. **Takasugi, N.,** Development of permanently proliferated and cornified vaginal epithelium in mice treated neonatally with steroid hormones and the implication in tumorigenesis, *Natl. Cancer Inst. Monogr.,* 51, 57, 1979.
24. **Ainslie, M. B. and Kohrman, A. F.,** The effect of 17α-hydroxyprogesterone caproate on vaginal development in mice: a preliminary report, *J. Toxicol. Environ. Health,* 3, 339, 1977.
25. **Takasugi, N.** Vaginal cornification in persistent-estrous mice, Endocrinology, 72, 607, 1963.
26. **Takasugi, N.,** Estrogen-independent proliferation of mouse vaginal epithelium transplanted from newborn into gonadectomized hosts, *Proc. Jpn. Acad.,* 47, 199, 1971.
27. **Terenius, L., Meyerson, B. J., and Palis, A.,** The effect of neonatal treatment with 17β-oestradiol or testosterone on the binding of 17β-oestradiol by mouse uterus and vagina, *Acta Endocrinol.,* 62, 671, 1969.
28. **Shyamala, G., Mori, T., and Bern, H. A.,** Nuclear and cytoplasmic oestrogen receptors in vaginal and uterine tissue of mice treated neonatally with steroids and prolactin, *J. Endocrinol.,* 63, 275, 1974.

29. **Kimura, T., Basu, S. L., and Nandi, S.,** Nature of induced persistent vaginal cornification in mice. III. Effects of estradiol and testosterone on vaginal epithelium *in vitro, J. Exp. Zool.*, 165, 497, 1967.
30. **Iguchi, T., Ohta, Y., and Takasugi, N.,** Mitotic activity of vaginal epithelial cells following neonatal injections of different doses of estrogen in mice, *Dev. Growth Differ.*, 18, 69, 1976.
31. **Kimura, T.,** Persistent vaginal cornification in mice treated with estrogen prenatally, *Endocrinol. Jpn.*, 22, 497, 1975.
32. **Shah, H. C. and McLachlan, J. A.,** The fate of diethylstilbestrol in the pregnant mice, *J. Pharmacol. Exp. Ther.*, 197, 687, 1976.
33. **McLachlan, J. A.,** Transplacental effects of diethylstilbestrol in mice, *Natl. Cancer Inst. Monogr.*, 51, 67, 1979.
34. **Mori, T.,** Effects of postpubertal oestrogen injections on mitotic activity of vaginal and uterine epithelial cells in mice treated neonatally with oestrogen, *J. Endocrinol.*, 64, 133, 1975.
35. **Kahn, R. H. and Bern, H. A.,** Antifolliculoid activity of vitamin A1, *Science*, 111, 516, 1950.
36. **Yasui, T., Iguchi, T., and Takasugi, N.,** Blockage of the occurrence of permanent vaginal changes in neonatally estrogen-treated mice by vitamin A; parabiosis and transplantation studies, *Endocrinol. Jpn.*, 24, 393, 1977.
37. **Yasui, T. and Takasugi, N.,** Prevention by vitamin A of the occurrence of permanent vaginal changes in neonatally estrogen-treated mice. An electron microscopic study, *Cell Tissue Res.*, 179, 475, 1977.
38. **Tachibana, H. and Takasugi, N.,** Restoration of normal responsiveness of vaginal and uterine epithelia to estrogen in neonatally estrogenized, A-vitaminized adult mice, *Proc. Jpn. Acad. Ser. B*, 56, 162, 1980.
39. **Takasugi, N.,** Morphogenesis of estrogen-independent proliferation and cornification of the vaginal epithelium in neonatally estrogenized mice, *Proc. Jpn. Acad.*, 47, 193, 1971.
40. **Takasugi, N. and Kamishima, Y.,** Development of vaginal epithelium showing irreversible proliferation and cornification in neonatally estrogenized mice: an electron microscope study, *Dev. Growth Differ.*, 15, 127, 1973.
41. **Mori, T., Iguchi, T., and Takasugi, N.,** Origin of permanently altered epithelial cells of the vagina in neonatally estrogen-treated mice, *J. Exp. Zool.*, 225, 99, 1983.
42. **Cunha, G. R. and Lee, A. K.,** The possible role of hemidesmosomes in neonatally estrogen-induced selection of a permanently altered abnormal vaginal epithelium, *J. Exp. Zool.*, 203, 361, 1978.
43. **Forsberg, J.-G.,** Mitotic rate and autoradiographic studies on the derivation and differentiation of the mouse vaginal anlage, *Acta Anat.*, 62, 266, 1965.
44. **Forsberg, J.-G.,** Studies on the cell degeneration rate during the differentiation of the epithelium in the uterine cervix and mullerian vagina of mouse, *J. Embryol. Exp. Morphol.*, 17, 433, 1967.
45. **Forsberg, J.-G. and Lannerstad, B.,** Chromosome aberrations during normal epithelial proliferation, *Nature (London)*, 217, 568, 1968.
46. **Forsberg, J.-G. and Källén, B.,** Cell death during embryogenesis, *Rev. Roum. Embryol. Cytol. Ser. Embryol.*, 5, 91, 1968.
47. **Källén, B.,** Overgrowth malformation and neoplasia in embryonic brain, *Confin. Neurol.*, 22, 40, 1962.
48. **Källén, B.,** Proliferation in the embryonic brain with special reference to the overgrowth phenomenon and its possible relationship to neoplasia, *Prog. Brain Res.*, 14, 263, 1965.
49. **Forsberg, J.-G.,** An estradiol mitotic rate inhibiting effect in the mullerian epithelium in neonatal mice, *J. Exp. Zool.*, 175, 369, 1970.
50. **Forsberg, J.-G. and Kalland, T.,** Neonatal estrogen treatment and epithelial abnormalities in the cervicovaginal epithelium of adult mice, *Cancer Res.*, 41, 721, 1981.
51. **Kimura, T., Kawashima, S., and Nishizuka, Y.,** Effects of prenatal treatment with estrogen on mitotic activity of vaginal anlage cells in mice, *Endocrinol. Jpn.*, 27, 739, 1980.
52. **Kohrman, A. F.,** The newborn mouse as a model for study of the effects of hormonal steroids in the young, *Pediatrics*, 62 (Suppl.), 1143, 1978.
53. **Kvinnsland, S.,** Estradiol-17β and cAMP: *in vitro* studies on the cervicovaginal epithelium of neonatal mice, *Cell Tissue Res.*, 175, 325, 1976.
54. **Kvinnsland, S.,** Adenylate cyclase activity in the uterine cervix of neonatal and immature mice. Influence of oestradiol-17β, *J. Endocrinol.*, 84, 255, 1980.
55. **Suzuki, Y. and Arai, Y.,** Induction of estrogen-independent persistent vaginal cornification in cyproterone acetate (CA)-induced feminized male mice, *Anat. Embryol.*, 151, 119, 1977.
56. **Forsberg, J.-G.,** The effect of estradiol-17β on the epithelium in the mouse vaginal anlage, *Acta Anat.*, 63, 71, 1966.
57. **Takasugi, N. and Bern, H. A.,** Tissue changes in mice with persistent vaginal cornification induced by early postnatal treatment with estrogen, *J. Natl. Cancer Inst.*, 33, 855, 1964.
58. **Kimura, T. and Nandi, S.,** Nature of induced persistent vaginal cornification in mice. IV. Changes in the vaginal epithelium of old mice treated neonatally with estradiol or testosterone, *J. Natl. Cancer Inst.*, 39, 75, 1967.

59. **McLachlan, J. A., Newbold, R. R., and Bullock, B. C.,** Long-term effects on the female mouse genital tract associated with prenatal exposure to diethylstilbestrol, *Cancer Res.*, 40, 3988, 1980.
60. **Mori, T. and Nishizuka, Y.,** Morphological alterations of basal cells of vaginal epithelium in neonatally oestrogenized mice, *Experientia*, 38, 389, 1982.
61. **Jones, L. A. and Pacillas-Verjan, R.,** Transplantability and sex steroid hormone responsiveness of cervicovaginal tumors derived from female BALB/cCrgl mice neonatally treated with ovarian steroids, *Cancer Res.*, 39, 2591, 1979.
62. **Dunn, T. B. and Green, A. W.,** Cysts of the epididymis, cancer of the cervix, granular cell myoblastoma and other lesions after estrogen injection in newborn mice, *J. Natl. Cancer Inst.*, 31, 425, 1963.
63. **Vorkerr, H., Menzer, R. H., Vorkerr, V. F., Jordan, S. W., and Kornfeld, M.,** Teratogenesis and carcinogenesis in rat offspring after transplacental and transmammary exposure to diethylstilbestrol, *Biochem. Pharmacol.*, 28, 1865, 1979.
64. **Bern, H. A. and Talamantes, F. J., Jr.,** Neonatal mouse models and their relation to disease in the human female, in *Developmental Effects of Diethylstilbestrol (DES) in Pregnancy*, Herbst, A. L. and Bern, H. A., Eds., Thieme-Stratton, New York, 1981, chap. 10.
65. **Takasugi, N.,** Carcinogenesis by vaginal transplants from ovariectomized neonatally estrogenized mice into ovariectomized normal hosts, *Gann*, 63, 73, 1972.
66. **Wong, L. M., Bern, H. A., Jones, L. A., and Mills, K. T.,** Effect of later treatment with estrogen on reproductive tract lesions in neonatally estrogenized mice, *Cancer Lett.*, 17, 115, 1982.
67. **Takasugi, N. and Bern, H. A.,** Crystals and concretions in the vaginae of persistent-estrous mice, *Proc. Soc. Exp. Biol. Med.*, 109, 622, 1962.
68. **Lamb, J. C., IV, Newbold, R. R., and McLachlan, J. A.,** Visualization by light and scanning electron microscopy of reproductive tract lesions in female mice treated transplacentally with diethylstilbestrol, *Cancer Res.*, 41, 4057, 1981.
69. **Forsberg, J.-G.,** The development of atypical epithelium in the mouse uterine cervix and vaginal fornix after neonatal estradiol treatment, *Br. J. Exp. Pathol.*, 50, 187, 1969.
70. **Forsberg, J.-G.,** Estrogen, vaginal cancer and vaginal development, *Am. J. Obstet. Gynecol.*, 113, 83, 1972.
71. **Plapinger, L. and Bern, H. A.,** Adenosis-like lesions and other cervicovaginal abnormalities in mice treated perinatally with estrogen, *J. Natl. Cancer Inst.*, 63, 507, 1979.
72. **Bern, H. A., Mills, K. T., Ostrander, P. I., Schoenrock, B., Graveline, B., and Plapinger, L.,** Cervicovaginal abnormalities in BALB/c mice treated neonatally with sex hormones, *Teratology*, 30, 267, 1984.
73. **Iguchi, T., Ostrander, P. L., Mills, K. T., and Bern, H. A.,** Induction of abnormal epithelial changes by estrogen in neonatal mouse vaginal transplants, *Cancer Res.*, 45, 5688, 1985.
74. **Chamness, G. C., Bannayan, G. A., Landy, L. A., Jr., Sheridan, P. J., and McGuire, W. L.,** Abnormal reproductive development in rats after neonatally administered antiestrogen (tamoxifen), *Biol. Reprod.*, 21, 1087, 1979.
75. **Greenman, D. L., Highman, B., Kodell, R. L., Morgan, K. T., and Norvell, M.,** Neoplastic and nonneoplastic responses to chronic feeding of diethylstilbestrol in C3H mice, *J. Toxicol. Environ. Health*, 14, 551, 1984.
76. **Eide, A.,** The effect of estradiol on the DNA synthesis in neonatal uterus and cervix, *Cell Tissue Res.*, 156, 551, 1975.
77. **Eide, A.,** The effect of oestradiol on the cell kinetics in the uterine and cervical epithelium of neonatal mice, *Cell Tissue Kinet.*, 8, 249, 1975.
78. **Forsberg, J.-G. and Lannerstad, B.,** The *in vitro* response of the mouse mullerian epithelium to estradiol, *Acta Embryol. Exp.*, p.45, 1970.
79. **Iguchi, T., Takase, M., and Takasugi, N.,** Development of vaginal adenosis-like lesions and uterine epithelial stratification in mice exposed perinatally to diethylstilbestrol, *Proc. Soc. Exp. Biol. Med.*, 181, 59, 1986.
80. **Forsberg, J.-G.,** unpublished data, 1986.
81. **Plapinger, L.,** Morphological effects of diethylstilbestrol on neonatal mouse uterus and vagina, *Cancer Res.*, 41, 4667, 1981.
82. **Åbro, A. and Kalland, T.,** The ultrastructure of a diethylstilbestrol-induced heterotopic epithelium and adenosis of the uterine cervix in mice, *Cell. Mol. Biol.*, 28, 319, 1982.
83. **Kvinnsland, S. and Forsberg, J.-G.,** Adenylate cyclase activity in mouse vaginal epithelium. Age related changes in basal activity and estradiol sensitivity, *Dev. Growth Differ.*, 19, 71, 1977.
84. **Døskeland, S.-O.,** unpublished data, 1985.
85. **Lee, A. E. and Rogers, L. A.,** Reduction of mitosis by an aqueous extract of uterine epithelium, *Cell Biol. Int. Rep.*, 5, 1093, 1981.
86. **Li, S. A., Klicka, J. K., and Li, J. J.,** Estrogen 2- and 4-hydroxylase activity, catechol estrogen formation, and implications for estrogen carcinogenesis in the hamster kidney, *Cancer Res.*, 45, 181, 1985.

87. **Purdy, R. H., Goldzieher, J. W., Le Quesne, P. W., Abdel-Baky, S., Durocher, C. K., Moore, P. H., Jr., and Rhim, J. S.,** Active intermediates and carcinogenesis, in *Catechol Estrogens,* Merriam, G. R. and Lipsett, M. B., Eds., Raven Press, New York, 1983.
88. **Degen, G. H. and McLachlan, J. A.,** Peroxidase-mediated in vitro metabolism of diethylstilbestrol and structural analogs with different biological activities, *Chem. Biol. Interact.,* 54, 363, 1985.
89. **Maydl, R., McLachlan, J. A., Newbold, R. R., and Metzler, M.,** Localization of diethylstilbestrol metabolites in the mouse genital tract, *Biochem. Pharmacol.,* 34, 710, 1985.
90. **Hillbertz-Nilsson, K. and Forsberg, J. G.,** Estrogen effects on sister chromatid exchanges in mouse uterine cervical and kidney cells, *J. Natl. Cancer Inst.,* 75, 575, 1985.
91. **Hillbertz-Nilsson, K. and Forsberg, J.-G.,** unpublished data, 1986.
92. **Cunha, G. R. and Fujii, H.,** Stromal-parenchymal interactions in normal and abnormal development of the genital tract, in *Developmental Effects of Diethylstilbestrol (DES) in Pregnancy,* Herbst, A. L. and Bern, H. A., Eds., Thieme-Stratton, New York, 1981, chap. 14.
93. **Yasuda, Y., Kihara, T., and Nishimura, H.,** Transplacental effect of ethinylestradiol on mouse vaginal epithelium, *Dev. Growth Differ.,* 19, 241, 1977.
94. **Rustia, M. and Shubik, P.,** Transplacental effects of diethylstilbestrol on the genital tract of hamster offspring, *Cancer Lett.,* 1, 139, 1976.
95. **Forsberg, J.-G.,** Late effects in the vaginal and cervical epithelia after injections of diethylstilbestrol into neonatal mice, *Am. J. Obstet. Gynecol.,* 121, 101, 1975.
96. **Newbold, R. R. and McLachlan, J. A.,** Vaginal adenosis and adenocarcinoma in mice exposed prenatally or neonatally to diethylstilbestrol, *Cancer Res.,* 42, 2003, 1982.
97. **Kalland, T.,** Long-term effects on the immune system of an early life exposure to diethylstilbestrol, *Branbury Rep.,* 11, 217, 1982.
98. **Adam, E., Kaufman, R. H., Adler-Storthz, K., Melnick, J. L., and Dreesman, G. R.,** A prospective study of association of herpes simplex virus and human papilloma virus infection with cervical neoplasia in women exposed to diethylstilbestrol in utero, *Int. J. Cancer,* 35, 19, 1985.
99. **Kalland, T. and Forsberg, J.-G.,** Natural killer cell activity and tumor susceptibility in female mice treated neonatally with diethylstilbestrol, *Cancer Res.,* 41, 5134, 1981.
100. **Rustia, M. and Shubik, P.,** Effects of transplacental exposure to diethylstilbestrol on carcinogenic susceptibility during postnatal life in hamster progeny, *Cancer Res.,* 39, 4636, 1979.
101. **Jordan, V. C., Clark, E. R., and Allen, K. E.,** Structure-activity relationships amongst nonsteroidal antioestrogens, in *Non-Steroidal Antioestrogens. Molecular Pharmacology and Antitumor Activity,* Sutherland, R. L. and Jordan, V. C., Eds., Academic Press, New York, 1981.
102. **Martin, L.,** Effects of antioestrogens on cell proliferation in the rodent reproductive tract, in *Non-Steroidal Antioestrogens. Molecular Pharmacology and Antitumor Activity,* Sutherland, R. L. and Jordan, V. C., Eds., Academic Press, New York, 1981, chap. 9.
103. **Forsberg, J.-G.,** Treatment with different antiestrogens in the neonatal period and effects in the cervicovaginal epithelium and ovaries of adult mice: a comparison to estrogen-induced changes, *Biol. Reprod.,* 32, 427, 1985.
104. **Taguchi, O. and Nishizuka, Y.,** Reproductive tract abnormalities in female mice treated neonatally with tamoxifen, *Am. J. Obstet. Gynecol.,* 151, 675, 1984.
105. **Adam, H. K.,** A review of the pharmacokinetics and metabolism of "Nolvadex" (Tamoxifen), in *Non-Steroidal Antioestrogens. Molecular Pharmacology and Antitumor Activity,* Sutherland, R. L. and Jordan, V. C., Eds., Academic Press, New York, 1981, chap. 4.
106. **Gorwill, R. H., Steele, H. D., and Sarda, I. R.,** Heterotopic columnar epithelium and adenosis in the vagina of the mouse after neonatal treatment with clomiphene citrate, *Am. J. Obstet. Gynecol.,* 144, 529, 1982.
107. **Clark, J. H. and McCormack, S. A.,** The effect of clomid and other triphenylethylene derivatives during pregnancy and the neonatal period, *J. Steroid Biochem.,* 12, 47, 1980.
108. **Johnson, L. D., Palmer, A. E., King, N. W., Jr., and Hertig, A. T.,** Vaginal adenosis in *Cebus apella* monkeys exposed to DES *in utero, Obstet. Gynecol.,* 57, 629, 1981.
109. **Hendrickx, A. G., Benirschke, K., Thompson, R. S., Ahern, J. K., Lucas, N. E., and Oi, R. H.,** The effects of prenatal diethylstilbestrol (DES) exposure on the genitalia of pubertal *Macaca mulatta.* I. Female offspring, *J. Reprod. Med.,* 22, 233, 1979.
110. **Taguchi, O., Cunha, G. R., and Robboy, S. J.,** Experimental study of the effect of diethylstilbestrol on the development of the human female reproductive tract, *Biol. Res. Pregnancy,* 4, 56, 1983.
111. **Robboy, S. J.,** A hypothetic mechanism of diethylstilbestrol (DES)-induced anomalies in exposed progeny, *Hum. Pathol.,* 14, 831, 1983.
112. **O'Brien, P. C., Noller, K. L., Robboy, S. J., Barnes, A. B., Kaufman, R. H., Tilley, B. C., and Townsend, D. E.,** Vaginal epithelial changes in young women enrolled in the national cooperative diethylstilbestrol adenosis (DESAD) project, *Obstet. Gynecol.,* 53, 300, 1979.

113. **Jefferies, J. A., Robboy, S. J., O'Brien, P. C., Bergstralh, E. J., Labarthe, D. R., Barnes, A. B., Noller, K. L., Hatab, P. A., Kaufman, R. H., and Townsend, D. E.,** Structural anomalies of the cervix and vagina in women enrolled in the diethylstilbestrol adenosis (DESAD) project, *Am. J. Obstet. Gynecol.*, 148, 59, 1984.
114. **Holt, L. H. and Herbst, A. L.,** DES-related female genital changes, *Semin. Oncol.*, 9, 341, 1982.
115. **Scully, R. E. and Welch, W. K.,** Pathology of the female genital tract after prenatal exposure to diethylstilbestrol, in *Developmental Effects of Diethylstilbestrol (DES) in Pregnancy*, Herbst, A. L. and Bern, H. A., Eds., Thieme-Stratton, New York, 1981, chap. 3.
116. **Robboy, S. J., Szyfelbein, W. M., Goellner, J. R., Kaufman, R. H., Taft, P. D., Richard, R. M., Gaffey, T. A., Prat, J., Virata, R., Hatab, P. A., McGorray, S. P., Noller, K. L., Townsend, D., Labarthe, D., and Barnes, A. B.,** Dysplasia and cytologic findings in 4,589 young women enrolled in diethylstilbestrol-adenosis (DESAD) project, *Am. J. Obstet. Gynecol.*, 140, 579, 1981.
117. **Burke, L., Antonioli, D., and Friedman, E. A.,** Evolution of diethylstilbestrol associated genital tract lesions, *Obstet. Gynecol.*, 57, 79, 1981.
118. **Robboy, S. J., Young, R. H., Welch, W. R., Truslow, G. Y., Prat, J., Herbst, A. L., and Scully, R. E.,** Atypical vaginal adenosis and cervical ectropion. Association with clear cell adenocarcinoma in diethylstilbestrol-exposed offspring, *Cancer*, 54, 869, 1984.
119. **Robboy, S. J., Welch, W. R., Young, R. H., Truslow, G. Y., Herbst, A. L., and Scully, R. E.,** Topographic relation of cervical ectropion and vaginal adenosis to clear cell adenocarcinoma, *Obstet. Gynecol.*, 60, 546, 1982b.
120. **Stafl, A. and Mattingly, R. F.,** Vaginal adenosis: a precancerous lesion?, *Am. J. Obstet. Gynecol.*, 120, 666, 1974.
121. **Herbst, A. L., Scully, R. E., and Robboy, S. J.,** Prenatal diethylstilbestrol exposure and human genital tract abnormalities, *Natl. Cancer Inst. Monogr.*, 51, 25, 1979.
122. **Yao, S. F., Reagan, J. W., and Richart, K. M.,** Cytologic diagnosis of diethylstilbestrol-related genital tract changes and evaluation of squamous cell neoplasia, in *Developmental Effects of Diethylstilbestrol (DES) in Pregnancy*, Herbst, A. L. and Bern, H. A., Eds., Thieme-Stratton, New York, 1981, chap. 4.
123. **Robboy, S. J., Noller, K. L., O'Brien, P., Kaufman, R. H., Townsend, D., Barnes, A. B., Gundersen, J., Lawrence, W. D., Bergstrahl, E., McGorray, S., Tilley, B. C., Anton, J., and Chazen, G.,** Increased incidence of cervical and vaginal dysplasia in 3,980 diethylstilbestrol-exposed young women. Experience of the national collaborative diethylstilbestrol adenosis project, *J. Am. Med. Assoc.*, 252, 2979, 1984b.
124. **Gilloteaux, J., Paul, R. J., and Steggles, A. W.,** Upper genital tract anomalies in the Syrian hamster as a result of *in utero* exposure to diethylstilbestrol, *Virchows Arch. A*, 398, 163, 1982.
125. **Viscomi, G. N., Gonzalez, K., and Taylor, K. J. W.,** Ultrasound detection of uterine abnormalities after diethylstilbestrol (DES) exposure, *Radiology*, 136, 733, 1980.
126. **Kent, J.,** Development of the infantile mouse uterus, *J. Reprod. Fertil.*, 43, 367, 1975.
127. **Ostrander, P. L., Mills, K. T., and Bern, H. A.,** Long-term responses of the mouse uterus to neonatal diethylstilbestrol treatment and to later sex hormone exposure, *J. Natl. Cancer Inst.*, 74, 121, 1985.
128. **Forsberg, J.-G., Tenenbaum, A., Rydberg, C., and Sernvi, C.,** Ovarian structure and function in neonatally estrogen-treated female mice, in *Estrogens in the Environment, Vol. 2*, McLachlan, J. A., Ed., Elsevier, New York, 1985, 327.
129. **Mori, T.,** Changes in alkaline phosphatase activity and mitotic rate in vaginal epithelium following estrogen injections in neonatally estrogenized mice, *Annot. Zool. Jpn.*, 40, 82, 1967.
130. **Takasugi, N. and Kimura, T.,** Estrogen sensitivity of vagina and uterus in neonatally estrogenized mice, *Gunma Symp. Endocrinol.*, 4, 185, 1967.
131. **Aihara, M., Kimura, T., and Kato, J.,** Dynamics of the estrogen receptor in the uteri of mice treated neonatally with estrogen, *Endocrinology*, 107, 224, 1980.
132. **Lerner, L. J., Vitale, A., and Oldani, C.,** Response of the immature, pubertal and mature rat to androgen and estrogen after treatment at birth with estradiol benzoate, testosterone propionate or an estrogen antagonist, in *Research on Steroids*, Vol. 7, Vermeulen, A., Klopper, A., Sciarra, F., Jungblut, P., and Lerner, L. J., Eds., North-Holland Press, Amsterdam, 1977.
133. **Hendry, N. J. and Leavitt, W. W.,** Binding and retention of estrogen in the uterus of hamsters treated neonatally with diethylstilbestrol, *J. Steroid Biochem.*, 17, 479, 1982.
134. **Walker, B. E.,** Reproductive tract anomalies in mice after prenatal exposure to DES, *Teratology*, 21, 313, 1980.
135. **Walker, B. E.,** Uterine tumors in old female mice exposed prenatally to diethylstilbestrol, *J. Natl. Cancer Inst.*, 70, 477, 1983.
136. **Baggs, R. B., Miller, R. K., Garman, R., and McKenzie, R. C.,** Teratogenic and neoplastic lesions of the genitourinary system in Wistar rats exposed *in utero* to diethylstilbestrol, *Teratology*, 21, 26A, 1980.
137. **Ennis, B. W. and Davies, J.,** Reproductive tract abnormalities in rats treated neonatally with DES, *Am. J. Anat.*, 164, 145, 1982.

138. **Mori, T.**, Ultrastructure of the uterine epithelium of mice treated neonatally with estrogen, *Acta Anat.*, 99, 462, 1977.
139. **Wordinger, R. J. and Morrill, A.**, Histology of the adult mouse oviduct and endometrium following a single prenatal exposure to diethylstilbestrol, *Virchows Arch. B*, 50, 71, 1985.
140. **Huseby, R. A. and Thurlow, S.**, Effects of prenatal exposure of mice to "low-dose" diethylstilbestrol and the development of adenomyosis associated with evidence of hyperprolactinemia, *Am. J. Obstet. Gynecol.*, 144, 939, 1982.
141. **Napalkov, N. P. and Anisimov, V. N.**, Transplacental effect of diethylstilbestrol in female rats, *Cancer Lett.*, 6, 107, 1979.
142. **Walker, B. E.**, Tumors of female offspring of mice exposed prenatally to diethylstilbestrol, *J. Natl. Cancer Inst.*, 73, 133, 1984.
143. **Greene, R. R., Burrill, M. W., and Ivy, A. C.**, Experimental intersexuality. The effects of estrogens on the antenatal sexual development of the rat, *Am. J. Anat.*, 67, 305, 1970.
144. **Newbold, R. R., Bullock, B. C., and McLachlan, J. A.**, Exposure to diethylstilbestrol during pregnancy permanently alters the ovary and oviduct, *Biol. Reprod.*, 28, 735, 1983.
145. **Newbold, R. R., Tyrey, S., Haney, A. F., and McLachlan, J. A.**, Developmentally arrested oviduct. A structural and functional defect in mice following prenatal exposure to diethylstilbestrol, *Teratology*, 27, 417, 1983.
146. **Henry, E. C., Miller, R. K., and Baggs, R. B.**, Direct fetal injections of diethylstilbestrol and 17β-estradiol: a method for investigating their teratogenicity, *Teratology*, 29, 297, 1984.
147. **Newbold, R. R., Bullock, B. C., and McLachlan, J. A.**, Progressive proliferative changes in the oviduct of mice following developmental exposure to diethylstilbestrol, *Teratogen. Carcinogen. Mutagen.*, 5, 473, 1985.
148. **Döhler, K. D., Hancke, J. L., Srivastava, S. S., Hofmann, C., Shryne, J. E., and Gorski, R. A.**, Participation of estrogens in female sexual differentiation of the brain; neuroanatomical, neuroendocrine and behavioral evidence, *Prog. Brain Res.*, 61, 99, 1984.
149. **Tenenbaum, A. and Forsberg, J.-G.**, Structural and functional changes in ovaries from adult mice treated with diethylstilboestrol in the neonatal period, *J. Reprod. Fertil.*, 73, 465, 1985.
150. **McLachlan, J. A., Newbold, R. R., Shah, H. C., Hogan, M. D., and Dixon, R. L.**, Reduced fertility in female mice exposed transplacentally to diethylstilbestrol (DES), *Fertil. Steril.*, 38, 364, 1982.
151. **Iguchi, T.**, Occurrence or polyovular follicles in ovaries of mice treated neonatally with diethylstilbestrol, *Proc. Jpn. Acad.*, 61(B), 288, 1985.
152. **Newbold, R. R., Carter, D. B., Harris, S. E., and McLachlan, J. A.**, Molecular differentiation of the mouse genital tract: altered protein synthesis following prenatal exposure to diethylstilbestrol, *Biol. Reprod.*, 30, 459, 1984.
153. **Csaba, G.**, Ontogeny and phylogeny of hormone receptors, *Monogr. Dev. Biol.*, 15, 1, 1981.
154. **Mukku, V. R. and Stancel, G. M.**, Regulation of epidermal growth factor receptor by estrogen, *J. Biol. Chem.*, 260, 9820, 1985.
155. **Cunha, G. R., Shannon, J. M., Vanderslice, K. D., Sekkingstad, M., and Robboy, S. J.**, Autoradiographic analysis of nuclear estrogen binding sites during postnatal development of the genital tract of female mice, *J. Steroid Biochem.*, 17, 281, 1982.
156. **Cunha, G. R., Brigsby, R. M., Cooke, P. S., and Sugimura, Y.**, Stromal epithelial interactions in the determination of hormonal responsiveness, in *Estrogens in the Environment, Vol. 2*, McLachlan, J. A., Ed., Elsevier, New York, 1985, 273.
157. **Døskeland, S.-O., Kalland, T., and Forsberg, J.-G.**, Studies on the differentiation pattern and hormonal sensitivity of an antigenic material specific for the cervicovaginal epithelium in fetal and neonatal mice, *Dev. Biol.*, 48, 184, 1976.
158. **Sheehan, D. M., Branham, W. S., Medloch, K. L., Olson, M. E., and Zehr, D. R.**, Uterine responses to estradiol in the neonatal rat, *Endocrinology*, 109, 76, 1981.
159. **Stock, G. and Gorski, J.**, The ontogeny of estrogen responsiveness reexamined: the differential effectiveness of diethylstilbestrol and estradiol on uterine deoxyribonucleic acid synthesis in neonatal rats, *Endocrinology*, 112, 2142, 1983.
160. **Rao, P. N. and Engelberg, J.**, Structural specificity of estrogens in the induction of mitotic chromatid non-disjunction in HeLa cells, *Exp. Cell Res.*, 48, 71, 1967.
161. **Rao, P. N.**, Estradiol induced mitotic delay in HeLa cells: reversal by calcium chloride and putrescine, *Exp. Cell Res.*, 57, 230, 1969.
162. **Sawada, M. and Ishidate, M.**, Colchicine-like effect of diethylstilbestrol (DES) on mammalian cells in vitro, *Mutat. Res.*, 57, 175, 1978.
163. **Hartley-Asp, B., Deinum, J., and Wallin, M.**, Diethylstilbestrol induces metaphase arrest and inhibits microtubule assembly, *Mutat. Res.*, 143, 231, 1985.
164. **Parry, E. M., Danford, N., and Parry, J. M.**, Differential staining of chromosomes and spindle and its use as an assay for determining the effect of diethylstilbestrol on cultured mammalian cells, *Mutat. Res.*, 105, 243, 1982.

165. **Sato, Y., Murai, T., Tsumuraya, M., Saito, H., and Kodama, M.,** Disruptive effect of diethylstilbestrol on microtubules, *Gann,* 75, 1046, 1984.
166. **Sharp, D. C. and Parry, J. M.,** Diethylstilboestrol: the binding and effects of diethylstilboestrol upon the polymerisation and depolymerisation of purified microtubule protein in vitro, *Carcinogenesis,* 6, 865, 1985.
167. **Chrisman, C. L. and Hinkle, L. L.,** Induction of polyploidy in mouse bone marrow cells with diethylstilbestrol-diphosphate, *Can. J. Genet. Cytol.,* 16, 831, 1974.
168. **Tsutsui, T., Maizumi, H., McLachlan, J. A., and Barrett, J. C.,** Aneuploidy induction and cell transformation by diethylstilbestrol: a possible chromosomal mechanism in carcinogenesis, *Cancer Res.,* 43, 3814, 1983.
169. **Birnboim, H. C.,** DNA clastogenic activity of diethylstilbestrol, *Biochem. Pharmacol.,* 34, 3251, 1985.
170. **Tsutsui, T., Degen, G. H., Schiffmann, D., Wong, A., Maizumi, H., McLachlan, J. A., and Barrett, J. C.,** Dependence on exogenous metabolic activation for induction of unscheduled DNA synthesis in Syrian hamster embryo cells by diethylstilbestrol and related compounds, *Cancer Res.,* 44, 184, 1984.
171. **Barrett, J. C., Wong, A., and McLachlan, J. A.,** Diethylstilbestrol induces neoplastic transformation without measurable gene mutation at two loci, *Science,* 212, 1402, 1981.
172. **Liehr, J. G.,** 2-Fluoroestradiol. Separation of estrogenicity from carcinogenicity, *Mol. Pharmacol.,* 23, 278, 1983.
173. **Metzler, M. and McLachlan, J. A.,** Oxidative metabolites of diethylstilbestrol in the fetal, neonatal and adult mouse, *Biochem. Pharmacol.,* 27, 1087, 1978.
174. **Miller, R. K., Hechmann, M. E., and McKenzie, R. C.,** Diethylstilbestrol: placental transfer, metabolism, covalent binding and fetal distribution in the Wistar rat, *J. Pharmacol. Exp. Ther.,* 220, 358, 1982.
175. **Maydl, R. and Metzler, M.,** Oxidative metabolites of diethylstilbestrol in the fetal Syrian golden hamster, *Teratology,* 30, 351, 1984.
176. **Haaf, H. and Metzler, M.,** *In vitro* metabolism of diethylstilbestrol by hepatic, renal and uterine microsomes of rats and hamsters. Effects of different inducers, *Biochem. Pharmacol.,* 34, 3107, 1985.
177. **Liehr, J. G., Randerath, K., and Randerath, E.,** Target organ-specific covalent DNA damage preceding diethylstilbestrol-induced carcinogenesis, *Carcinogenesis,* 6, 1067, 1985.
178. **Liehr, J. G., DaGue, B. B., and Ballafore, A. M.,** Reactivity of 4′4′-diethylstilbestrol quinone, a metabolic intermediate of diethylstilbestrol, *Carcinogenesis,* 6, 829, 1985.

Chapter 5

LONG-TERM EFFECTS OF PERINATAL TREATMENT WITH SEX STEROIDS AND RELATED SUBSTANCES ON REPRODUCTIVE ORGANS OF FEMALE MICE

Takao Mori and Taisen Iguchi

TABLE OF CONTENTS

I. Introduction 64

II. Changes Occurring in the Reproductive Organs and Adrenal Gland 64
- A. Vagina 64
- B. Clitoris 69
- C. Uterus 69
- D. Oviduct 70
- E. Ovary 71
- F. Adrenal 72

III. Evaluation of Animal Models 73

IV. Conclusions 74

Acknowledgments 74

References 74

I. INTRODUCTION

In laboratory rodents, the effects of neonatal treatments with androgens or estrogens on the sex differentiation of the brain have been extensively studied. When administered appropriately to female neonates, the hormones induce permanent functional changes in the hypothalamo-hypophysial axis leading to the anovulatory syndrome associated with persistent vaginal estrous state (see Chapter 1). After the findings of Takasugi et al.[1] that persistent, irreversible cornification of the vaginal epithelium could be induced by neonatal treatment of mice with adequate amounts of estrogen, it was demonstrated that neonatal treatment of mice with estradiol diethylstilbestrol (DES) produces hyperplastic proliferation of the vaginal epithelium, which could result in precancerous or cancerous lesions after more than 10 months.[2-4] Approximately 10 years later, Herbst et al.[5] pointed out that intrauterine exposure of human fetuses to DES (for the purpose of maintaining pregnancy) is frequently associated with a wide range of reproductive tract abnormalities including tumorigenesis in female offspring (see Chapter 10). Since then, neonatal treatment of mice with estrogen has attracted renewed attention as a useful experimental model for the analysis of tumorigenesis induced by transplacental exposure of human fetuses to DES.

The intent of this paper is to survey pathological alterations induced in the reproductive organs by exposure of perinatal female rodents, mice in particular, to sex steroids and related substances and to discuss at some length their relation to human lesions occurring under similar circumstances.

II. CHANGES OCCURRING IN THE REPRODUCTIVE ORGANS AND ADRENAL GLAND

A. Vagina

After perinatal treatment of female mice with adequate doses of estrogens by direct injection or by administration to pregnant mother (transplacental exposure), the most marked changes occurring in the vagina are hyperplastic proliferation of the epithelium associated with extensive cornification. The vaginal changes are either ovary-dependent, due to continuous production of ovarian estrogen as a consequence of permanent functional alteration of the hypothalamo-hypophysial-ovarian system, or ovary- or estrogen-independent autonomous proliferation of the epithelium affected by estrogens given perinatally.[6-10] Differences in vaginal response seem to be caused by a difference in mode of hormone administration, dose of the hormone, or, sometimes, the strain of mice used. In general, higher doses induce the ovary-independent and lower doses lead to the ovary-dependent permanent proliferation-cornification of the vaginal epithelium. There are no marked differences in the histology of the vaginal epithelium according to age, regardless of whether the proliferation-cornification of the epithelium is ovary dependent or independent, provided the ovaries are left *in situ* in the mice with the former type vagina.[7]

The most important finding in mice in the biomedical field is that the mice given perinatal treatment with estrogens show a higher incidence of vaginal lesions at advanced ages as compared with intact controls.[6-13] The vaginal epithelium exhibits marked hyperplastic proliferation and extensive downgrowths accompanied by cysts with a cornified core or with masses of degenerating epithelial cells and cellular debris together with leukocytes and erythrocytes (Figure 1). The epithelial hyperplasias develop into precancerous lesions with epithelial projections and downgrowths extending into the deep stroma and muscle layer through the disrupted basal lamina (Figure 2). The lesions mainly culminate in squamous-cell carcinomas with advance of age (Figure 3). At first, it was thought that there are some differences in histological characteristics of the vaginal lesions between estrogen-treated mice and DES-exposed human females. The human lesions involve the retention of glandular

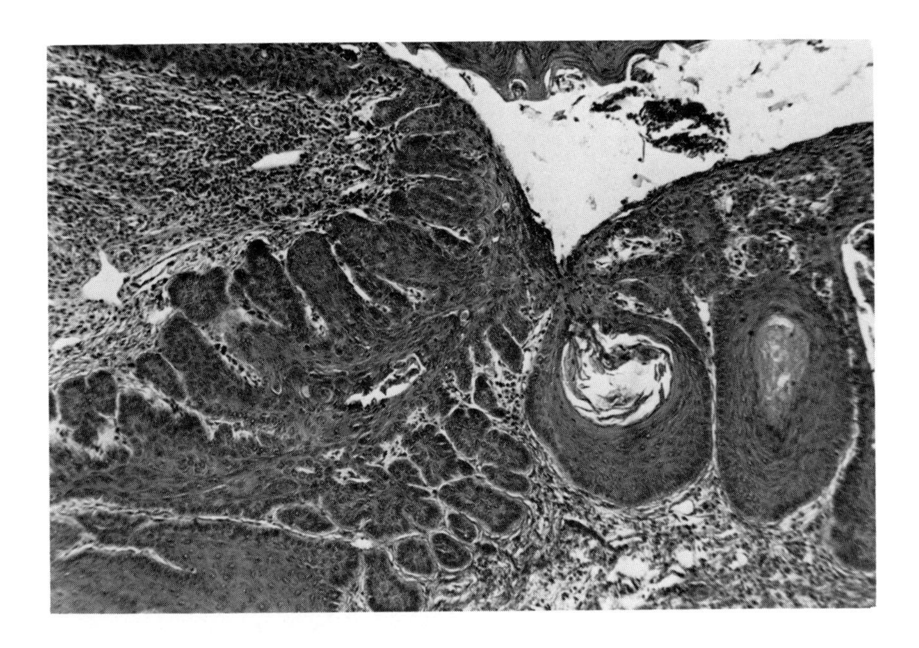

FIGURE 1. Vaginal lesions consisting of marked hyperplastic proliferation and downgrowths of epithelial cells in a 10-month-old BALB/c mouse treated with 20 μg estradiol-17β for 5 days from the day of birth. Cysts bearing pearls, masses of epithelial cells, cellular debris, and accumulation of leukocytes and erythrocytes are visible. (Magnification × 140.)

Müllerian epithelium characteristic of adenosis in which columnar cells are arranged as a single layer in areas of the fornical stratified epithelium or as gland-like structures in the subepithelial stroma. Furthermore, the advanced human lesions are largely classifiable as clear-cell adenocarcinomas, squamous-cell carcinomas being rarely found.[11,14-17] However, accumulating evidence shows that although adenosis occurs in mice treated with estrogens perinatally, the lesioned areas are relatively smaller in mice than in humans.[18-31] In addition, adenocarcinomas, frequently associated with squamous-cell lesions, develop in the mouse vagina, especially in the cervical region.[22,24,27,28,31] In aged mice of the NMRI strain receiving neonatal DES treatment, considerable adenosis-like proliferations occur in the vagina and some of them exhibit histopathological characteristics of adenocarcinoma.[19,27] According to Mori,[31-34] in mice of the C57BL and C3H strains, treatment with 20 μg estradiol for 5 days from the day of birth invariably results in the development of squamous-cell lesions, whereas in BALB/c mice, similar treatment induces adenosis and adenocarcinomas as well (Figures 4 and 5). These findings clearly demonstrate the strain difference in the induction of vaginal lesions of squamous-cell or columnar-cell type. Recently, it has been reported that ovarian estrogen may play an important role in the development of adenosis in prenatally DES-exposed ICR mice.[35] Thus, it is now well established that mouse vaginal lesions more or less resemble human lesions, depending on the dose of estrogen given perinatally, the duration of perinatal exposure to estrogen, the age of sacrifice, and the strain of mice used. Clear-cell adenocarcinoma has not yet been reported in the mouse. On the other hand, it has been reported that atypical squamous epithelium comes to appear in DES-exposed women with benign vaginal adenosis, suggesting that the glandular epithelium in the vagina is gradually replaced by a squamous metaplastic epithelium with the advance of age.[36-38]

Adenosis-like lesions are found in NMRI mice at 2 to 6 months of age and in C57BL mice at 35 days of age, after neonatal treatment with antiestrogens, e.g., tamoxifen.[39-41]

Several lines of evidence provided by mouse experiments demonstrate that estradiol and DES have qualitatively the same effects on the induction of genital tract lesions regardless

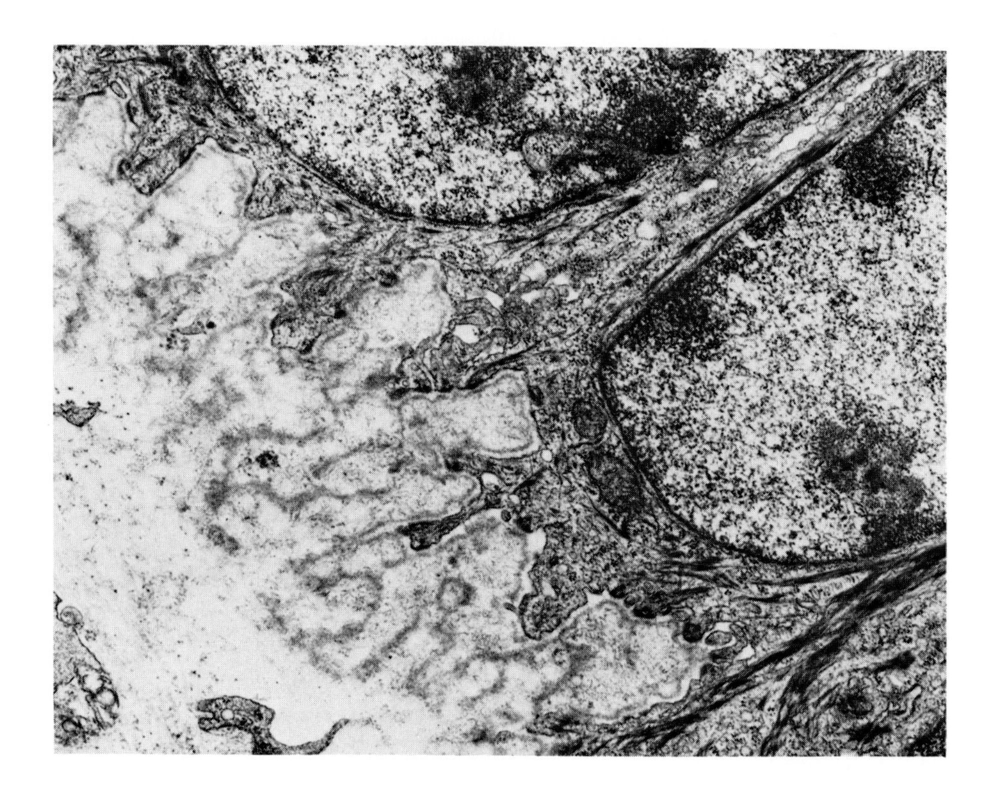

FIGURE 2. Basal region of vagina of a 10-month-old C57BL mouse treated neonatally with 20 μg estradiol-17β for the first 5 days after birth. Note foldings and disruption of basal lamina and cytoplasmic projections of basal cells. (Magnification × 11,800.)

of whether they are given prenatally or neonatally.[42] Administration of estradiol directly to fetuses can also induce persistent vaginal epithelial cornification associated with hyperplastic proliferation.[43]

In mice, the effects of androgens, testosterone, and testosterone propionate appear to be similar in effect to estradiol-17β.[44-48] Treatment with 5α-dihydrotestosterone, a nonaromatizable androgen, has been shown to be effective in this respect[49,50] and 5β-dihydrotestosterone which is biologically inactive on adult reproductive organs also induces persistent vaginal cornification.[51] After transplantation into syngeneic ovariectomized hosts, a majority of neonatal mouse vaginas cultured in a medium containing either estrogens or 5α-dihydrotestosterone exhibited estrogen-independent proliferation and cornification of the epithelium, whereas only 15% of neonatal vaginas transplanted after prior exposure to testosterone in vitro showed a proliferation of the epithelium.[52] These findings suggest that estrogens and nonaromatizable androgens are able to act directly on the vaginal epithelium of newborn mice to cause some permanent changes, while aromatizable androgens such as testosterone act on the epithelium after conversion into estradiol or some related estrogens. Recent studies have demonstrated that aromatase and/or 5α-reductase inhibitors can slightly decrease the incidence of the vaginal changes induced by testosterone.[138] At present, it is not known whether or not different mechanisms are involved in the induction of the irreversible vaginal changes caused by different hormones. Androgen treatment often produces a persistent mucification of the vaginal epithelium associated with irreversible proliferation (Figure 6). In addition, direct administration of testosterone propionate to fetuses is especially effective in the production of vaginal changes.[53]

Neonatal administration of progesterone[22,54,55] and 17α-hydroxyprogesterone caproate[50,56]

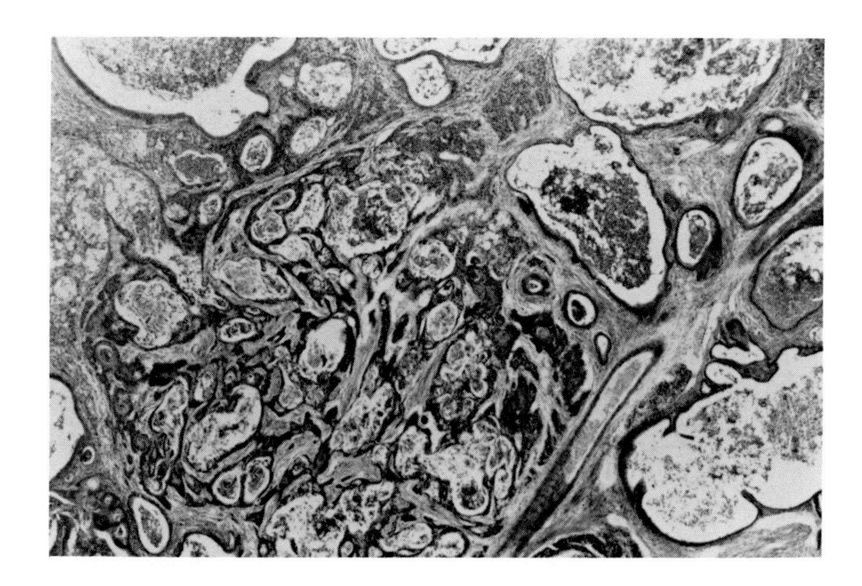

FIGURE 3. Squamous-cell carcinoma of vagina in a 20-month-old BALB/c mouse treated neonatally with 20 μg estradiol-17β for 5 days from the day of birth. (Magnification × 100.)

affects the vagina in a manner comparable to estrogen. If treated neonatally with progesterone alone or progesterone plus estradiol, vaginal lesions including adenocarcinoma may occur in BALB/c mice when they reach about 2 years of age.[22]

It is of interest that treatment with coumestrol, estrogen isolated from the ladino clover, also induces irreversible changes in the mouse vaginal epithelium[57] since this finding suggests the possible involvement of dietary estrogen in the production of human vaginal lesions.[58,59] The predominant dietary estrogens to which humans are exposed are naturally occurring phytoestrogens, rather than synthetic estrogens, e.g., DES.[60] Under certain circumstances, phytoestrogens may reach circulating levels high enough to exert effects on the vaginas in animals ingesting them. A well-known example is hyperestrogenization and subsequent infertility in sheep grazing clover.[60-62] Some fungi produce estrogenic mycotoxins, e.g., zearalenone produced by *Fusarium* fungi infesting stored corn. The "moldy corn syndrome" observed in female pigs fed corn contaminated with *Fusarium* involves vaginal prolapse and infertility.[59] It is worth noting that the presence of coumestrol and other naturally occurring estrogenic substances in food could contribute to potential hazard to humans.[57,59]

The ovary plays a role in the manifestation of the effects of neonatal steroid treatment on the vagina, since neonatal ovariectomy reduces the incidence of vaginal changes.[63,64] Moreover, in neonatally estrogenized, ovariectomized mice, treatment with estradiol enhances vaginal changes.[46,65-69] These findings appear to suggest that girls born of DES-treated mothers had better avoid unnecessary exposure to exogenous estrogen. However, some workers have reported that continuous exposure of neonatally estrogenized mice to further estrogenic stimulation when adult induces no appreciable effect on the vagina or even reduces vaginal changes.[25,70]

It has also been shown that neonatal treatment with estrogen of female mice causes reduction of the vaginal responsiveness to short-term estrogenic stimulation given after maturation,[71] cytosolic and nuclear estrogen binding sites being markedly reduced in the vagina of the mice.[72-74]

Ultrastructural studies on the vaginal epithelium of mice neonatally treated with estrogen have been extensively performed.[67,75-78] During the development of cancerous changes in the vagina, the first sign of morphological disorder is an invasion of affected epithelial cells

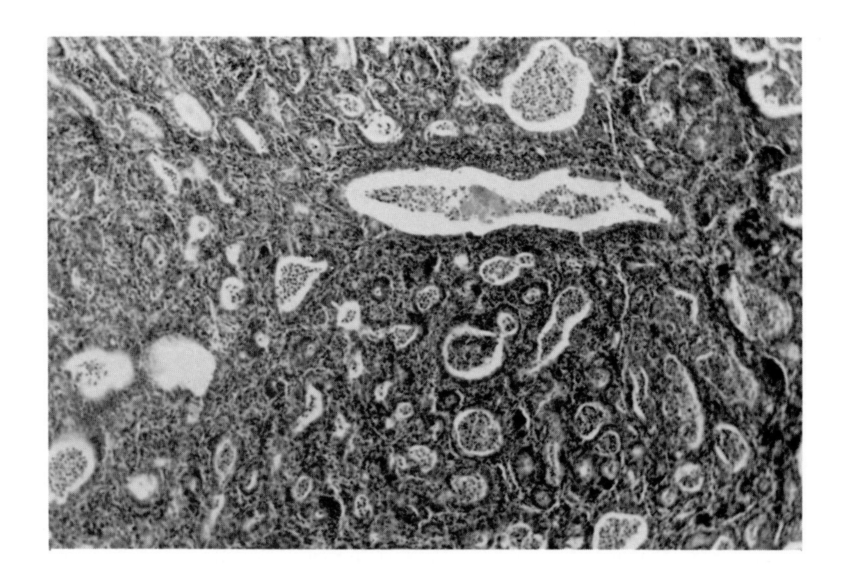

FIGURE 4. Adenocarcinoma of vagina in a 20-month-old BALB/c mouse treated neonatally with 20 μg estradiol-17β for the first 5 days after birth. (Magnification × 100.)

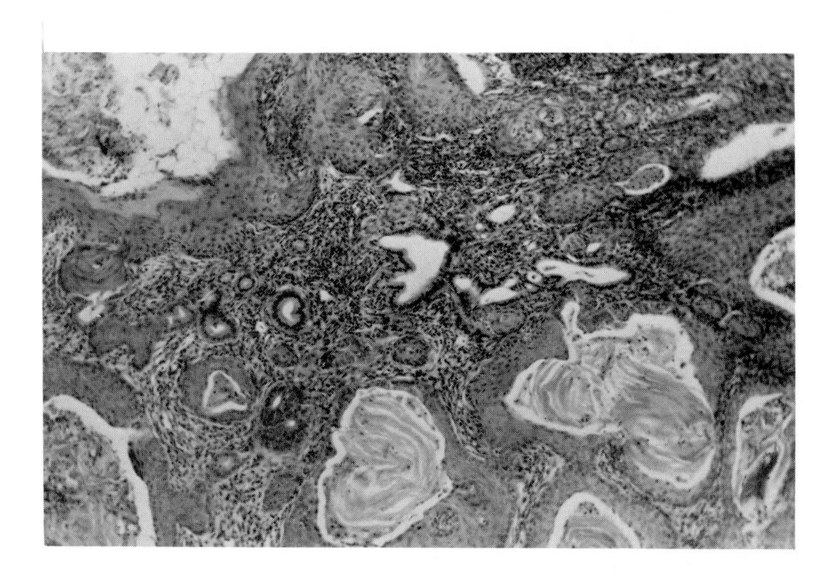

FIGURE 5. Adenosis in hyperplastic lesion of cervix in a 20-month-old BALB/c mouse treated neonatally with 20 μg estradiol-17β for 5 days from the day of birth. Note gland-like structures in the subepithelial stroma. (Magnification × 100.)

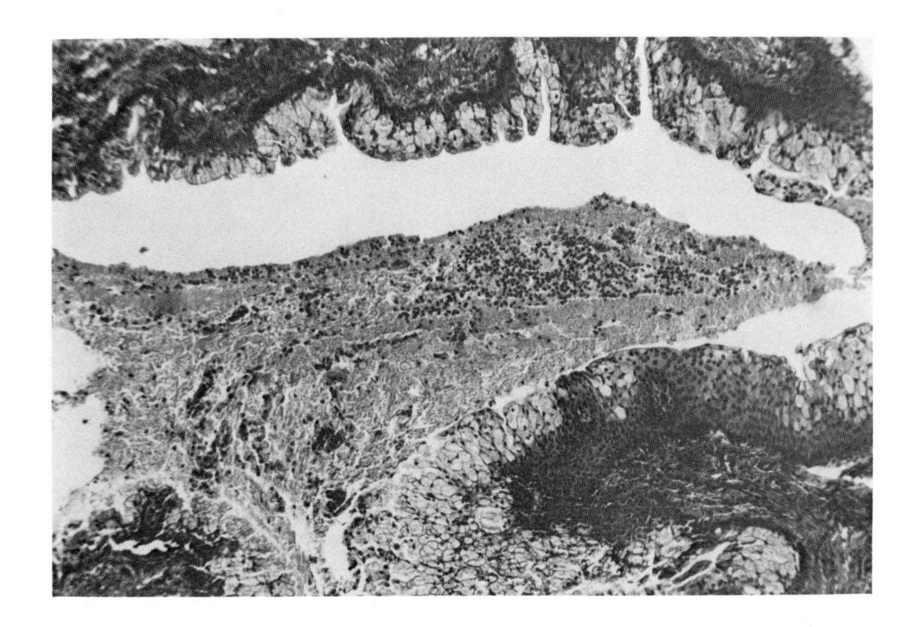

FIGURE 6. Mucification of vaginal epithelium in a 10-month-old C57BL mouse treated neonatally with 20 μg testosterone for 10 days from the day of birth. (Magnification × 100.)

into the stroma through gaps in the basal lamina.[76,78] Marked foldings and disruption of basal lamina become evident with the advance of age (Figure 2). Half desmosomes between epithelial cells and basal lamina and desmosomes between epithelial cells decrease in number by 3 months after treatment with estrogen. Mitochondria are also reduced in number in epithelial cells as they are at advanced ages.[78]

Although primary vaginal cancers rarely cause deaths in women, they still contribute to a significant rise in the death rate of middle-aged females.[11,12] In view of the association of DES treatment of pregnant women and the development of reproductive tract abnormalities in their female offspring and various information obtained from animal models, it is highly desirable that clinical hormone usage, especially perinatal and continuous treatment, be done with great care.

B. Clitoris

The occurrence of clitorial abnormalities such as hypospadia has been reported in mice given prenatal[26,79] or postnatal[4,30,33,80] treatment with estrogens and postnatal treatment with tamoxifen.[40,41] Os clitoridis as well as hypospadia has also been found in female mice treated neonatally with androgen.[49]

C. Uterus

Uteri of mice treated perinatally with estrogens or androgens frequently show cystic endometrial hyperplasia.[6,13] Perinatal treatment with estradiol also results in a high incidence of localized or overall squamous stratification of the uterine epithelium (Figure 7). In the epidermized areas of the endometrium ultrastructural studies reveal that basal cells of the epithelium resemble those of the vaginal epithelium in mice treated neonatally with estrogen.[81]

Decreased responsiveness to the growth-promoting effects of estrogens and the consequent functional disturbances have been reported in the mouse uterus exposed perinatally to natural estrogens or DES.[71,82-85] Uterine hypoplasia associated with suppression of gland formation in mice exposed to DES prenatally is found after maturation of the animals.[24,30,83,85] Similar

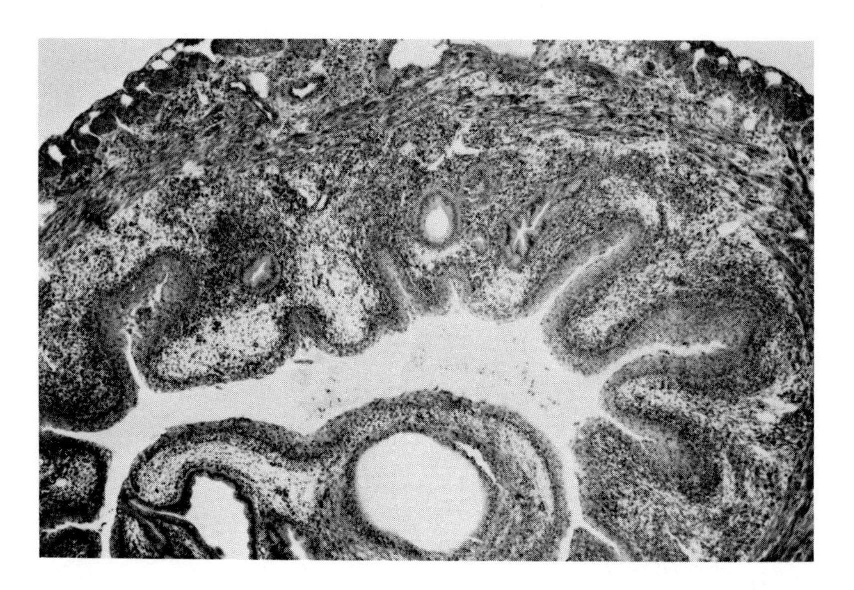

FIGURE 7. Metaplasia of uterine epithelium of a 15-month-old BALB/c mouse treated neonatally with 20 μg estradiol-17β for the first 10 days after birth. Note stratification of epithelium. (Magnification × 100.)

lesions, involving suppression of uterine growth, secretory function, and cellular differentiation, are noted in women prenatally exposed to DES.[86] Tensile strength (rigidity) of the uterus is also decreased in DES-exposed mice. Cytosolic and nuclear estrogen receptor sites are markedly reduced.[72,87] Since the binding affinity of estrogen receptors in such animals is not different from that in the controls, this reduction must be due to quantitative rather than qualitative changes in estrogen receptor protein.[87] Uterine growth associated with genesis of uterine glands is also suppressed in mice treated neonatally with tamoxifen.[41]

Prenatal DES exposure of CD-1 mice has been reported to induce a high incidence of uterine adenocarcinoma.[28] However, it has recently been shown that the development of adenomatous uterine hyperplasia characterized by an increase of uterine size and number of glands is suppressed in BALB/c mice neonatally treated with estrogen when compared to the intact controls.[31]

On the other hand, it has been reported that prenatal exposure of mice to estrogens induces uterine adenomyosis,[24,28,88] in which the endometrial glands accompanying stromal fibroblasts invade the myometrium and connective tissue space between the inner circular and outer longitudinal smooth muscle layers. In contrast, it is also stated that neonatal treatment of female mice with estradiol inhibits the induction of adenomyosis, which is of common occurrence in the age-matched old normal controls.[31] Therefore, it seems likely that impaired responsiveness of the uterus to the action of estrogens may be responsible for the decreased incidence of adenomatous hyperplasia and adenomyosis. There are two possible explanations for these different results of perinatal treatment with estrogens: (1) responsiveness of the uterine endometrium to estrogens is different between neonatal and prenatal mice (neonatal treatment with estrogen directly affects the endometrium leading to squamous metaplasia, while prenatal treatment results in the continuous exposure of the endometrium to ovarian estrogen due to altered functions of the hypothalamo-hypophysial-ovarian axis) and (2) strain difference in sensitivity to estrogen.

D. Oviduct

The most interesting histological change in the oviduct of old mice treated perinatally

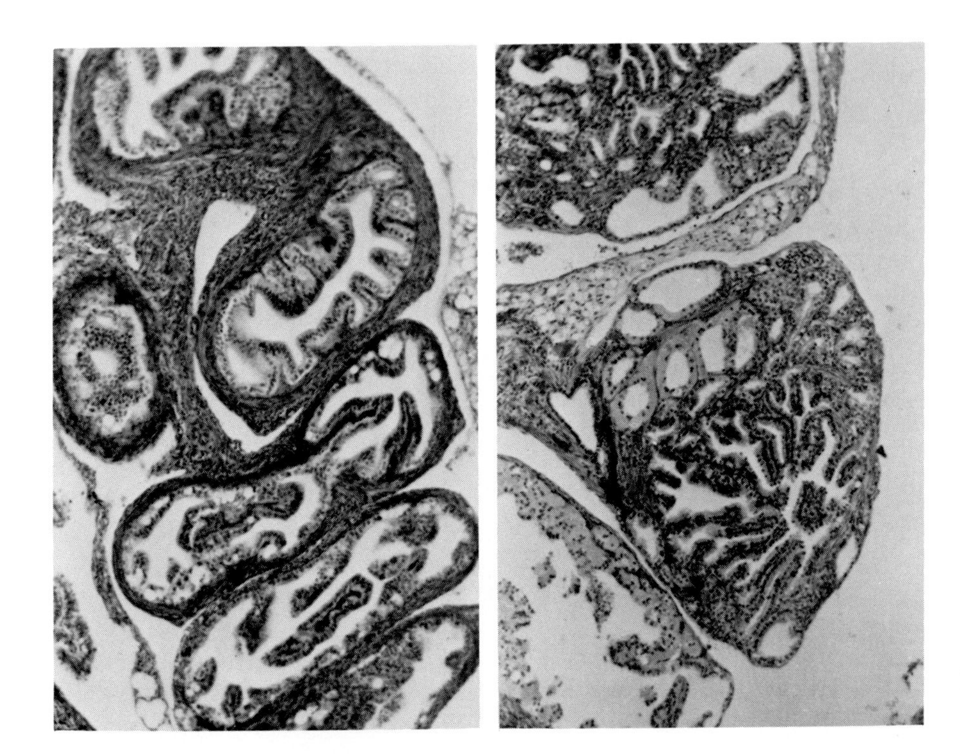

FIGURE 8. Oviduct of a 20-month-old BALB/c mouse. (Left) Mouse treated neonatally with oil vehicle for 5 days from the day of birth. (Right) Mouse treated neonatally with 20 μg estradiol-17β for 5 days from the day of birth. Note epithelial hyperplasia and growth of mucosal epithelium through the muscularis. (Magnification × 230.)

with estrogens is a marked epithelial hyperplasia with gland-like structures extending into the thin muscle wall of the oviducts.[31,89-91] The serosal surface of the oviducts is irregular shaped with polyps containing tubular epithelial projections (Figure 8). Mesonephric remnants are also incorporated into the stroma of the oviduct. Although adenomatous folds and projections from the ducts are markedly developed in and through the wall of the oviductal muscularis, the lesions do not spread to other organs with no metastases being encountered.[89,90]

Oviductal malformations, similar to those found in mice receiving perinatal estrogen treatment, have been reported in young women exposed to DES prenatally. Clinical studies also reveal a high incidence of sterility and an increased frequency of ectopic pregnancy in DES-exposed women.[92-95]

E. Ovary

The ovaries of mice treated perinatally with estrogens or androgens contain vesicular follicles of varying sizes but lack corpora lutea (anovulatory syndrome). The interstitial tissue undergoes marked hypertrophy, showing medullary tubule-like structures and a large amount of ceroid deposition. In addition, ovarian inflammations and intraovarian and paraovarian cysts derived from mesonephric remnants are occasionally observed.[4,7,90] Polyovular follicles frequently occur in the ovaries of mice exposed neonatally to estrogen, aromatizable androgen (but not nonaromatizable androgen), or progestin (Figure 9).[91,96-98] Natural and synthetic estrogen is able to induce a higher incidence of polyovular follicles than is aromatizable androgen, suggesting the role of the conversion of aromatizable androgen to estrogen in the formation of polyovular follicles. Polyovular follicles are also found in

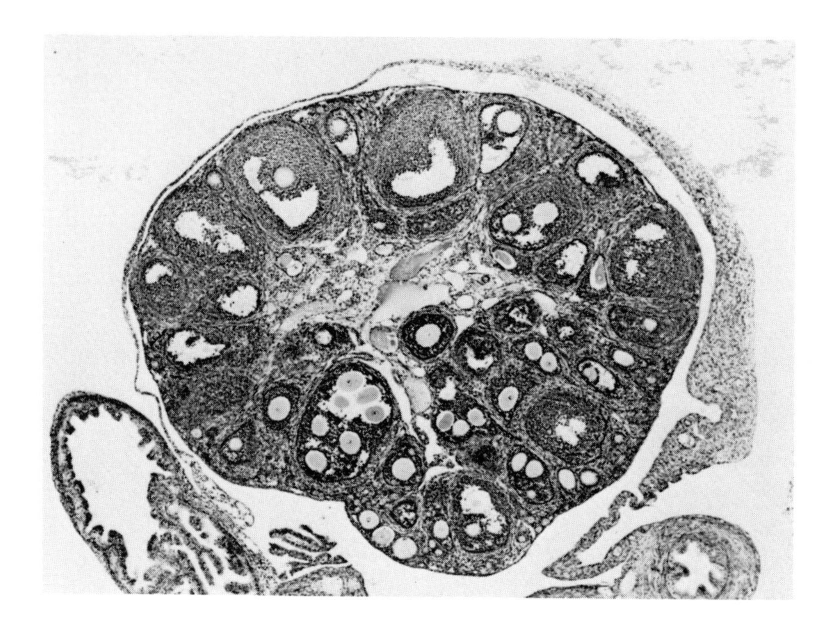

FIGURE 9. Polyovular follicle of ovary of 30-day-old ICR mouse whose mother was given 2000 μg DES for 4 days from day 15 of gestation. (Magnification × 205.)

ovaries of mice given neonatal injections of coumestrol or zearalenone.[99] Prenatal exposure of mice to DES also induces a higher incidence of polyovular follicles than in untreated controls.[100] Primordial and mature polyovular follicles with unequal sized oocytes are found in ovaries of DES-exposed mice. Polyovular follicles gradually increase in number from 5 to 30 days of age. It seems likely that polyovular follicles may result from the fusion of uniovular primordial follicles and/or from enclosure of more than one oocyte during the process of formation of primordial follicles.

Recently it has been reported that more than 80% of oocytes in small follicles undergo degeneration in mice given neonatal injections of antiestrogen, tamoxifen, much like well-known oocyte degeneration in the ovaries of hypophysectomized animals. In addition, follicles of tamoxifen-exposed mice fail to luteinize in response to human chorionic gonadotropin.[41] These findings may imply that the hypothalamo-hypophysial-ovarian system is deteriorated by neonatal tamoxifen treatment.

It is postulated that the development of medullary tubule-like structures in the interstitial tissue of the ovaries of aged DES-exposed mice may be an outcome of the direct effect of DES on ovarian development.[91] In contrast, signs of ovarian aging, e.g., a decrease in responsiveness to circulating gonadotropins, occur later in neonatally estrogen-treated mice than in the normal intact controls.[101] These may be less direct effects of neonatal estrogen treatment on the ovaries. Ovarian tumors, e.g., granulosa cell tumors, are frequently encountered in aged mice perinatally exposed to estrogens (Figure 10).[3,24,31] It has not yet been determined whether the ovarian tumorigenesis results from the abnormal endocrine environment caused by hypothalamic lesions or by direct effects of estrogen given neonatally on the ovaries. Incidence of ovarian tumors in DES-exposed women should be studied and compared to that encountered in unexposed women.

F. Adrenal

The adrenal cortex also seems to be affected by neonatal hormone treatment. The most marked change is earlier ceroidogenesis in the cortex in neonatally estrogenized mice than in the controls.[102] However, in neonatally estrogenized mice incidence of adrenocortical

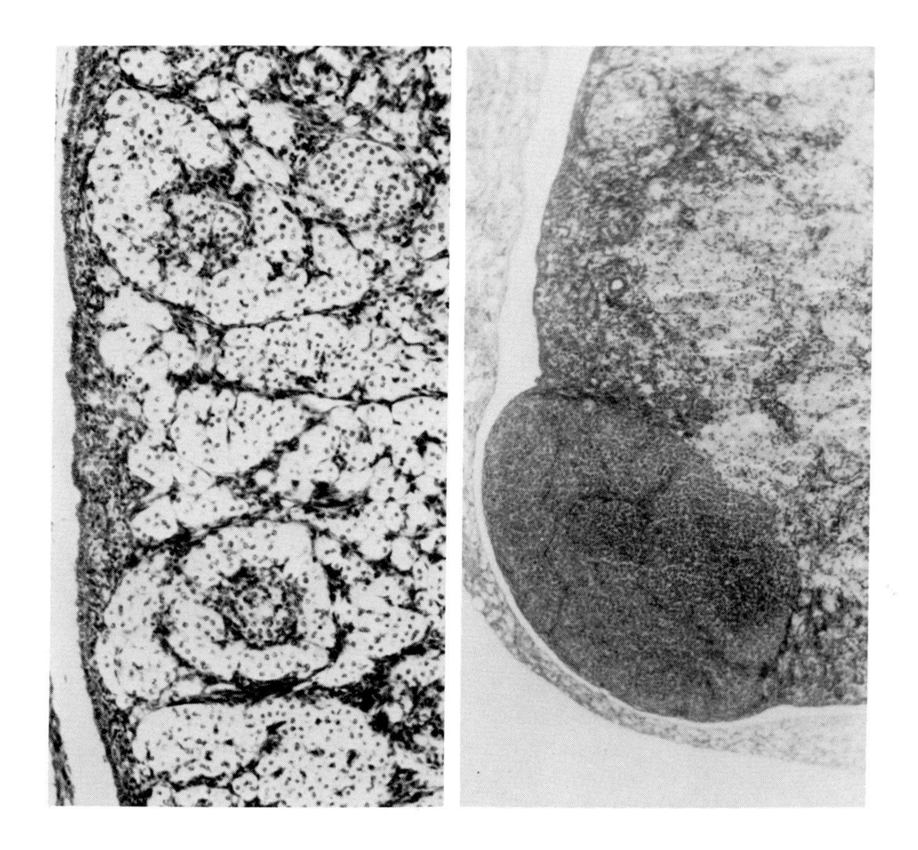

FIGURE 10. Ovaries of mice treated neonatally with estradiol-17β. (Left) Ovary of a 15-month-old C3H mouse treated neonatally with 0.5 μg estradiol-17β for 7 days from the day of birth. Note downgrowths from surface epithelium and markedly hypertrophied interstitial tissue. (Magnification × 140.). (Right) Ovary of a 20-month-old BALB/c mouse treated neonatally with 20 μg estradiol-17β for 5 days from the day of birth. Note small granulosa cell tumor (bottom of the figure) and hypertrophy of interstitial tissue with ceroid deposition. (Magnification × 100.)

nodular hyperplasia after ovariectomy is lower than in the intact controls.[25] Whether the neonatally administered hormones primarily act on the hypothalamo-hypophysial-adrenal axis or on the adrenocortical level is not known. Circulating levels of corticosterone are found to be in the normal range,[103] even if pathological changes have occurred in adrenocortical structure (ceroid deposition, etc.).[32,33,102,104]

III. EVALUATION OF ANIMAL MODELS

It is well established that neonatal treatments with estrogens or androgens also induce pathological changes in the reproductive tracts of rats[105-111] and hamsters.[112-114] Cherry and Glucksmann[107] have demonstrated in rats of the Lister strain that one or two daily injections of 70 μg estradiol monobenzoate or 1.25 mg of testosterone propionate from the day of birth resulted in sarcomas, papillomas, or carcinomas in the cervico-vaginal tract when the rats reached more than 7 months of age. Rustia and Shubik[112] have shown that pregnant hamsters treated with 20 or 40 mg DES per kg body weight by intragastric intubation on days 14 and 15 of gestation developed hyperplastic and neoplastic lesions in the reproductive tract in a majority of their female progeny. The females bearing the tumors were in a state of continuous estrus. Vaginal ridging and vaginal adenosis were observed in some pubertal female rhesus monkeys exposed prenatally to 1 mg/day DES.[115]

On the other hand, long-term intermittent or continuous post-pubertal treatment with estrogens causes various types of hyperplastic lesions and cancers in the reproductive organs of laboratory animals. In this respect, the duration of estrogen treatment is more important than the range of dosages used.[116-118] In addition, a number of recent studies have demonstrated a consistent, strongly positive association between exposure to estrogenic substances and development of climacteric symptoms as well as endometrial cancer, providing evidence for a positive dose-response relationship.[119-121] Furthermore, it is reported that the sequential, cyclic use of estrogens and progestins as oral contraceptives may induce an increase in the incidence of breast[122] and endometrial cancer.[123-127] Although the hormonal mechanisms involved in the development of human cancers are not yet well understood,[128,129] it is thought that long-term estrogenic stimulation may be an important etiological factor for the development of some cancers in human reproductive organs.

In mice, neonatal administration of vitamin A together with estrogens or androgens can prevent irreversible ovary-independent vaginal lesions.[130-136] The development of ovary-dependent changes of the vaginal epithelium is suppressed by early ovariectomy. In addition, instillment of progesterone into the vaginal lumen of neonatally estrogenized mice can inhibit the occurrence of permanent cornification and lesions of the vaginal epithelium.[137] Similar local therapies may be applicable for preventing lesions in humans.

IV. CONCLUSIONS

Although the underlying mechanisms for the genesis of lesions and cancers in the reproductive organs in animals perinatally exposed to estrogens or androgens have not been thoroughly worked out, accumulating evidence undoubtedly contributes to the understanding of the causation of lesions occurring in the reproductive organs of women prenatally exposed to DES.

ACKNOWLEDGMENTS

The authors wish to express their cordial thanks to Emeritus Professor K. Takewaki of the University of Tokyo for his kind criticism and help in preparing the manuscript. This work was supported by Grants-in-Aid for Fundamental Scientific Research from the Ministry of Education, Science and Culture, Japan.

REFERENCES

1. **Takasugi, N., Bern, H. A., and DeOme, K. B.,** Nature of the vaginal cornification in persistent-estrous mice, *Science,* 138, 438, 1962.
2. **Takasugi, N.,** Vaginal cornification in persistent-estrous mice, *Endocrinology,* 72, 607, 1963.
3. **Dunn, T. B. and Green, A. W.,** Cyst of the epididymis, cancer of the cervix, granular cell myoblastoma, and other lesions after estrogen injection in newborn mice, *J. Natl. Cancer Inst.,* 31, 425, 1963.
4. **Takasugi, N. and Bern, H. A.,** Tissue changes in mice with persistent vaginal cornification induced by early postnatal treatment with estrogen, *J. Natl. Cancer Inst.,* 33, 855, 1964.
5. **Herbst, A. L., Ulfelder, H., and Poskanzer, D. C.,** Adenocarcinoma of the vagina, association of maternal stilbestrol therapy with tumor appearance in young women, *N. Engl. J. Med.,* 284, 878, 1971.
6. **Takasugi, N., Kimura, T., and Mori, T.,** Irreversible changes in mouse vaginal epithelium induced by early postnatal treatment with steroid hormones, in *The Post-Natal Development of Phenotype,* Kazda, S. and Denenberg, V. H., Eds., Academia, Prague, 1970, 229.
7. **Takasugi, N.,** Cytological basis for permanent vaginal changes in mice treated neonatally with steroid hormones, *Int. Rev. Cytol.,* 44, 193, 1976.

8. **Bern, H. A.,** The neonatal mouse — tumorigenesis after short-term exposure to hormones and its possible relevance to human syndrome, in Proc. Symp. Endocrine-Induced Neoplasia, Omaha, October 21 to 22, 1976, 31.
9. **Kohrman, A. F.,** The newborn mouse as a model for study of the effects of hormonal steroids in the young, *Pediatrics,* 62 (Suppl.), 1143, 1978.
10. **Forsberg, J.-G.,** Induction of conditions leading to cancer in the genital tract by estrogen during the differentiation phase of the genital epithelium, in *Hormones and Embryonic Development, Advances in the Biosciences,* Vol. 13, Raspe, G., Ed., Pergamon Press, Oxford, 1973, 139.
11. **Herbst, A. L. and Bern, H. A.,** *Developmental Effects of Diethylstilbestrol (DES) in Pregnancy,* ThiemeStratton, New York, 1981.
12. **Mori, T. and Takasugi, N.,** Cervico-vaginal tumors in humans and laboratory animals, in *Hormone Related Tumors,* Nagasawa, H. and Abe, K., Eds., Springer-Verlag, Berlin, 1981, 227.
13. **McLachlan, J. A. and Fabro, S. E.,** Altered postnatal development following intrauterine exposure to hormonally active chemicals, in *Advances in Pharmacology and Therapeutics II,* Vol. 5, Yoshida, H., Hagihara, Y., and Ebashi, S., Eds., Pergamon Press, Oxford, 1982, 211.
14. **Robboy, S. J., Scully, R. E., Welch, W. R., and Herbst, A. L.,** Intrauterine diethylstilbestrol exposure and its consequences, *Arch. Pathol. Lab. Med.,* 101, 1, 1977.
15. **Herbst, A. L., Scully, R. E., and Robboy, S. J.,** Prenatal diethylstilbestrol exposure and human genital tract abnormalities, *Natl. Cancer Inst. Monogr.,* 51, 25, 1979.
16. **Bibbo, M.,** Transplacental effects of diethylstilbestrol, in *Current Topics in Pathology,* Vol. 66, Grundmann, E. and Kirsten, W. H., Eds., Springer-Verlag, Berlin, 1979, 191.
17. **Shapiro, S. and Slone, D.,** The effects of exogenous female hormones on the fetus, *Epidemiol. Rev.,* 1, 110, 1979.
18. **Forsberg, J.-G.,** Estrogen, vaginal cancer, and vaginal development, *Am. J. Obstet. Gynecol.,* 113, 83, 1972.
19. **Forsberg, J.-G.,** Late effects in the vaginal and cervical epithelia after injections of diethylstilbestrol into neonatal mice, *Am. J. Obstet. Gynecol.,* 121, 101, 1975.
20. **Yasuda, Y., Kihara, T., and Nishimura, H.,** Transplacental effect of ethinyl estradiol on mouse vaginal epithelium, *Dev. Growth Differ.,* 19, 241, 1977.
21. **Kalland, T., Fossberg, T. M., and Forsberg, J.-G.,** Localization of ^{3}H-estradiol-17β in diethylstilbestrol-induced adenosis, *Obstet. Gynecol.,* 51, 464, 1978.
22. **Jones, L. A. and Bern, H. A.,** Cervicovaginal and mammary gland abnormalities in BALB/cCrgl mice treated neonatally with progesterone and estrogen, alone or in combination, *Cancer Res.,* 39, 2560, 1979.
23. **Plapinger, L. and Bern, H. A.,** Adenosis-like lesions and other cervicovaginal abnormalities in mice treated perinatally with estrogen, *J. Natl. Cancer Inst.,* 63, 507, 1979.
24. **McLachlan, J. A., Newbold, R. R., and Bullock, B. C.,** Long-term effects on the female mouse genital tract associated with prenatal exposure to diethylstilbestrol, *Cancer Res.,* 40, 3988, 1980.
25. **Kawashima, S., Mori, T., Kimura, T., Arai, Y., and Nishizuka, Y.,** Effects of estrogen treatment on persistent hyperplastic lesions of the vagina in neonatally estrogenized mice, *Endocrinol. Jpn.,* 27, 533, 1980.
26. **Walker, B. E.,** Reproductive tract anomalies in mice after prenatal exposure to DES, *Teratology,* 21, 313, 1980.
27. **Forsberg, J.-G. and Kalland, T.,** Neonatal estrogen treatment and epithelial abnormalities in the cervicovaginal epithelium of adult mice, *Cancer Res.,* 41, 721, 1981.
28. **Walker, B. E.,** Uterine tumors in old female mice exposed prenatally to diethylstilbestrol, *J. Natl. Cancer Inst.,* 70, 477, 1983.
29. **Iguchi, T., Ostrander, P. L., Mills, K. T., and Bern, H. A.,** Induction of abnormal epithelial changes by estrogen in neonatal mouse vaginal transplants, *Cancer Res.,* 45, 5688, 1985.
30. **Iguchi, T., Takase, M., and Takasugi, N.,** Development of vaginal adenosis-like lesions and uterine epithelial stratification in mice exposed perinatally to diethylstilbestrol, *Proc. Soc. Exp. Biol. Med.,* 181, 59, 1986.
31. **Mori, T.,** Abnormalities in the reproductive system of aged mice after neonatal estradiol exposure, *J. Endocrinol. Invest.,* 9, 397, 1986.
32. **Mori, T.,** Changes in reproductive organs and some other glands in old C3H/MS mice treated neonatally with low doses of estrogen, *Annot. Zool. Jpn.,* 41, 43, 1968.
33. **Mori, T.,** Changes in the reproductive and some other organs in old C3H/MS mice given high dose estrogen injections during neonatal life, *Annot. Zool. Jpn.,* 41, 85, 1968.
34. **Mori, T.,** Development of hyperplastic lesions and changes in phosphatase activity in vaginal epithelium in neonatally estrogenized mice, *Annot. Zool. Jpn.,* 42, 133, 1969.
35. **Iguchi, T., Takei, T., Takase, M., and Takasugi, N.,** Estrogen participation in induction of cervicovaginal adenosis-like lesions in immature mice exposed prenatally to diethylstilbestrol, *Acta Anat.,* 127, 110, 1986.

36. **Stafl, A. and Mattingly, R. F.,** Vaginal adenosis: a precancerous lesions? *Am. J. Obstet. Gynecol.*, 120, 666, 1974.
37. **Fetherston, W. C.,** Squamous neoplasia of vagina related to DES syndrome, *Am. J. Obstet. Gynecol.*, 122, 176, 1975.
38. **Mattingly, R. F. and Stafl, A.,** Cancer risk in diethylstilbestrol-exposed offspring, *Am. J. Obstet. Gynecol.*, 126, 543, 1976.
39. **Forsberg, J.-G.,** Treatment with different antiestrogens in the neonatal period and effects in the cervicovaginal epithelium and ovaries of adult mice; a comparison to estrogen-induced changes, *Biol. Reprod.*, 32, 427, 1985.
40. **Taguchi, O. and Nishizuka, Y.,** Reproductive tract abnormalities in female mice treated neonatally with tamoxifen, *Am. J. Obstet. Gynecol.*, 151, 675, 1985.
41. **Iguchi, T., Hirokawa, M., and Takasugi, N.,** Occurrence of genital tract abnormalities and bladder hernia in female mice exposed neonatally to tamoxifen, *Toxicology*, 42, 1, 1986.
42. **McLachlan, J. A.,** Prenatal exposure to diethylstilbestrol in mice: toxicological studies, *J. Toxicol. Environ. Health*, 2, 527, 1977.
43. **Kimura, T.,** Persistent vaginal cornification in mice treated with estrogen prenatally, *Endocrinol. Jpn.*, 22, 497, 1975.
44. **Takasugi, N.,** Persistent changes in vaginal epithelium of ovariectomized mice induced by early postnatal injections of testosterone or progesterone, *J. Fac. Sci. Univ. Tokyo Sec. 4*, 10, 397, 1964.
45. **Kimura, T. and Nandi, S.,** Nature of induced persistent vaginal cornification in mice. IV. Changes in the vaginal epithelium of old mice treated neonatally with estradiol or testosterone, *J. Natl. Cancer Inst.*, 39, 75, 1967.
46. **Bern, H. A., Jones, L. A., Mills, K. T., Kohrman, A., and Mori, T.,** Use of the neonatal mouse in studying long-term effects of early exposure to hormones and other agents, *J. Toxicol. Environ. Health*, Suppl. 1, 103, 1976.
47. **Ohta, Y. and Iguchi, T.,** Development of the vaginal epithelium showing estrogen-independent proliferation and cornification in neonatally androgenized mice, *Endocrinol. Jpn.*, 23, 333, 1976.
48. **Iguchi, T. and Ohta, Y.,** Electron-microscopic study on the development of permanently proliferated and cornified vaginal epithelium in mice treated neonatally with androgen, *Acta Anat.*, 108, 469, 1980.
49. **Iguchi, T. and Takasugi, N.,** Occurrence of permanent changes in vaginal and uterine epithelia in mice treated neonatally with progestin, estrogen and aromatizable or non-aromatizable androgens, *Endocrinol. Jpn.*, 23, 327, 1976.
50. **Bern, H. A., Mills, K. T., Ostrander, P. L., Schoenrock, B., Graveline, B., and Plapinger, L.,** Cervicovaginal abnormalities in BALB/c mice treated neonatally with sex hormones, *Teratology*, 30, 267, 1984.
51. **Yanai, R., Mori, T., and Nagasawa, H.,** Long-term effects of prenatal and neonatal administration of 5β-dihydrotestosterone on normal and neoplastic mammary development in mice, *Cancer Res.*, 37, 4456, 1977.
52. **Iguchi, T.,** Effects of sex hormones on neonatal mouse vaginal epithelium *in vitro*, *Proc. Jpn. Acad.*, B60, 414, 1984.
53. **Taguchi, O., Nishizuka, Y., and Takasugi, N.,** Irreversible lesions in female reproductive tracts of mice after prenatal exposure to testosterone propionate, *Endocrinol. Jpn.*, 24, 385, 1977.
54. **Jones, L. A. and Bern, H. A.,** Long-term effects of neonatal treatment with progesterone, alone and in combination with estrogen, on the mammary gland and reproductive tract of female BALB/cfC3H mice, *Cancer Res.*, 37, 67, 1977.
55. **Jones, L. A. and Pacillas-Verjan, R.,** Transplantability and sex steroid hormone responsiveness of cervicovaginal tumors derived from female BALB/cCrgl mice neonatally treated with ovarian steroids, *Cancer Res.*, 39, 2591, 1979.
56. **Ainslie, M. and Kohrman, A. F.,** The effect of 17β-hydroxyprogesterone caproate on vaginal development in mice: a preliminary report, *J. Toxicol. Environ. Health*, 3, 339, 1977.
57. **Burroughs, C. D., Bern, H. A., and Robert Stokstad, E. L.,** Prolonged vaginal cornification and other changes in mice treated neonatally with coumestrol, a plant estrogen, *J. Toxicol. Environ. Health*, 15, 51, 1985.
58. **Verdeal, K. and Ryan, D. S.,** Naturally-occurring estrogens in food stuffs, *J. Food Prot.*, 44, 577, 1979.
59. **McLachlan, J. A., Korach, K. S., Newbold, R. R., and Degen, G. H.,** Diethylstilbestrol and other estrogens in the environment, *Fund. Appl. Toxicol.*, 4, 686, 1984.
60. **Stob, M.,** Naturally occurring food toxicants: estrogens, in *CRC Handbook of Naturally Occurring Food Toxicants*, Recheige, M., Ed., CRC Press, Boca Raton, Fla., 1983, 81.
61. **Adams, N. R.,** Pathological changes in the tissues of infertile ewes with clover disease, *J. Comp. Pathol.*, 86, 29, 1976.
62. **Adams, N. R.,** A changed responsiveness to oestrogen in ewes with clover disease, *J. Reprod. Fertil. Suppl.*, 30, 223, 1981.

63. **Takasugi, N.**, Effect of estrogen on vaginal epithelium of ovariectomized adult mice receiving early postnatal injections of estrogen, *J. Fac. Sci. Univ. Tokyo Sec. 4,* 10, 403, 1964.
64. **Iguchi, T.**, Ovarian influence on mitosis and alkaline phosphatase activity in mouse vaginal epithelium permanently affected by neonatal injections of 5α-dihydrotestosterone, *Proc. Jpn. Acad.*, 52, 579, 1976.
65. **Kohrman, A. F. and Greenberg, R. E.**, Permanent effects of estradiol on cellular metabolism of the developing mouse vagina, *Dev. Biol.*, 18, 632, 1968.
66. **Mori, T.**, Long-lasting effect of estrogen injections on alkaline phosphatase activity in the vaginal epithelium of neonatally estrogenized adult mice, *Proc. Jpn. Acad.*, 44, 516, 1968.
67. **Mori, T.**, Ultrastructural characteristics of the vaginal epithelium of neonatally estrogenized mice in response to subsequent estrogen treatment, *Endocrinol. Jpn.*, 23, 341, 1976.
68. **Iguchi, T.**, Mitotic activity in vaginal epithelium in neonatally androgenized mice following estrogen administration, *Proc. Jpn. Acad.*, 53, 113, 1977.
69. **Mori, T. and Nishizuka, M.**, Additional effects of postpuberal estrogen injections on the vaginal epithelium in neonatally estrogenized mice, *Acta Anat.*, 100, 369, 1978.
70. **Wong, L. M., Bern, H. A., Jones, L. A., and Mills, K. T.**, Effect of later treatment with estrogen on reproductive tract lesions in neonatally estrogenized female mice, *Cancer Lett.*, 17, 115, 1982.
71. **Takasugi, N. and Kimura, T.**, Estrogen sensitivity of vagina and uterus in neonatally estrogenized mice, *Gunma Symp. Endocrinol.*, 4, 185, 1967.
72. **Shyamala, G., Mori, T., and Bern, H. A.**, Nuclear and cytoplasmic oestrogen receptors in vaginal and uterine tissue of mice treated neonatally with steroids and prolactin, *J. Endocrinol.*, 63, 275, 1974.
73. **Carlton, B. D.**, Ontogeny of Estrogen and Progesterone Receptors in Mouse Uterus and Vagina: Influence of Diethylstilbestrol Treatment During the Neonatal Period, Ph.D. thesis, University of Cincinnati, Cincinnati, 1977.
74. **Bern, H. A., Edery, M., Mills, T. M., Kohrman, A. F., Mori, T., and Larson, L.**, Longterm alterations in histology and steroid receptor levels of the genital tract and mammary gland following neonatal exposure of female BALB/cCrgl mice to various doses of diethylstilbestrol, *Cancer Res.*, 47, 4165, 1987.
75. **Takasugi, N. and Kamishima, Y.**, Development of vaginal epithelium showing irreversible proliferation and cornification in neonatally estrogenized mice: an electron microscope study, *Dev. Growth Differ.*, 15, 127, 1973.
76. **Kvinnsland, S. and Åbro, A.**, Epithelial ultrastructure and the distribution of an estradiol sensitive antigen in the vagina of adult mice, *Z. Zellforsch. Mikrosk. Anat.*, 136, 263, 1973.
77. **Mori, T., Nagahama, Y., Bern, H. A., and Young, P. N.**, Ultrastructural changes in vaginal epithelium of mice neonatally treated with estrogen and prolactin, *Anat. Rec.*, 179, 225, 1974.
78. **Mori, T. and Nishizuka, Y.**, Morphological alterations of basal cells of vaginal epithelium in neonatally oestrogenized mice, *Experientia*, 38, 389, 1982.
79. **Nomura, T. and Kanzaki, T.**, Induction of urogenital anomalies and some tumors in the progeny of mice receiving diethylstilbestrol during pregnancy, *Cancer Res.*, 37, 1099, 1977.
80. **Warner, M. R., Warner, R. L., and Clinton, C. W.**, Reproductive tract calculi, their induction, age incidence, composition and biological effects in Balb/cCrgl mice injected as newborns with estradiol-17β, *Biol. Reprod.*, 20, 310, 1979.
81. **Mori, T.**, Ultrastructure of the uterine epithelium of mice treated neonatally with estrogen, *Acta Anat.*, 99, 462, 1977.
82. **Mori, T.**, Changes in alkaline phosphatase activity and mitotic rate in vaginal epithelium following estrogen injections in neonatally estrogenized mice, *Annot. Zool. Jpn.*, 40, 82, 1967.
83. **Maier, D. B., Newbold, R. R., and McLachlan, J. A.**, Prenatal diethylstilbestrol exposure alters murine uterine responses to prepubertal estrogen stimulation, *Endocrinology,* 116, 1878, 1985.
84. **Ostrander, P. L., Mills, K. T., and Bern, H. A.**, Long-term responses of the mouse uterus to neonatal diethylstilbestrol treatment and to later sex hormone exposure, *J. Natl. Cancer Inst.*, 74, 121, 1985.
85. **Iguchi, T. and Takasugi, N.**, Postnatal development of uterine abnormalities in mice exposed to DES *in utero, Biol. Neonate,* 52, 97, 1987.
86. **Haney, A. F., Hammond, C. B., Soules, M. R., and Creasman, W. T.**, Diethylstilbestrol-induced upper genital tract abnormalities, *Fertil. Steril.*, 31, 142, 1979.
87. **Aihara, M., Kimura, T., and Kato, J.**, Dynamics of the estrogen receptor in the uteri of mice treated neonatally with estrogen, *Endocrinology,* 107, 224, 1980.
88. **Güttner, J.**, Adenomyosis in mice, *Z. Versuchstierkd.*, 22, 249, 1980.
89. **Newbold, R. R., Bullock, B. C., and McLachlan, J. A.**, Exposure to diethylstilbestrol during pregnancy permanently alters the ovary and oviduct, *Biol. Reprod.*, 28, 735, 1983.
90. **Newbold, R. R., Tyrey, S., Haney, A. F., and McLachlan, J. A.**, Developmentally arrested oviduct: a structural and functional defect in mice following prenatal exposure to diethylstilbestrol, *Teratology,* 27, 417, 1983.
91. **Haney, A. F., Newbold, R. R., and McLachlan, J. A.**, Prenatal diethylstilbestrol exposure in the mouse: effects on ovarian histology and steroidogenesis in vitro, *Biol. Reprod.*, 30, 471, 1984.

92. **Benjamin, C. L. and Beaver, D. C.,** The pathogenesis of salpingitis isthmica modosa, *Am. J. Clin. Pathol.,* 21, 212, 1951.
93. **Kaufman, R. H., Adam, E., Binder, G. L., and Gerthoffer, E.,** Upper genital tract changes and pregnancy outcome in offspring exposed *in utero* to diethylstilbestrol, *Am. J. Obstet. Gynecol.,* 137, 299, 1980.
94. **DeCherney, A. H., Cholst, I., and Naftolin, F.,** Structure and function of the fallopian tubes following exposure to diethylstilbestrol (DES) during gestation, *Fertil. Steril.,* 36, 741, 1981.
95. **Shen, S. C., Bansal, M., Purrazzella, R., Malviya, V., and Strauss, L.,** Benign glandular inclusions in lymph nodes, endosalpingitis and salpingitis isthmica nodosa in a young girl with clear cell adenocarcinoma of the cervix, *Am. J. Surg. Pathol.,* 7, 293, 1983.
96. **Iguchi, T.,** Occurrence of polyovular follicles in ovaries of mice treated neonatally with diethylstilbestrol, *Proc. Jpn. Acad.,* B61, 288, 1985.
97. **Forsberg, J.-G., Tenenbaum, A., Rydberg, C., and Sernvi, S.,** Ovarian structure and function in neonatally estrogen treated female mice, in *Estrogens in the Environment,* Vol. II, McLachlan, J. A., Ed., Elsevier, New York, 1985, 327.
98. **Iguchi, T., Takasugi, N., Bern, H. A., and Mills, K. T.,** Frequent occurrence of polyovular follicles in ovaries of mice exposed neonatally to diethylstilbestrol, *Teratology,* 34, 29, 1986.
99. **Burroughs, C. D., Williams, B. A., Mills, K. T., and Bern, H. A.,** Genital tract abnormalities in female C57BL/Crgl mice exposed neonatally to phytoestrogens (coumestrol and zearalenone), *Cancer Res.,* 27, 220, 1986.
100. **Iguchi, T. and Takasugi, N.,** Polyovular follicles in the ovary of immature mice exposed prenatally to diethylstilbestrol, *Anat. Embryol.,* 175, 53, 1986.
101. **Mori, T.,** Age-related changes in ovarian responsiveness to gonadotropins in normal and neonatally estrogenized mice, *J. Exp. Zool.,* 207, 451, 1979.
102. **Mori, T.,** Ceroid deposition in the adrenal cortices of ddY mice treated neonatally with estrogen, *J. Fac. Sci. Univ. Tokyo Sec. 4,* 13, 139, 1974.
103. **Hawkins, E. F., Young, P. N., Hawkins, A. M. C., and Bern, H. A.,** Adrenocortical function: corticosterone levels in female BALB/c and C3H mice under various conditions, *J. Exp. Zool.,* 194, 479, 1975.
104. **Mori, T., Bern, H. A., Mills, K. T., and Young, P. N.,** Long-term effects of neonatal steroid exposure on mammary gland development and tumorigenesis in mice, *J. Natl. Cancer Inst.,* 57, 1057, 1976.
105. **Takasugi, N. and Kimura, T.,** Hyperplasia of vaginal epithelium unaffected by ovariectomy in rats receiving early postnatal injections of estrogen, *J. Fac. Sci. Univ. Tokyo Sec. 4,* 10, 381, 1964.
106. **Kimura, T. and Takasugi, N.,** Persistent hyperplasia of vaginal epithelium following ovariectomy in rats given injections of androgen in early postnatal life, *J. Fac. Sci. Univ. Tokyo Sec. 4,* 10, 391, 1964.
107. **Cherry, C. P. and Glucksmann, A.,** The induction of cervico-vaginal tumors in oestrogenised and androgenised rats, *Br. J. Cancer,* 22, 728, 1968.
108. **Takewaki, K.,** Reproductive organs and anterior hypophysis of neonatally androgenized female rats, *Sci. Rep. Tokyo Woman's Christian Coll.,* 1, 31, 1968.
109. **Arai, Y., Suzuki, Y., and Nishizuka, Y.,** Hyperplastic and metaplastic lesions in the reproductive tract of male rats induced by neonatal treatment with diethylstilbestrol, *Virchows Arch. A,* 376, 21, 1977.
110. **Boylan, E. S.,** Morphological and functional consequences of prenatal exposure to diethylstilbestrol in the rat, *Biol. Reprod.,* 19, 854, 1978.
111. **Napalkov, N. P. and Anisimov, V. N.,** Transplacental effect of diethylstilbestrol in female rats, *Cancer Lett.,* 6, 107, 1979.
112. **Rustia, M. and Shubik, P.,** Effects of transplacental exposure to diethylstilbestrol on carcinogenic susceptibility during postnatal life in hamster progeny, *Cancer Res.,* 39, 4636, 1979.
113. **Rustia, M.,** Role of hormonal imbalance in transplacental carcinogenesis induced in Syrian golden hamsters by sex hormones, *Natl. Cancer Inst. Monogr.,* 51, 77, 1979.
114. **Gilloteaux, J., Paul, R. J., and Steggles, A. W.,** Upper genital tract abnormalities in the Syrian hamster as a result of *in utero* exposure to diethylstilbestrol, *Virchows Arch. A,* 398, 163, 1982.
115. **Hendrickx, A. G., Benirschke, K., Thompson, R. S., Ahern, J., Lucas, W. E., and Oi, R.,** The effects of prenatal diethylstilbestrol (DES) exposure on the genitalia of pubertal *Macaca mulatta, Teratology,* 17, 23A, 1978.
116. **Pan, S. C. and Gardner, W. U.,** Carcinomas of the uterine cervix and vagina in estrogen- and androgen-treated hybrid mice, *Cancer Res.,* 8, 337, 1948.
117. **Gilmour, M. D.,** An investigation into the influence of oestrone on the growth and on the genesis of malignant cells, *J. Pathol. Bacteriol.,* 45, 179, 1937.
118. **Heywood, R. and Wadsworth, P. F.,** The experimental toxicology of estrogens, *Pharmacol. Ther.,* 8, 125, 1980.
119. **Smith, D. C., Prentice, R., Thompson, D. J., and Herrmann, W. L.,** Association of exogenous estrogen and endometrial carcinoma, *N. Engl. J. Med.,* 293, 1164, 1975.

120. **Ziel, H. K. and Finkle, W. D.,** Increased risk of endometrial carcinoma among users of conjugated estrogens, *N. Engl. J. Med.,* 293, 1167, 1975.
121. **Mack, T. M., Pike, M. C., Henderson, B. E., Pfeffer, R. I., Gerkins, V. R., Arthur, M., and Brown, S. E.,** Estrogens and endometrial cancer in a retirement community, *N. Engl. J. Med.,* 294, 1262, 1976.
122. **Gondos, B.,** Histologic changes associated with oral contraceptive usage, *Ann. Clin. Lab. Sci.,* 6, 291, 1976.
123. **Lyon, F. A.,** The development of adenocarcinoma of the endometrium in young women receiving long-term sequential oral contraception, *Am. J. Obstet. Gynecol.,* 123, 299, 1975.
124. **Lyon, F. A. and Frisch, M. J.,** Endometrial abnormalities occurring in young women on long-term sequential oral contraception, *Obstet. Gynecol.,* 47, 639, 1976.
125. **Stern, E., Forsythe, A. B., Youkeles, L., and Coffelt, C. F.,** Steroid contraceptive use and cervical dysplasia: increased risk of progression, *Science,* 196, 1460, 1977.
126. **Antunes, C. M. F., Stolley, P. D., Rosenshein, N. B., Davies, J. L., Tonascia, J. A., Brown, C., Burnett, L., Rutledge, A., Pokempher, M., and Garcia, R.,** Endometrial cancer and estrogen use, *N. Engl. J. Med.,* 300, 9, 1979.
127. **Weiss, N. S. and Sayvetz, T. A.,** Incidence of endometrial cancer in relation to the use of oral contraceptives, *N. Engl. J. Med.,* 302, 551, 1980.
128. **Bush, R. S.,** *Malignancies of the Ovary, Uterus and Cervix,* Edward Arnold, London, 1979.
129. **James, V. H. and Reed, M. J.,** Steroid hormones and human cancer, *Prog. Cancer Res. Ther.,* 14, 471, 1980.
130. **Mori, T.,** Effects of neonatal injections of estrogen in combination with vitamin A on the vaginal epithelium of adult mice, *Annot. Zool. Jpn.,* 41, 113, 1968.
131. **Mori, T.,** Further studies on the inhibitory effect of vitamin A on the development of ovary-independent vaginal cornification in neonatally estrogenized mice, *Proc. Jpn. Acad.,* 45, 115, 1969.
132. **Yasui, T. and Takasugi, N.,** Prevention of vitamin A of the occurrence of permanent vaginal changes in neonatally estrogen-treated mice, *Cell Tissue Res.,* 179, 475, 1977.
133. **Yasui, T., Iguchi, T., and Takasugi, N.,** Blockage of the occurrence of permanent vaginal changes in neonatally estrogen-treated mice by vitamin A; parabiosis and transplantation studies, *Endocrinol. Jpn.,* 24, 393, 1977.
134. **Iguchi, T. and Takasugi, N.,** Blockade by vitamin A of the occurrence of permanent vaginal changes in mice treated neonatally with 5α-dihydrotestosterone, *Anat. Histol. Embryol.,* 155, 127, 1979.
135. **Tachibana, H., Iguchi, T., and Takasugi, N.,** Different perinatal periods of vitamin A administration for prevention of the occurrence of permanent vaginal changes in mice treated neonatally with estrogen, *Zool. Sci.,* 1, 777, 1984.
136. **Iguchi, T., Iwase, H., Kato, H., and Takasugi, N.,** Prevention by vitamin A of the occurrence of permanent vaginal and uterine changes in ovariectomized adult mice treated neonatally with diethylstilbestrol and its nullification in the presence of ovaries, *Exp. Clin. Endocrinol.,* 85, 129, 1985.
137. **Jones, L. A., Verjan, R. P., Mills, K. T., and Bern, H. A.,** Prevention by progesterone of cervicovaginal lesions in neonatally estrogenized BALB/c mice, *Cancer Lett.,* 23, 123, 1984.
138. **Iguchi, T., Todoroki, R., Takasugi, N., and Petrow, V.,** unpublished observations.

Chapter 6

LONG-TERM EFFECTS OF PERINATAL EXPOSURE TO HORMONES AND RELATED SUBSTANCES ON NORMAL AND NEOPLASTIC GROWTH OF MURINE MAMMARY GLANDS

Hiroshi Nagasawa and Takao Mori

TABLE OF CONTENTS

I. Introduction 82

II. Effects of Hormones and Related Substances on Normal Mammary Gland Growth 82

III. Effects of Hormones and Related Substances on Neoplastic Mammary Gland Growth 82
- A. Effects of Hormones and Related Substances on Spontaneous Mammary Tumorigenesis 82
- B. Effects of Hormones and Related Substances on Carcinogen-Induced Mammary Tumorigenesis 83

IV. Effects of Hormones and Related Substances on Mammotropic Hormone Secretion 84

V. Effects of Hormones and Related Substances on Mammary Gland Responsiveness to Hormones 85

VI. Summary 85

References 85

I. INTRODUCTION

Clinical studies beginning with that of Herbst and Scully[1] indicated the relationship between the ingestion of diethylstilbestrol (DES) by women during the first trimester of pregnancy and exceptionally earlier and higher incidence of vaginal and cervical cancers in their female offspring. In mice and rats, perinatal exposure to DES, steroid hormones, and related substances induces abnormalities and malignancies in the reproductive tracts.[2-4] Mammary glands and reproductive tracts are controlled by similar hormones and, therefore, the exposure to hormones and synthetic analogs shows profound effects on mammary glands in experimental animals,[5-7] while it is far from conclusive in humans. Mori et al.[6] have reviewed the long-term effects of perinatal exposure to hormones on normal and neoplastic mammary gland growth in rodents. This chapter is an extension of that review.

II. EFFECTS OF HORMONES AND RELATED SUBSTANCES ON NORMAL MAMMARY GLAND GROWTH

The long-term effects of perinatal treatments with hormones on normal mammary gland growth vary according to the doses, types, and treatment periods of hormones and the ages, sexes, and strains of animals, etc.[6,7]

Recently, Tomooka and Bern[8] found that in female BALB/c mice daily injections of estradiol or DES for the first 5 days of postnatal life inhibited mammary gland growth on the 6th day, while the treatment stimulated growth after 4 weeks. Testosterone and 5α-dihydrotestosterone (5α-DHT) also enhanced mammary gland growth 4 weeks after treatment, whereas they showed no effects at 6 days of age. No immediate effects were evident with progesterone and 5β-dihydrotestosterone (5β-DHT). Furthermore, similar estradiol administration between 1 to 5 days of age induced the most mammary abnormalities at 12 months of age, such as dilated ducts, hyperplastic alveolar nodule (HAN)-like lesions, or aberrant secretory state.[9] The incidence of abnormalities declined markedly when the treatment was begun after day 5.[9]

Single subcutaneous injection of 4 mg monosodium glutamate (MSG) to female mice on the day of birth inhibited normal and preneoplastic mammary gland growth (C3H/He and SHN)[10,11] and pregnancy-dependent mammary tumor development (GR/A).[12]

Daily subcutaneous injections of 100 μg (290.4 IU) vitamin A for the first 5 days of postnatal life to GR/A female mice resulted in a substantial increase in the incidence of pregnancy-dependent mammary tumors.[13]

III. EFFECTS OF HORMONES AND RELATED SUBSTANCES ON NEOPLASTIC MAMMARY GLAND GROWTH

A. Effects of Hormones and Related Substances on Spontaneous Mammary Tumorigenesis

Neonatal treatment with estrogen, progesterone, or androgen induced an increase in mammary tumor development in several strains of mice bearing mammary tumor virus (MTV),[6] which is an essential factor for neoplastic mammary response to perinatal hormone treatment as well as for spontaneous mammary tumor development in mice. However, Mori[14] and Jones and Bern[15] have recently reported stimulation by similar treatments of both preneoplastic and neoplastic mammary gland growth in MTV-unexpressed female BALB/c mice. Daily subcutaneous injections of 5α-DHT for the first 5 days of postnatal life resulted in a stimulation of neoplastic mammary gland growth associated with the ovarian anovulatory syndrome and the stimulated pituitary prolactin secretion, while prenatal treatment with the hormone showed no effects.[16] 5β-DHT, which is biologically inactive and does not bind to androgen receptors in adult mice, could also enhance mammary tumorigenesis and the effects were more marked than with 5α-DHT.[17] Similar treatment with vitamin A stimulated autonomous mammary tumor development in GR/A mice.[13] These findings strongly suggest that the effects of perinatal exposure to some agents are often quite different both quantitatively and qualitatively from those observed after exposure to adults.

Temporary pituitary grafting, from which only prolactin is secreted predominantly, during 1 to 23 days of age was also effective on the induction of mammary tumors in mice.[6] Neonatal single injection of MSG, which inhibited normal and preneoplastic mammary gland growth in mice (Section II), showed no effects on spontaneous mammary tumorigenesis.[11]

CB-154, a potent prolactin release suppressor, is widely used clinically for the therapy of amenorrhea-galactorrhea, suppression of pubertal lactation infertility, postpill anovulation, etc. In view of the use of CB-154 during pregnancy, the long-term effects of perinatal exposure to this drug on mammary tumorigenesis in mice were studied.[18] Neither prenatal (0.3 mg for 4 days from day 12 to day 15 of pregnancy) nor neonatal (0.06 mg for the first 5 days of postnatal life) treatment significantly affected mammary tumorigenesis in SHN mice.

While perinatal hormone treatments can modulate spontaneous mammary tumorigenesis in mice, data on the effects of the different times of exposure are scanty. Moreover, estrogen participation after maturity in this process would also be of much importance since most women will generally experience steroid hormones, especially through contraceptive pills and environmental naturally occurring photoestrogens such as coumestrol from ladino clover and zearalenone from moldy corn.[19] In this respect, Nagasawa et al.[20] examined in SLN mice the long-term effects of prenatal (12 or 17 days of fetal life) or neonatal (on the day of birth) treatment with DES (5 or 0.1 μg) or progesterone (1000 or 100 μg) with or without additional treatment with estradiol benzoate (EB) in the form of a subcutaneous pellet implantation for 2 to 5 months of age. Spontaneous mammary tumorigenesis of females given DES on perinatal 12 days was significantly lower than that of mice receiving progesterone or oil (control) on the same day. This difference in mammary tumorigenesis was not altered by EB after maturity. Exposure to progesterone at prenatal 17 days with or without EB after maturity also inhibited mammary tumorigenesis at advanced ages. Mammary tumorigenesis was enhanced by neonatal treatment with DES or progesterone; however, it was arrested by EB after maturity. These results indicate that long-term effects of perinatal exposure to DES or steroids on mammary tumorigenesis at advanced ages are largely dependent upon perinatal age of the subjects and the fact that sex hormones and related substances after maturity can modulate these perinatal hormone effects. Neonatal exposure to steroid hormones had little effect on spontaneous mammary tumorigenesis in rats.[6]

B. Effects of Hormones and Related Substances on Carcinogen-Induced Mammary Tumorigenesis

There are some papers on the long-term effects of neonatal hormone treatments on carcinogen-induced mammary tumorigenesis in MTV unexpressed mice, all of which were negative.[6] However, long-term effects of perinatal hormone exposure on carcinogen-induced mammary tumorigenesis in rats vary according to the experimental conditions which are either inhibitory, stimulatory, or slightly effective. Administration of testosterone propionate (500 μg) or EB (100 μg) immediately after ovariectomy on the day of birth did not significantly influence the incidence, histopathology, or estrogen responsiveness of 7,12-dimethylbenz(*a*)anthracene (DMBA)-induced mammary tumors of Sprague-Dawley rats.[21] This suggests the importance of ovaries in this process. It has been observed that prenatal hormone exposure induced the alteration in the characteristics of carcinogen-induced mammary tumors of rats. Boylan and Calhoon[22,23] found that the combination of prenatal exposure to DES and postnatal treatment with DMBA resulted in a significant increase in the number of mammary tumors per rat compared to rats treated with DMBA alone. They[24] also observed that, when compared to controls, a higher number of these DMBA-induced mammary tumors in female rats given DES neonatally overcame the initial inhibitory effects of ovariectomy and began to grow again. DMBA administration at 6 months of age to female hamsters, whose mothers received a single dose of 10 mg DES per kg body weight on day 14 of pregnancy, resulted in a higher incidence of mammary tumors as well as ovarian and uterine tumors and adrenal melanomas than did DMBA alone.[25]

IV. EFFECTS OF HORMONES AND RELATED SUBSTANCES ON MAMMOTROPIC HORMONE SECRETION

The amount of hormone acting on the target cells is one of the essential factors for cell growth. Accumulated data strongly suggest considerable alteration in hormone secretion of anterior pituitary and ovaries by perinatal hormone treatments.[6] Kalland et al.[26] reported that in female NMRI mice 5 μg DES treatment for the first 5 days of postnatal life enhanced pituitary response to estrogen for prolactin secretion in later life. In estrogenized and androgenized rats, Vaticón et al.[27] found the difference between estrogen and androgen to be the ability to alter the prolactin control system and higher sensitivity in females than males in the hypothalamic response to the treatments. Loss of cyclicity of hormone secretion was also induced by neonatal hormone treatments. Plasma prolactin level was higher in neonatally estrogenized or prolactinized mice than in the control at metaestrus/diestrus; however, no difference was observed in the level at proestrus/estrus.[28] Continued estrus is usually seen in neonatally hormone-treated animals. The increased prolactin secretion in these animals is dependent upon ovaries,[29] implying a critical role for sustained secretion of estrogen in this process.

Meanwhile, no difference was found in plasma prolactin levels at autopsy between tumorous mice receiving different schedules of perinatal treatments with DES or progesterone.[18] Lopez et al.[30] found that plasma prolactin level was significantly lower in C3H/MTV+ female mice given 2.5 μg DES for the first 5 days of life than in the control at 10 weeks of age, but not at 5 and 10 months of age. Furthermore, modulation of pituitary response to dopamine by neonatal DES treatments was observed.[30] Treatments of mice with DES significantly decreased plasma prolactin levels at 6 weeks of age and neonatal DES and 17α-hydroxyprogesterone caproate also declined prolactin synthesis at 10 weeks of age; however, the percent of prolactin release did not appear affected by these treatments.[31] Thus, neonatal hormone treatment affected mechanisms that regulate prolactin synthesis, but not those that regulate release. Boylan et al.[32] observed that in female Sprague-Dawley rats given two injections of 0.6 μg DES on days 15 and 18 of prenatal life, serum prolactin level was lower than in the control at 2 months of age and was not associated with differences in serum levels of estrogen and progesterone. On the other hand, prolactin level at 9 months of age was comparable to that of the control.

Both normal and neonatally androgenized female rats responded in a similar manner to blinding, olfactory bulbectomy, and pinealectomy, exhibiting diurnal plasma prolactin pattern. However, there was an apparent difference in the effects of estrogen on diurnal circulating prolactin pattern between these normal and neonatally androgenized rats.[36]

Neonatal single injection of 4 mg MSG to female mice on the day of birth resulted in the chronically lower plasma growth hormone (GH) levels.[11] The treatment had little effect on the plasma prolactin level,[11] estrous cycles,[11] and reproduction[12] (Section II). MSG and related substances are known to induce acute degeneration of hypothalamic neurons in perinatal animals.[37]

Recently, several hormones and related substances have been identified in milk; however, their physiological significance is obscure.[38,39] Nagasawa et al.[40] found a relationship between milk levels of prolactin, GH, or progesterone and the growth or the puberty of offspring.

V. EFFECTS OF HORMONES AND RELATED SUBSTANCES ON MAMMARY GLAND RESPONSIVENESS TO HORMONES

Cell responsiveness or susceptibility to hormones is another important factor for cell growth. The effects of perinatal hormones and related substances on mammary gland responsiveness to mammotropic hormones are not conclusive. The responsiveness is generally little changed or increased by perinatal hormone treatments; however, a few experimental results infer the decrease of responsiveness in rats.[6] In this respect, Heideman et al.[41] found that progesterone-binding activity of DMBA-induced mammary tumors of rats was elevated by prenatal exposure to DES during the third trimester, but not by the exposure during the second trimester and no alteration was observed in estrogen binding capacities between treatments. Mammary tissue of adult (about 75 days of age) male Holtsman rats whose mothers received 10 mg cyproterone acetate, an antiandrogen, daily for 10 days beginning day 11 of pregnancy showed presence of estrogen-binding sites comparable to the levels in the normal females, while it was undetectable in normal males.[41] These rats also had the higher mammary gland activity of conversion of ^{3}H-androstenedione to estrogens than did normal males.[42] Mammary gland response to growth factors was lower in the neonatally estrogenized mice than in the control.[43]

VI. SUMMARY

In this chapter, the experimental results of the long-term effects of perinatal exposure to hormones and related substances on normal (Section II) and neoplastic (Section III) mammary gland growth were reviewed. In addition, alteration by the treatments of mammotropic hormone secretion (Section IV) and mammary gland responsiveness to hormones (Section V) as essential factors for mammary growth were discussed. All would be important in determining the possible human consequences of antenatal exposure to hormones and other agents.

REFERENCES

1. **Herbst, A. L. and Scully, R. D.,** Adenocarcinoma of the vagina in adolescence. A report of 7 cases including 6 clear-cell carcinoma (so-called mesonephromas), *Cancer (Philadelphia),* 25, 745, 1970.
2. **Takasugi, N.,** Cytological basis for permanent vaginal changes in mice treated neonatally with steroid hormones, *Int. Rev. Cytol.,* 44, 193, 1975.
3. **Bern, H. A., Jones, L. A., Mori, T., and Young, P.,** Exposure of neonatal mice to steroids: long-term effects on the mammary gland and other reproductive structures, *J. Steroid Biochem.,* 6, 673, 1975.
4. **Burroughs, C. D., Bern, H. A., and Stockstad, E. L. R.,** Prolonged vaginal cornification and other changes in mice treated neonatally with coumestrol, a potent estrogen, *J. Toxicol. Environ. Health,* 15, 51, 1985.
5. **Bern, H. A., Jones, L. A., Mills, K. T., Kohrman, A., and Mori, T.,** Use of the neonatal mouse in studying longterm effects of early exposure to hormones and other agents, *J. Toxicol. Environ. Health,* Suppl. 1, 103, 1976.
6. **Mori, T., Nagasawa, H., and Bern, H. A.,** Long-term effects of perinatal exposure to hormones on normal and neoplastic mammary growth in rodents: a review, *J. Environ. Pathol. Toxicol.,* 3, 191, 1980.
7. **Hoshino, K.,** Hormonal teratogenesis in mammary glands of the mouse, in *Advances in the Study of Birth Defects,* Vol. 2, Persaud, T. V. N., Ed., MTP Press Ltd., Lancaster, 1979, 139.
8. **Tomooka, Y. and Bern, H. A.,** Growth of mouse mammary glands after neonatal sex hormone treatment, *J. Natl. Cancer Inst.,* 69, 1347, 1982.
9. **Bern, H. A., Mills, K. T., and Jones, L. A.,** Critical period for neonatal estrogen exposure in occurrence of mammary gland abnormalities in adult mice, *Proc. Soc. Exp. Biol. Med.,* 172, 239, 1983.
10. **Nagasawa, H., Yanai, R., and Kikuyama, S.,** Irreversible inhibition of pituitary prolactin and growth hormone secretion and of mammary gland development in mice by monosodium glutamate administered neonatally, *Acta Endocrinol. (Copenhagen),* 75, 249, 1974.

11. **Nagasawa, H., Noguchi, Y., Mori, T., Niki, K., and Namiki, H.**, Suppression of normal and neoplastic mammary growth and uterine adenomyosis with reduced growth hormone level in SHN mice given monosodium glutamate neonatally, *Eur. J. Cancer*, 21, 1547, 1985.
12. **Nagasawa, H. and Yanai, R.**, Inhibition of pregnancy-dependent mammary tumorigenesis by a single treatment of neonatal GR/A mice with monosodium glutamate, *Proc. Soc. Exp. Biol. Med.*, 158, 128, 1978.
13. **Nagasawa, H.**, Stimulation by neonatal treatment with vitamin A of spontaneous mammary tumor development in GRS/A mice, *Breast Cancer Res. Treat.*, 4, 205, 1984.
14. **Mori, T.**, Abnormalities in the reproductive system of aged mice after neonatal estradiol exposure, *J. Endocrinol. Invest.*, 9, 397, 1986.
15. **Jones, L. A. and Bern, H. A.**, Cervicovaginal and mammary gland abnormalities in BALB/cCrgl mice treated neonatally with progesterone and estrogen, alone or in combination, *Cancer Res.*, 39, 2560, 1979.
16. **Yanai, R., Nagasawa, H., Mori, T., and Nakajima, Y.**, Long-term effects of perinatal exposure to 5α-dihydrotestosterone on normal and neoplastic mammary development in mice, *Endocrinol. Jpn.*, 28, 231, 1981.
17. **Yanai, R., Mori, T., and Nagasawa, H.**, Long-term effects of prenatal and neonatal administration of 5β-dihydrotestosterone on normal and neoplastic mammary development in mice, *Cancer Res.*, 37, 4456, 1977.
18. **Nagasawa, H. and Yanai, R.**, The influence of prenatal or neonatal administration of 2-bromo-α-ergocriptine on pituitary prolactin secretion and normal and neoplastic mammary growth in adult mice, *J. Endocrinol. Invest.*, 1, 273, 1978.
19. **McLachlan, J. A., Korach, K. S., Newbold, R. R., and Degen, G. H.**, Diethylstilbestrol and other estrogens in the environment, *Fund. Appl. Toxicol.*, 4, 686, 1984.
20. **Nagasawa, H., Mori, T., and Nakajima, Y.**, Long-term effects of progesterone or diethylstilbestrol with or without estrogen after maturity on mammary tumorigenesis, *Eur. J. Cancer*, 16, 1583, 1980.
21. **Verhoeven, G., Vandoren, G., Heyns, W., Kuhn, E. R., Janssens, J. P., Teuwen, D., Goddeeris, P., Lesaffre, E., and De Moor, P.**, Incidence, growth and oestradiol-receptor levels of 7,12-dimethylbenz(a)anthracene-induced mammary tumours in rats: effects of neonatal sex steroids and oestradiol implants, *J. Endocrinol.*, 95, 357, 1982.
22. **Boylan, E. S. and Calhoon, R. E.**, Mammary tumorigenesis in the rat following prenatal exposure to diethylstilbestrol and postnatal treatment with 7,12-dimethylbenz(a)anthracene, *J. Toxicol. Environ. Health*, 5, 1059, 1979.
23. **Boylan, E. S. and Calhoon, R. E.**, Transplacental action of diethylstilbestrol on mammary carcinogenesis in female rats given one or two doses of 7,12-dimethylbenz(a)anthracene, *Cancer Res.*, 43, 4879, 1983.
24. **Boylan, E. S. and Calhoon, R. E.**, Prenatal exposure to diethylstilbestrol: ovarian-independent growth of mammary tumors induced by 7,12-dimethylbenz(a)anthracene, *J. Natl. Cancer Inst.*, 66, 649, 1981.
25. **Rustia, M. and Shubik, P.**, Effects of transplacental exposure to diethylstilbestrol on carcinogen susceptibility during postnatal life in hamster progeny, *Cancer Res.*, 39, 4636, 1979.
26. **Kalland, T., Forsberg, J.-G., and Sinha, Y. N.**, Long-term effects of neonatal DES treatment on plasma prolactin in female mice, *Endocr. Res. Commun.*, 7, 157, 1980.
27. **Vaticón, M. D., Galaz, M. C. F., Tejero, A., and Aguilar, E.**, Alteration of prolactin control in adult rats treated neonatally with sex steroids, *J. Endocrinol.*, 105, 429, 1985.
28. **Nagasawa, H., Mori, T., Yanai, R., Bern, H. A., and Mills, K. T.**, Longterm effects of neonatal hormone treatments on plasma prolactin levels in BALB/cfC3H and BALB/c female mice, *Cancer Res.*, 38, 942, 1978.
29. **Nagasawa, H., Yanai, R., Jones, L. A., Bern, H. A., and Mills, K. T.**, Ovarian dependence of the stimulating effect of neonatal hormone treatment on plasma levels of prolactin in female mice, *J. Endocrinol.*, 79, 391, 1978.
30. **Lopez, J., Ogren, L., and Talamantes, F.**, Neonatal diethylstilbestrol treatment: response of prolactin to dopamine or estradiol in adult mice, *Endocrinology*, 119, 1020, 1986.
31. **Lopez, J., Ogren, L., and Talamantes, F.**, Effects of neonatal treatment with diethylstilbestrol and 17α-hydroxyprogesterone caproate on *in vitro* pituitary prolactin secretion, *Life Sci.*, 34, 2303, 1984.
32. **Boylan, E. S., Calhoon, R. E., and Vonderhaar, B. K.**, Transplacental action of diethylstilbestrol on reproductive and endocrine organs, mammary glands, and serum hormone levels in two- and nine-month-old female rats, *Cancer Res.*, 43, 4872, 1983.
33. **Boylan, E. S. and Calhoon, R. E.**, Transplacental action of diethylstilbestrol on mammary carcinogenesis in female rats given one or two doses of 7,12-dimethylbenz(a)anthracene, *Cancer Res.*, 43, 4879, 1983.
34. **Gala, R. R., Clarke, W. P., Haisenleder, D. J., Pan, J.-T., and Pieper, D. R.**, The influence of blinding, olfactory bulbectomy, and pinealectomy on twenty four-hour plasma prolactin levels in normal and neonatally androgenized female rats, *Endocrinology*, 115, 1256, 1984.

35. **Gala, R. R., Clarke, W. P., Haisenleder, D. J., Pan, J.-T., and Pieper, D. R.,** The influence of blinding, olfactory bulbectomy and pinealectomy on plasma and anterior pituitary prolactin levels and on uterine and anterior pituitary weights in normal and neonatally androgenized rats, *Life Sci.*, 36, 1617, 1985.
36. **Gala, R. R.,** The influence of estrogen administration on plasma prolactin levels in the neonatally androgenized (NA) rat, *Proc. Soc. Exp. Biol. Med.*, 166, 216, 1981.
37. **Olney, J. W., Sharpe, L. G., and Feigin, R. D.,** Glutamate-induced brain damage in infant primates, *J. Neuropathol. Exp. Neurol.*, 31, 464, 1972.
38. **Koldowsky, O.,** Hormones in milk: minireview, *Life Sci.*, 26, 1838, 1980.
39. **Pope, G. S. and Swinburne, J. K.,** Reviews of the progress of dairy science: hormones in milk: their physiological significance and values as diagnostic aids, *J. Dairy Res.*, 47, 427, 1980.
40. **Nagasawa, H., Naito, T., Namiki, H., Inaba, T., and Mori, J.,** Relationship between milk levels of hormones and growth or puberty of offspring in mice, *Exp. Clin. Endocrinol.*, in press.
41. **Heideman, P. H., Wittliff, J. L., Calhoon, R. E., and Boylan, E. S.,** Influence of prenatal exposure to diethylstilbestrol on estrogen and progesterone binding proteins in uteri and dimethylbenzanthracene-induced mammary tumors of the rat, *J. Toxicol. Environ. Health,* 8, 667, 1981.
42. **Rajendran, K. G., Shah, P. N., Dubey, A. K., and Begli, N. P.,** Mammary gland differentiation in adult male rat — effect of prenatal exposure to cyproterone acetate, *Endocr. Res. Commun.*, 4, 267, 1977.
43. **Tomooka, Y., Bern, H. A., and Nandi, S.,** Growth of mammary epithelial cells from neonatally sex hormone-exposed mice in serum free collagen gel culture, *Cancer Lett.*, 20, 255, 1983.

Chapter 7

NEOPLASTIC AND NON–NEOPLASTIC LESIONS IN MALE REPRODUCTIVE ORGANS FOLLOWING PERINATAL EXPOSURE TO HORMONES AND RELATED SUBSTANCES

Retha R. Newbold and John A. McLachlan

TABLE OF CONTENTS

I. Introduction 90

II. Lesions of the Rete Testis 90

III. Lesions of Müllerian Duct Remnants 94

IV. Lesions of the Corpus Testis 100

V. Embryological Considerations 105

VI. Conclusion 106

Acknowledgments 106

References 107

I. INTRODUCTION

Normal differentiation and morphogenesis of developing tissues can be altered by a variety of compounds during the perinatal period. In particular, it has been known for decades that estrogens can cause teratogenic and carcinogenic changes in the reproductive tracts of experimental animals following exposure during critical periods of development.[1-4] Since many chemicals such as insecticides, drugs, and many naturally occurring compounds have estrogenic activity,[5] interest has developed regarding the biological effects of these estrogenic compounds on the developing fetus. The dramatic effects in women who were exposed *in utero* to diethylstilbestrol (DES), a potent synthetic estrogen, illustrate this concern and are discussed by Herbst and Rotmensch in a subsequent chapter. Alterations in the genital tract of men exposed prenatally to DES are covered by Gill in Chapter 11. Although these chapters summarize the reported DES effects in humans, little is known about the long-lasting changes simply because of the age of the patients. In particular, to date there are no reported indications of neoplastic alterations in men exposed to DES *in utero*[6-11] except the reports of testicular neoplasia by Gill et al.[12-13] and Conley et al.[14]

In this chapter, recent findings will be presented describing neoplastic and non-neoplastic changes in male mice following prenatal exposure to DES, and the embryological considerations that may result in these abnormalities will be explored. Lesions such as rete testis adenocarcinoma and Müllerian duct lesions will be discussed. Other abnormalities such as undescended and hypoplastic testes, infertility, epididymal cysts, sperm abnormalities, prostatic inflammation, squamous metaplasia of the prostatic utricle, etc. have been covered in a previous review.[15] We consider this rodent model to be a system for understanding the influence of perinatal DES exposure in the human male.

II. LESIONS OF THE RETE TESTIS

Various studies in our laboratory have pointed to the fact that the mesonephric duct, as well as the Müllerian duct, is a target for DES.[16-18] In fact, in adult female mice[18] and humans[16] exposed prenatally to DES, hyperplastic mesonephric remnants are a common finding. The relationship of these dysmorphogenic mesonephric structures to the pathogenesis of hyperplastic or neoplastic disease in the female is still being studied. The findings, however, raise the possibility that structures derived from the mesonephric ducts or tubules may be dysplastic in male offspring. Byskov[19,20] attributes the rete system to mesonephric tubular origin. Therefore, the rete testis, a clearly definable adult structure apparently derived from mesonephric tubules, was evaluated for hyperplastic or neoplastic changes in male CD-1 mice exposed prenatally to DES (100 μg/kg) on days 9 through 16 of gestation.

In a group of animals ranging from 10 to 18 months of age, 56% (130 out of 233) of the male mice exposed to DES during gestation had various degrees of papillary proliferation and hyperplasia of the epithelium of the rete testis. The simplest form of this lesion consisted of knob-like overgrowths of the cuboidal epithelium and diffuse hyperplasia of the epithelium (Figure 1). These overgrowths had coalesced to form papillomas consisting of small cuboidal cells with increased hyperchromatism and vacuolated cells in other DES-treated animals. Control males, 10 to 18 months of age, had mild focal epithelial hyperplasia of the rete testis in 23 out of 96 (24%) mice, but papillary proliferation of the epithelium was never observed.

A more severe lesion also was observed in mice after *in utero* exposure to DES. This lesion, resembling adenocarcinoma of the rete testis, was not found in any controls (0 out of 96) but was found in 5% of the prenatally DES-exposed mice (11 out of 233). Although distant metastases could not be found, the tumors often infiltrated into the seminiferous tubules; the histological pattern of these tumors was suggestive of either papillary adeno-

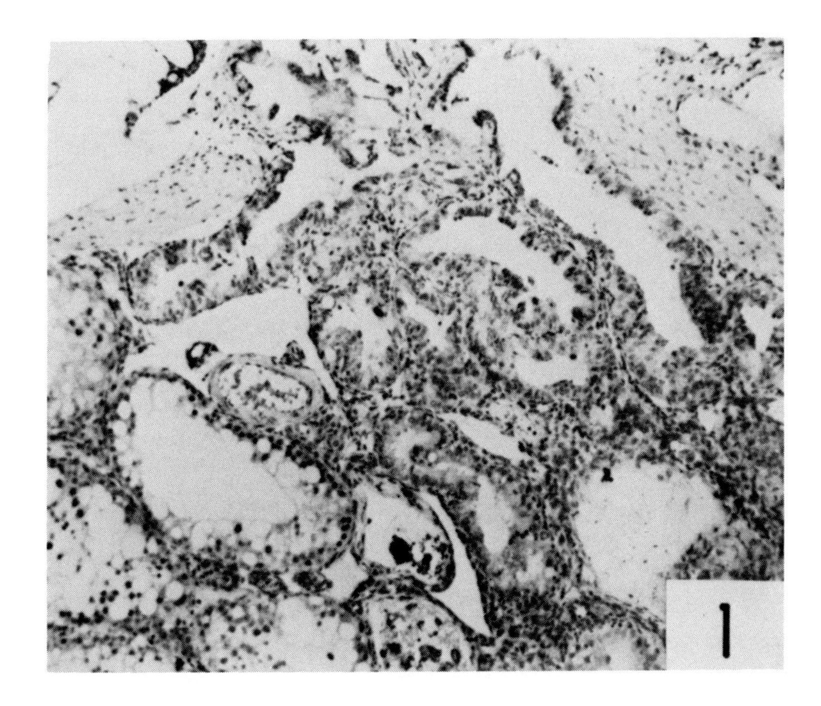

FIGURE 1. Diffuse hyperplasia of the epithelium of the rete testis from a 14-month-old prenatally DES-exposed mouse. The rete joins the efferent duct at the top of the photomicrograph. (H.E.; magnification × 25.) (From Newbold, R. R., Bullock, B. C., and McLachlan, J. A., *Cancer Res.*, 45, 5145, 1985. With permission.)

carcinoma (Figure 2) or tubulopapillary adenocarcinoma (Figure 3). This lesion occupied approximately 10% of the section surface of the testis, and there was an area of local invasion (Figure 4). There were increased cellularity and papillary infoldings in the epithelium. The epithelial cells were mainly cuboidal and appeared to be enlarged. There was considerable pleomorphism of nuclei. Mitotic figures were not numerous but often were abnormal.

The incidence of the lesion resembling adenocarcinoma of the rete testis in the DES males was 1 of 31 (3%), 10 months; 1 of 21 (8%), 13 months; 4 of 49 (8%), 14 months; 1 of 52 (2%), 15 months; 1 of 18 (6%), 16 months; and 3 of 34 (9%), 18 months. No tumors were found in the control series of mice. A more detailed report of these animals studied is given in Newbold et al.[21,22]

Adenocarcinoma of the rete testis in humans and experimental animals is extremely rare. Criteria were first formulated in 1945 by Felk and Hunter[23] and later by others,[24-27] and have been accepted by most investigators as the basis for identifying this neoplasm in humans. In summary, the diagnostic criteria of tumor of the rete testis are (1) involvement centering on the mediastinum testis rather than in the testis proper, (2) lack of direct extension through the parietal tunica, (3) transition from normal epithelial structure to neoplastic structures in the rete testis, (4) no evidence of teratoma, and (5) lack of any other primary tumor.

In the mouse, as in the human, tumors were confined to the mediastinum testis. There was maximal involvement of the rete testis but minimal involvement of the corpus. Tumor cells also appeared to be transformed from cytologically normal rete epithelium to neoplastic epithelium. Sections were examined from the mouse testis containing the lesion and from the opposite testis, but no teratomatous elements were found. In one mouse, both an interstitial cell tumor in the corpus testis and an adenocarcinoma of the rete testis were observed. Thus, the lesions in mice conform closely to the criteria established for rete adenocarcinoma in humans.

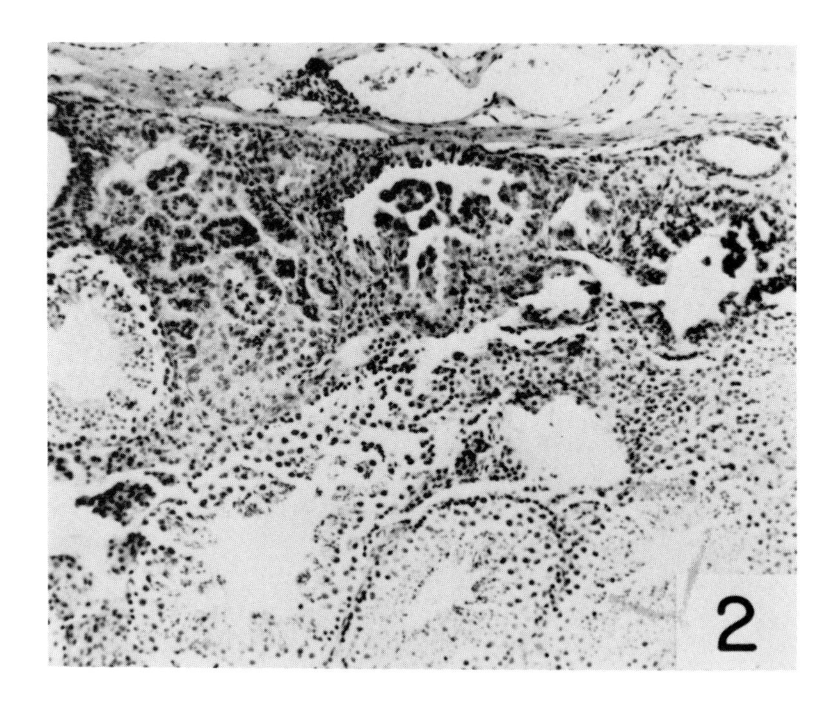

FIGURE 2. Papillary adenocarcinoma of the rete testis from a 13-month-old prenatally DES-treated mouse. The rete is filled by papillary projections covered by pleomorphic epithelium. (H.E.; magnification × 25.) (From Newbold, R. R., Bullock, B. C., and McLachlan, J. A., *Cancer Res.*, 45, 5145, 1985. With permission.)

In addition, the rare occurrence of adenocarcinoma of the rete testis in humans shows no age preference (age range from 20 to 80 years), and tumors were reported almost equally divided between the right and left side. Likewise, in the animal model there was no statistical difference between the prevalence of rete testis lesions in mice age 10 and 18 months of age. In humans, the tumors usually were present as a testicular mass often associated with a hydrocele.[28] It is of special interest that, in the group of human cases reported, there are at least three tumors in maldescended testes. Cryptorchidism has been implicated as a predisposing factor for testicular cancer such as seminoma. The previously reported high incidence of retained testes in mice following prenatal DES exposure[15,29] and the occurrence of this specific rare lesion of the testis resembling rete adenocarcinoma, described in the present report, raise the possibility of an association between cryptorchidism, prenatal DES exposure, and testicular cancer. In fact, researchers studying risk factors for cancer of the testis have evidence supporting this association.[30-35] Although cryptorchidism results in decreased or lack of spermatogenesis in male mice,[15] inactivity cannot solely account for the higher prevalence of rete lesions since 4 of the 11 animals diagnosed with the rete lesion had spermatogenesis occurring in the testis.

The demonstration of an extremely rare lesion such as rete testis adenocarcinoma is unique in its prevalence alone since it appears in 5% of the prenatal DES animals. Recently, Yoshitomi and Morii[36] reported the spontaneous occurrence of a rete adenocarcinoma in a 23-month-old mouse from a colony of 500 aged mice (0.2% incidence). Therefore, our data suggest prenatal exposure to DES results in at least a 20-fold increase in the prevalence of adenocarcinoma. In fact, if these prenatal DES males were allowed to age beyond 18 months, they might have a higher incidence of abnormalities of the rete.

In an attempt to increase the incidence of rete tumors in the prenatally DES-exposed male

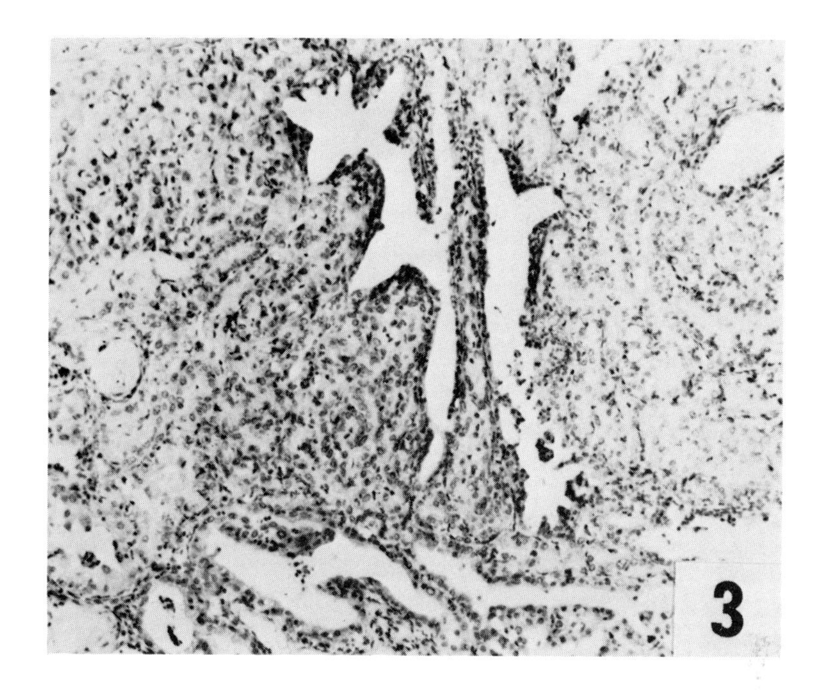

FIGURE 3. Adenocarcinoma of the rete testis from a 16-month-old prenatally DES-treated mouse. While the papillary pattern is focally apparent, the majority of the tumor has a tubular arrangement of pleomorphic epithelial cells. Rete epithelium in channels at the bottom of the photomicrograph is hyperplastic. This tumor occupies approximately 10% of the section surface of the testis. (H.E.; magnification × 25.) (From Newbold, R. R., Bullock, B. C., and McLachlan, J. A., *Cancer Res.*, 45, 5145, 1985. With permission.)

mice, estradiol (5 mg) pellets were subcutaneously implanted for 2 months prior to sacrifice. Secondary exposure to an estrogen in adult life did not increase the tumor incidence at this site. The rete lesions did not appear to be responsive to estrogen stimulation. These data suggest that the rete tumors were a result of exposure to DES during a critical stage early in development.

Since 1943,[37] the induction of interstitial cell tumors in adult mice with DES has been well studied; however, no studies of estrogen-treated adult mice[38] have reported abnormalities of the rete. The increased incidence of lesions in the rete testis reported in this paper thus suggests that this lesion may be associated with developmental exposure to DES. In addition, early exposure in development may be important since there are no reports of this lesion in other rodent models treated late prenatally or neonatally with DES or other estrogenic compounds.[39-42] An increased incidence of a rare tumor, vaginal adenocarcinoma, focused attention on the adverse effects on female offspring of women given DES while pregnant. The occurrence of rete testis tumors in the male offspring of mice given DES during pregnancy suggests that this may be an analogous situation since naturally occurring rete testis lesions are extremely rare among mice.

While no reports of rete hyperplasia or adenocarcinoma in humans have been attributed to prenatal exposure to DES, three cases of seminomas have been described in prenatally DES-exposed men, suggesting an association of prenatal DES treatment with subsequent development of testicular tumors.[12-14] A recent report[36] states that rete adenocarcinoma can be misdiagnosed as seminoma and seminoma must be ruled out before a diagnosis of rete adenocarcinoma can be made; thus, caution should be taken in diagnosing any testicular lesions associated with prenatal DES exposure.

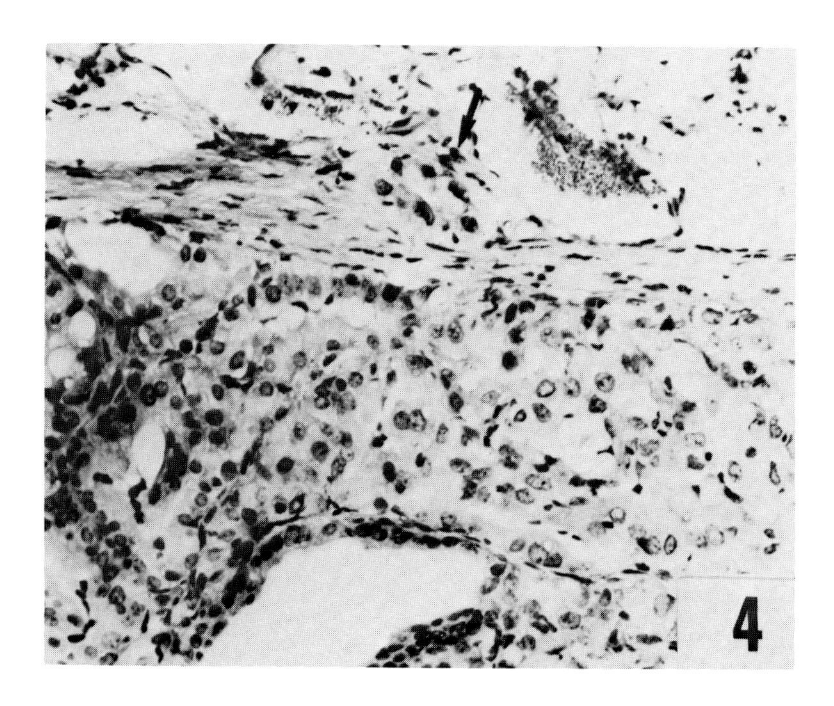

FIGURE 4. Higher magnification of the lesion suggestive of tubolopapillary adenocarcinoma in Figure 5a; there is an area of local invasion (↑). (H.E.; magnification × 60.) (From Newbold, R. R., Bullock, B. C., and McLachlan, J. A., *Cancer Res.*, 45, 5145, 1985. With permission.)

III. LESIONS OF MÜLLERIAN DUCT REMNANTS

Early in genital tract development, the Müllerian (paramesonephric) ducts in the male fetus regress in response to Müllerian-inhibiting substance (MIS) produced by the fetal testes.[43,44] The mechanism by which this substance acts is uncertain; however, MIS or an interaction of MIS and testicular secretions seems to be essential for Müllerian duct regression during the critical phase of sexual development. Since 1939, it has been known that exogenous estrogens interfere with the regression of male Müllerian ducts.[45] Studies from our laboratory have also shown that prenatal treatment of mice with DES results in the persistence of Müllerian duct derivatives which are homologous to the female duct structures such as oviduct, uterus, and upper vagina in adult male offspring.[15,29,46] Since a similar report by Driscol and Taylor[47] has described persistent Müllerian duct remnants in humans exposed *in utero* to DES, retention of Müllerian tissue in adults may be a general DES-induced biological phenomenon. Recently, several studies in our laboratory have suggested that DES exerts its effect mainly by altering the response of the Müllerian duct tissue itself rather than by suppression of MIS production by the fetal testes or a structural defect in the MIS.[48,49]

By using an organ culture assay system for MIS activity, we were able to study the mechanisms of DES inhibition of MIS-induced regression of the Müllerian ducts. In this culture system, DES-treated or control-indifferent ducts (embryonic reproductive tracts) were cocultured along with treated or control embryonic testes. Prenatal DES exposure was by subcutaneous injection to the mother (100 μg/kg body weight) on days 9 through 12 of gestation. Embryonic tissues were removed on day 13 of gestation and cultured for 72 hr. In organ culture, Müllerian duct regression, comparable to that seen in vivo, occurred when control reproductive tracts were associated with control testes. However, maintenance of the Müllerian ducts was observed in all the tissues when DES-treated testes and DES-treated

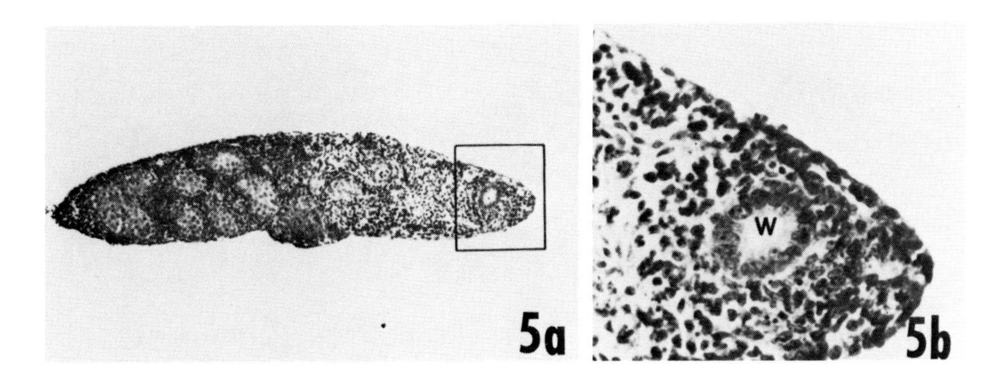

FIGURE 5. Photomicrograph of the recombination of tissues cultured for 72 hr. Reproductive $tract_{control}$ + $testis_{control}$. The Müllerian duct has regressed in the cranial portion of the reproductive tract, while the Wolffian duct (W) is maintained. (H.E.; magnification: a, × 30; b, × 160.) (From Newbold, R. R., Suzuki, Y., and McLachlan, J. A., *Endocrinology,* 115, 1863, 1984. With permission.)

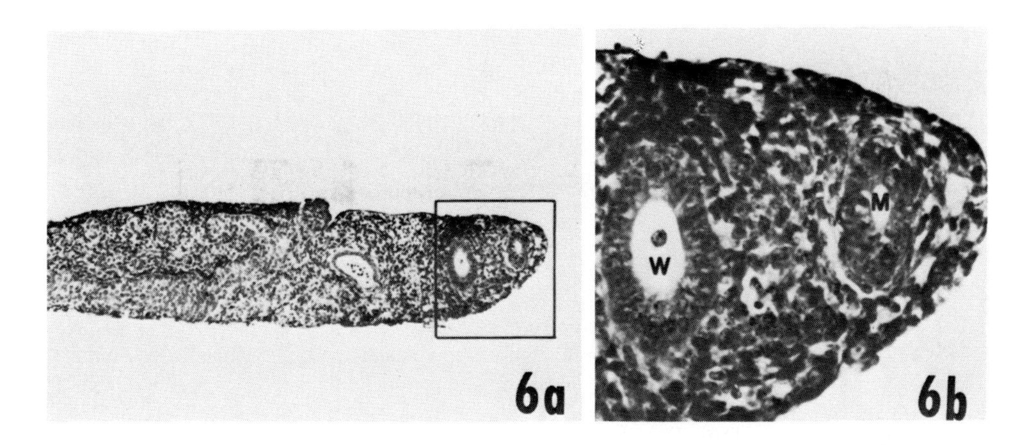

FIGURE 6. Photomicrograph of the recombination of tissues cultured for 72 hr. Reproductive $tract_{DES}$ + $testis_{DES}$. The Müllerian duct (M) persists and appears as it did at the original time of explant. The Wolffian duct (W) is also identified in this section and appears stimulated compared to control ducts. (H.E.; magnification: a, × 30; b, × 160.) (From Newbold, R. R., Suzuki, Y., and McLachlan, J. A., *Endocrinology,* 115, 1863, 1984. With permission.)

reproductive tracts were cultured together. When recombinations of control reproductive tracts and DES-treated testes were formed, there was regression of the Müllerian ducts in 87%. But, in the combinations of control testes and DES-treated reproductive tracts, 41% of the cultured tissue showed partial regression of the Müllerian duct and 59% showed no regression (Figures 5 to 8). These data support previous in vivo results that prenatal exposure to DES has an inhibitory effect on Müllerian duct regression and also further suggest that this inhibitory effect is mainly due to a decrease in responsiveness of the treated embryonic Müllerian duct.[49]

Another report from our laboratory described altered protein synthesis in reproductive tract tissues exposed prenatally to DES. Results of culturing DES-exposed fetal genital tracts with [^{35}S]methionine-containing media have revealed alterations in the protein synthetic patterns of the tissue. Analysis of approximately 280 spots on the fluorograms of 2-D gel electrophoretic samples showed a reproducible subset of proteins in the region spanned by pI 5.5 to 6.0 and molecular weight 45,000 to 90,000 daltons. In this constellation of spots, there is a protein at pI 5.8/70,000 (5.8/70) which is decreased or missing in reproductive

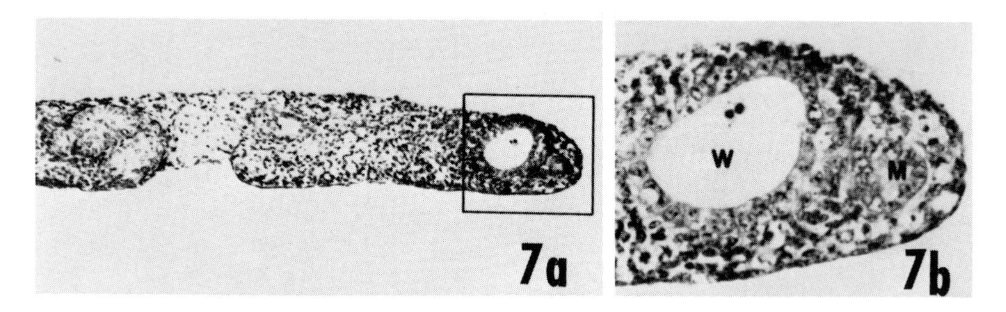

FIGURE 7. Photomicrographs of the recombination of tissues cultured for 72 hr. Reproductive $tract_{control}$ + $testis_{DES}$. The Müllerian duct (M) lumen is significantly reduced, and the mesenchyme has started to condense around the duct. This photograph demonstrates partial regression of the Müllerian duct. The Wolffian (W) is apparently enlarged in this section. (H.E.; magnification: a, × 30, b, × 160.) (From Newbold, R. R., Suzuki, Y., and McLachlan, J. A., *Endocrinology,* 115, 1863, 1984. With permission.)

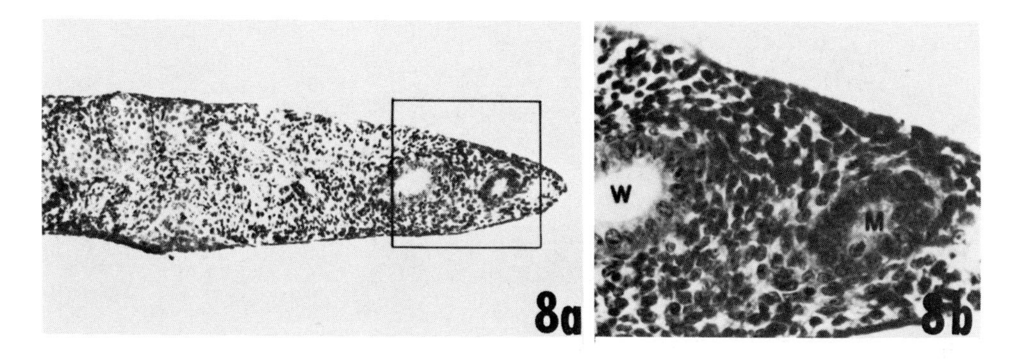

FIGURE 8. Photomicrographs of the recombination of tissues cultured for 72 hr. Reproductive $tract_{DES}$ + $testis_{control}$. There is little regression of the Müllerian duct (M). It persists as it did at the original time of culture and as it is in Figure 6. The Wolffian duct (W) is present and appears stimulated compared to control duct. (H.E.; magnification: a, × 30; b, × 160.) (From Newbold, R. R., Suzuki, Y., and McLachlan, J. A., *Endocrinology,* 115, 1863, 1984. With permission.)

tract tissues exposed to DES during fetal development. The fetal genital tract tissues at 14 days of gestation synthesize this protein, but at later stages the synthesis of this protein is greatly diminished or missing as a result of DES exposure. For example, on day 2 after birth, protein 5.8/70 identified in the control animal (Figure 9a) is missing in the DES-exposed animals (Figure 9b). The identity of this missing protein has not been established, but it appears to be a reliable and reproducible marker for a biochemical change in reproductive tract tissues exposed to DES *in utero*.[48] The relationship between altered protein synthesis and subsequent reproductive tract alterations including tumors is unclear, but it does suggest another direct effect of DES on the developing fetal genital tract.

In order to study the long-term effects on retained Müllerian duct remnants in male offspring, we again used the prenatally DES-exposed mouse model. Persistent Müllerian remnants in the cranial portion of the male reproductive tract were studied; caudal persistent Müllerian remnants (prostatic utricle) have been previously described[15], as well as other genital abnormalities induced in other male rodent models treated prenatally or neonatally with DES or other estrogenic compounds.[1,2,4,50-57] Pregnant outbred CD-1 mice were subcutaneously injected with daily doses of DES (100 μg/kg) on days 9 through 16 of gestation. DES-exposed male offspring and age-matched control males were sacrificed at 10 to 18 months of age and examined for reproductive tract abnormalities in the area of the testis, epididymis, and vas deferens. Prominent Müllerian remnants were observed in 268 out of

FIGURE 9 (a and b). Full fluorograms of the region spanned by pI 5.0 to 7.0 on the abscissa and molecular weight 16,000 to 90,000 daltons on the ordinate of proteins from the reproductive tract of 2-day-old neonatal mice; (a) control and (b) exposed to DES 100 μg/kg on days 9 through 16 of gestation. These tissues were labeled with ^{35}S-methionine for 2 hr before they were processed for isoelectric focusing. The pH range scale and the molecular weight scale are not precise but give a ± 5% map location of major proteins. In each case, 200,000 cpm were loaded onto the isoelectric gels and were exposed (fluorography) for 8 days. The arrow identifies the location of spot 5.8/70. Note that this spot is missing in the DES-treated tissue (b). (From Newbold, R. R., Carter, D. B., Harris, S. E., and McLachlan, J. A., *Biol. Reprod.*, 30, 459, 1984. With permission.)

277 (97%) of the DES-exposed males but not in any of the control animals (Table 1). These remnants demonstrated an ability to differentiate into "female-like structures" homologous to oviduct and uterus. The Müllerian remnants were enlarged and cystic and shared supporting connective tissue with adjacent male structures. Previously reported lesions,[29] termed "epididymal cysts", were histologically determined to be cystic oviduct-like structures and were, therefore, considered a Müllerian duct abnormality. Pathological changes in these male oviductal homologs included diverticuli similar to those previously reported in DES-exposed female mice.[58,59] Squamous metaplasia in the uterine homolog was observed in DES-exposed males at 13 and 17 months of age (Figure 10). Adenomas were seen in 2 animals at 15 months of age (Figure 11), and cystadenomas (Figure 12) were seen in the Müllerian remnants of 15 animals ranging from 14 to 18 months of age. In addition, there were some malignant lesions such as a carcinoma (Figure 13), two stromal tumors (Figure 14), one stromal sarcoma (Figure 15), one complex stromal sarcoma (Figure 16), and one adenocarcinoma (Figure 17).

Epididymal structures were also altered by prenatal exposure to DES. Inflammation and sperm granulomas were prevalent in DES-treated animals as young as 10 months but were

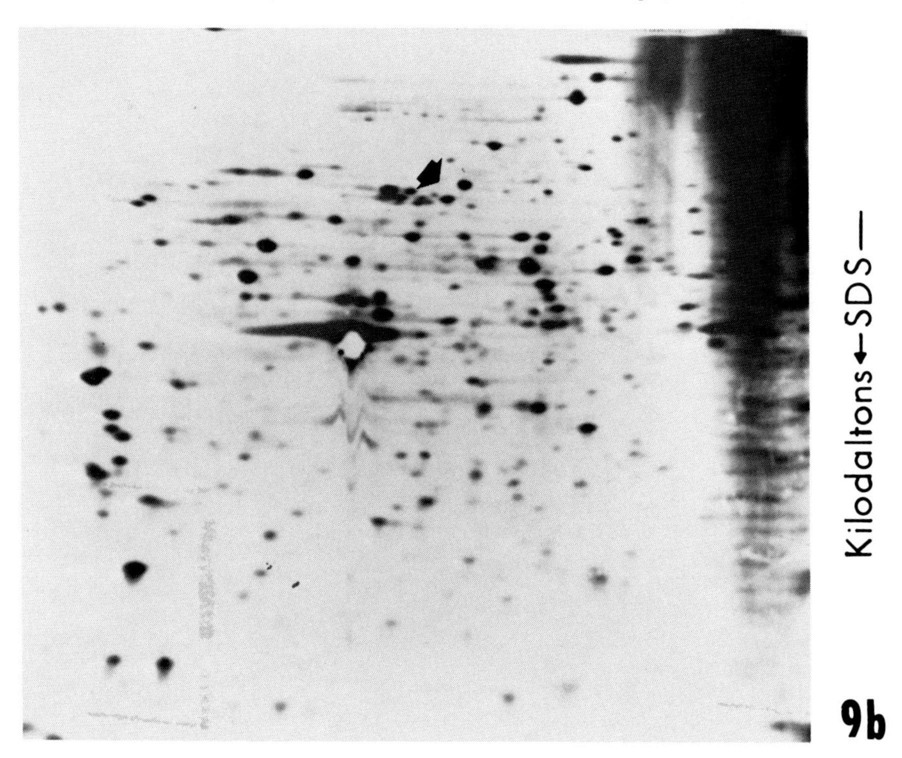

FIGURE 9 (cont'd)

observed only in control animals at 18 months of age. Cysts specifically of epididymal duct origin were seen in 32 out of 277 (12%) DES-treated animals. Hyperplasia (1 out of 55, 15 months) and adenoma (1 out of 55, 15 months) of the epididymal duct in separate animals were also observed. No comparable abnormalities were noted in 122 control males of corresponding ages (Table 1).

These data demonstrate that transplacental exposure to DES affects the differentiation and normal development of the male genital tract involving both the Müllerian (paramesonephric) and Wolffian (mesonephric) ducts. The long-term changes in both of these tissues include lesions, some of which are neoplastic, although the natural history of the lesions is not known.

It was not clear whether the male Müllerian remnants developed tumors spontaneously rather than as a direct consequence of DES treatment. Also, the possibility of factors from the testis stimulating tumor formation needs to be determined. The fact that DES-treated males had higher levels of circulating estrogens may also increase the incidence of tumors, although attempts at implanting estrogen pellets in 10-month-old prenatally DES-exposed male mice did not increase the incidence of Müllerian duct tumors.[60] Alternatively, the retained Müllerian remnants may produce toxic factors which stimulate tumor formation in the male. These possibilities must be investigated.

Certainly, the retained Müllerian duct remnants appear to have hyperplastic and dysplastic potential and appear to be more susceptible to tumor formation than adjacent male structures, although a higher incidence of epididymal cysts was also observed in DES-treated males. In fact, accumulating experimental data support the contention that DES-treated male mice are at higher risk for tumor or cyst formation than normal males. Also, there was some

Table 1
ABNORMALITIES IN MALE MICE EXPOSED PRENATALLY TO DIETHYLSTILBESTROL[a]

	Number of animals	
	Out of 277	Percent
Müllerian remnants[b]		
Prominent structure	268	97[c]
Cystic structure	121	44[c]
Cystic hyperplasia	18	6[c]
Hyperplasia	85	31[c]
Inflammation	27	10[c]
Tumors[d]	23	8[c]
Wolffian derivatives[e]		
Cystic epididymis	32	12[c]
Inflammation	87	29[c]
Sperm granuloma of epididymis	16	6[c]
Hyperplasia of epididymal duct	1	0
Tumors[f]	1	0

[a] Males were the 10- to 18-month-old offspring of CD-1 mice treated subcutaneously with DES (100 μg/kg) on days 9 through 16 of gestation.

[b] Squamous metaplasia and gland formation in male Müllerian remnants are not included in the table because of low incidence.

[c] Significantly different from corresponding controls: Fisher Exact Test $p < 0.05$.

[d] Tumors of the Müllerian remnants were distributed and cytologically characterized as follows: 2 adenomas, 15 cystadenomas, 1 carcinoma, 2 stromal tumors, 1 stromal sarcoma, 1 complex stromal sarcoma, and 1 adenocarcinoma.

[e] At 18 months of age, the control animals had minor inflammatory changes in Wolffian derivatives. One 18-month-old control animal had severe inflammation of the epididymis and another animal had a sperm granuloma. No other lesions were observed in the control animals at any age in this study.

[f] Tumor in the Wolffian derivatives was adenoma of the epididymis.

From Newbold et al., *Teratogen. Carcinogen. Mutagen.*, submitted. With permission.

suggestion that the uterine homolog may be more susceptible to tumor formation than the oviductal homolog.

A clear direct relationship between prenatal DES exposure and neoplasia of the Müllerian remnants in men has not been established. Reports of increased numbers of abnormalities in the genital and urinary tracts were found among sons of women treated with DES during pregnancy, but no evidence was detected to suggest a positive link to cancer[6-11] except for the reports from two laboratories suggesting an association with testicular seminoma.[12-14] Although any definitive judgment would be premature since these sons are still quite young, this report raises the possibility that remnants of Müllerian duct structures may pose a risk of developing benign or malignant changes in humans as they age.

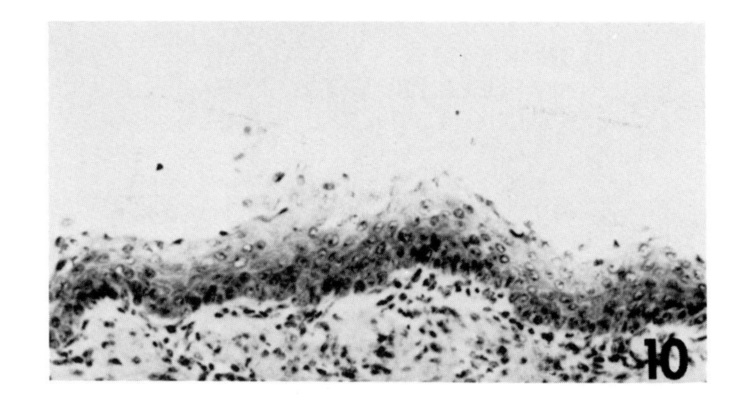

FIGURE 10. Squamous metaplasia of a Müllerian remnant in a 17-month-old DES-100 male associated with severe epididymitis. (H.E.; magnification × 30.) (From Newbold, R. R., Bullock, B. C., and McLachlan, J. A., *Tert. Ca. Mut.*, 7, 377, 1987. With permission.)

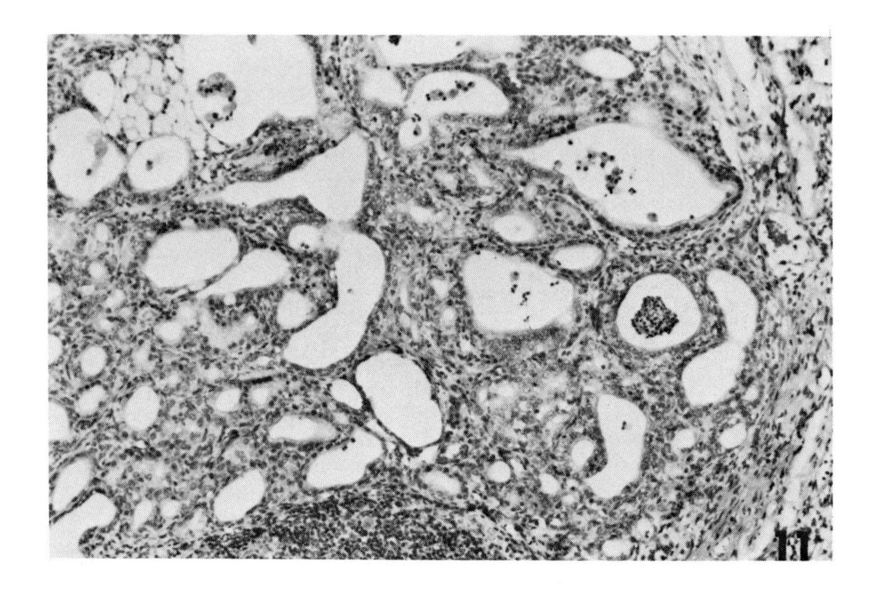

FIGURE 11. Adenoma of Müllerian duct origin. A collection of gland-like structures have oviductal-type epithelium. Although cytologically benign, there is local invasion. 15-month-old DES-100 male. (H.E.; magnification × 40.) (From Newbold, R. R., Bullock, B. C., and McLachlan, J. A., *Tert. Ca. Mut.*, 7, 377, 1987. With permission.)

IV. LESIONS OF THE CORPUS TESTIS

To follow up on two reports suggesting an association of prenatal DES exposure and later development of testicular seminoma,[12-14] we investigated testicular lesions in the prenatal DES mouse model. Animals were exposed to 100 μg/kg on days 9 to 16 of gestation and examined at 10 to 18 months of age. In addition to nonmalignant abnormalities such as retained testes, which have been previously reported in this mouse model[15,29] and described in men exposed prenatally to DES,[7-13] there were degenerative changes, ranging from mild to severe, observed in the retained and descended testes of the prenatal DES-exposed animals

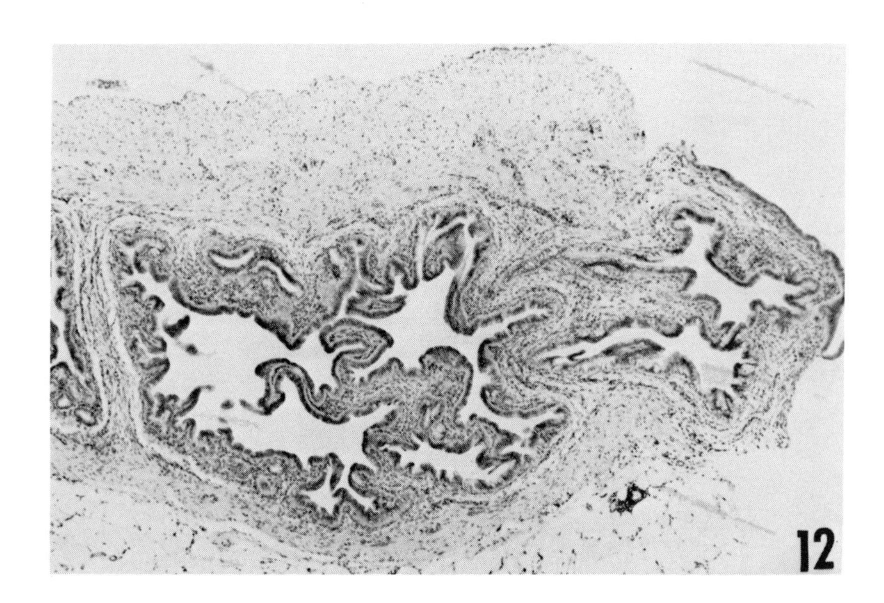

FIGURE 12. Papillary cystadenoma of the Müllerian duct remnant. Partially collapsed cystic structure has projecting stalks of epithelium similar to oviductal folds but in a more complex pattern. 15-month-old DES-100 male. (H.E.; magnification × 6.) (From Newbold, R. R., Bullock, B. C., and McLachlan, J. A., *Tert. Ca. Mut.*, 7, 377, 1987. With permission.)

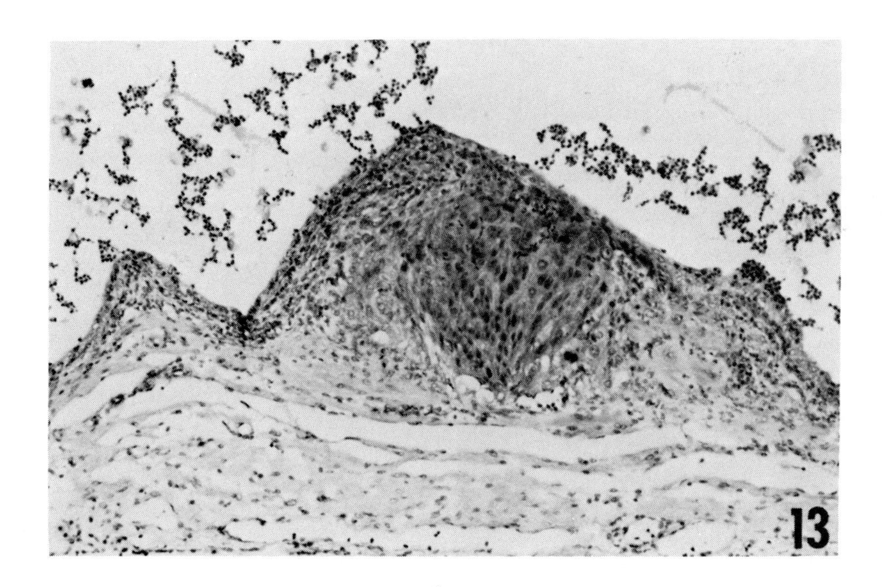

FIGURE 13. Lesion resembling intraepithelial carcinoma of the Müllerian remnant shows lack of progression of differentiation in large cells with hyperchromatic nuclei. 14-month-old DES-100 male. (H.E.; magnification × 60.) (From Newbold, R. R., Bullock, B. C., and McLachlan, J. A., *Tert. Ca. Mut.*, 7, 377, 1987. With permission.)

(Table 2). Degenerative changes were described by four categories of increasing severity: (1) relatively normal-looking testis but smaller than control, (2) marked reduced spermatogenesis with giant cells within the lumen of the seminiferous tubules, (3) atrophied seminiferous tubules with hylanized basement membrane and thickened arterioles, and (4) mostly necrotic or scar tissue comprising the testis. Categories 2, 3, and 4 are included in Table 2

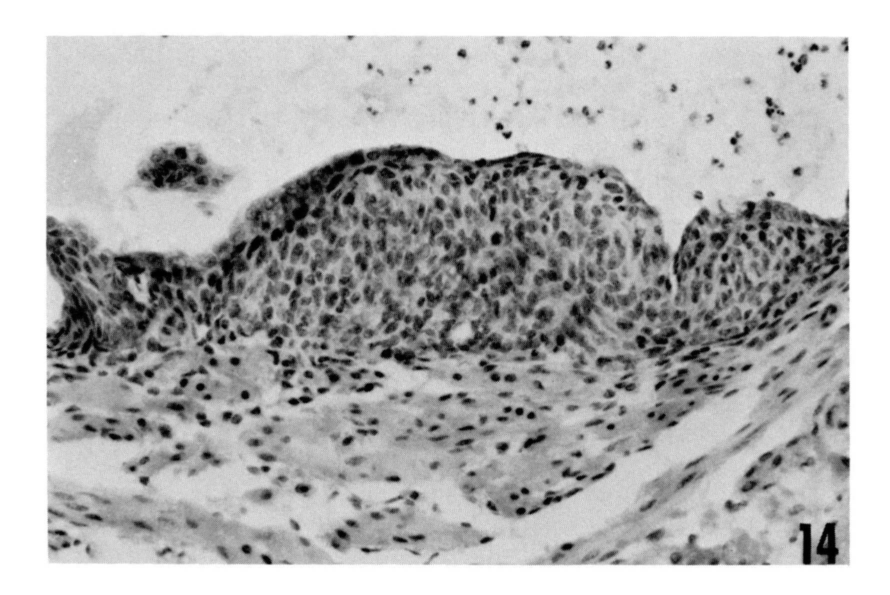

FIGURE 14. Small stromal tumor in a cystic Müllerian duct structure. 10-month-old DES-100 male. (H.E.; magnification × 30.) (From Newbold, R. R., Bullock, B. C., and McLachlan, J. A., *Tert. Ca. Mut.*, 7, 377, 1987. With permission.)

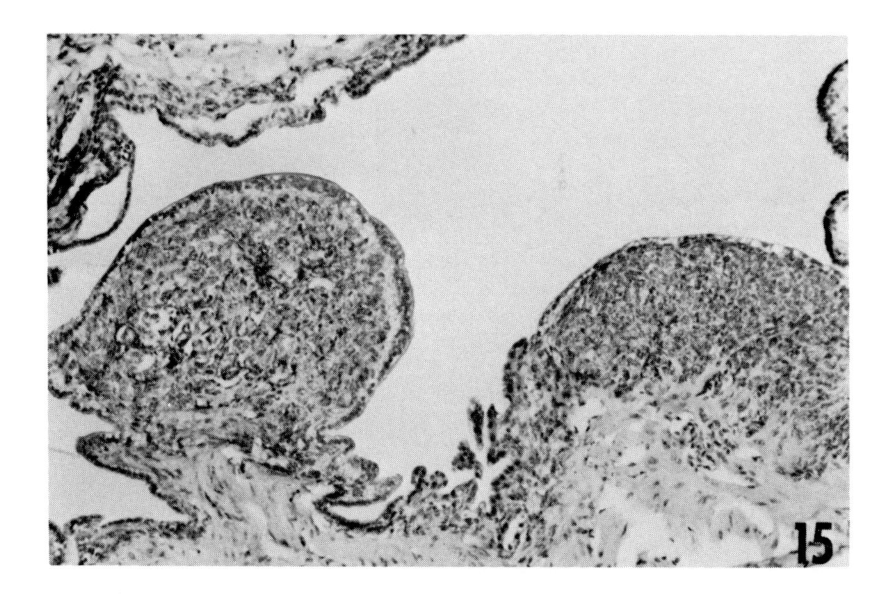

FIGURE 15. Lesion of the Müllerian remnant resembling a stromal sarcoma in an 18-month-old DES-100 male. In the two polypoid structures, there is a highly cellular area beneath an intact vacuolated surface epithelium. (H.E.; magnification × 30.) (From Newbold, R. R., Bullock, B. C., and McLachlan, J. A., *Tert. Ca. Mut.*, 7, 377, 1987. With permission.)

as degenerative changes. These types of changes were not observed in any of the control animals.

Mineralization was seen in 9% of the DES-exposed mice. This was primarily in degenerative retained testes and often within reminiscent seminiferous tubules.

Marked inflammation was seen in 8% of the DES animals (Table 2). Inflammatory changes

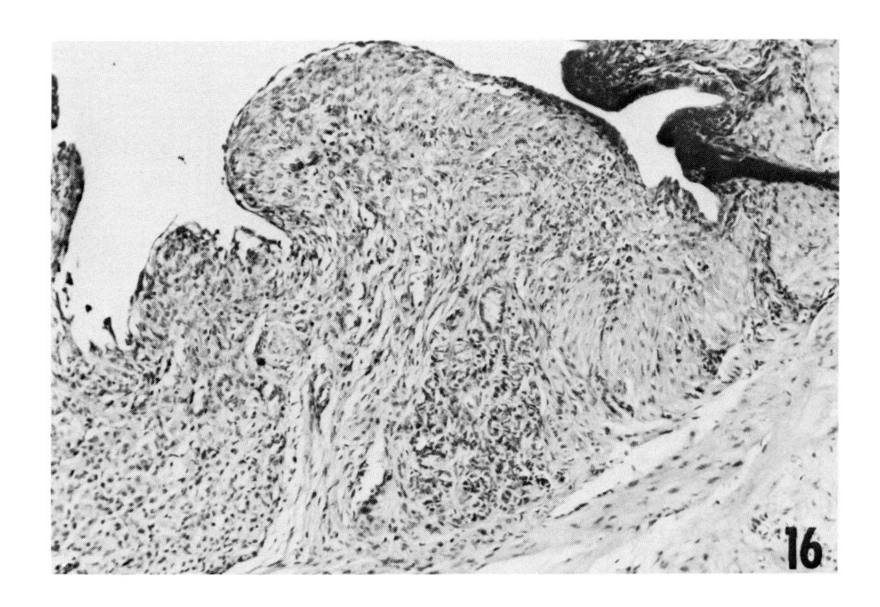

FIGURE 16. Lesion resembling a complex stromal sarcoma in the Müllerian remnant of an 18-month-old DES-100 male. (H.E.; magnification × 30.) (From Newbold, R. R., Bullock, B. C., and McLachlan, J. A., *Tert. Ca. Mut.*, 7, 377, 1987. With permission.)

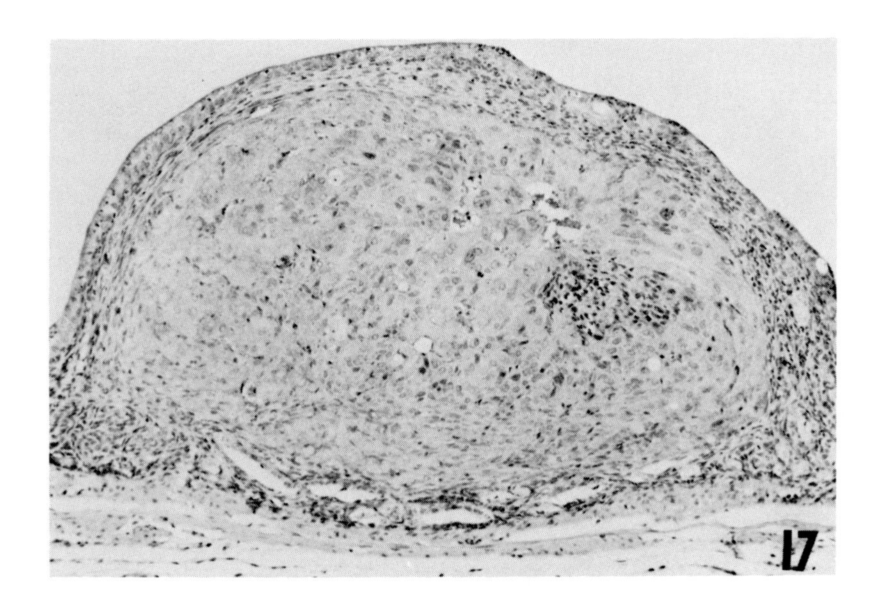

FIGURE 17. Adenocarcinoma of Müllerian remnant found within the wall of a large Müllerian cyst. These cells secrete a small amount of PAS-positive material. Adjacent glands within the cyst were PAS-positive. 18-month-old DES-100 male. (H.E.; magnification × 30.) (From Newbold, R. R., Bullock, B. C., and McLachlan, J. A., *Tert. Ca. Mut.*, 7, 377, 1987. With permission.)

Table 2
TESTICULAR LESIONS FOLLOWING PRENATAL EXPOSURE TO DIETHYLSTILBESTROL[a]

Lesion	10 month	11 month	12 month	13 month	14 month	15 month	16 month	17 month	18 month	Total
Retained[b]	45/47 (96)	17/18 (94)	9/9 (100)	22/23 (96)	49/52 (94)	45/55 (87)	16/20 (80)	9/9 (100)	40/44 (91)	252/277 (91)
Inflammation	7/47 (15)	2/18 (11)	0/9 (0)	1/23 (4)	7/52 (13)	4/55 (7)	0/20 (0)	2/9 (22)	0/44 (0)	23/277 (8)
Degenerative changes[c]	39/47 (83)	15/18 (83)	8/9 (89)	20/23 (87)	40/52 (77)	51/55 (93)	18/20 (90)	7/9 (78)	28/44 (64)	226/277 (82)
Interstitial cell tumor	0/47 (0)	0/18 (0)	0/9 (0)	0/23 (0)	0/52 (0)	0/55 (0)	0/20 (0)	0/9 (0)	2/44 (5)	2/277 (1)
Interstitial cell carcinoma	1/47 (2)	0/18 (0)	0/9 (0)	0/23 (0)	3/52 (6)	0/55 (0)	0/20 (0)	0/9 (0)	1/44 (2)	5/277 (2)

Note: Numbers in parentheses are percentages.

[a] Lesions in the reproductive tract of male mice exposed prenatally to DES. Males were the 10- to 18-month-old offspring of CD-1 mice treated with DES (100 μg/kg, subcutaneously) on days 9 through 16 of gestation. Similar abnormalities were not observed in the control animals in this study.

[b] This number represents animals with at least one retained testis.

[c] These data include only categories 2, 3, and 4 (see text).

appeared to be secondary to either escape of sperm from seminiferous tubules into interstitial tissues and/or necrosis. There did not appear to be primary inflammatory disease in the testes.

Sperm granulomas were frequently seen in degenerating testes. These were typical granulomatous reactions with recognizable sperm mixed with or surrounded by neutrophils in a collar of large foamy macrophages.

Interstitial cell tumors were seen in 2 out of 277 DES-exposed mice. These lesions were small tumors occurring in 18-month-old animals. In addition to the two benign tumors, there were five malignant interstitial cell tumors (2%).

These data suggest that mice exposed to DES have an increased risk of developing testicular abnormalities including hypotropic testes and interstitial cell tumors. Although the incidence of testicular tumors is low in the corpus testis, the additional finding of rete testis adenocarcinoma raises the combined incidence of testicular tumors to 8% in the DES-exposed animals. Prior to these recent findings, most of the testicular abnormalities have been subtle, but the new data suggest that the changes in the testis, including neoplasia, are significant.

Although seminomas have been experimentally induced in dogs,[61] germ cell tumors of the testis are very rare in rodents.[62] Cryptorchidism is considered to be a predisposing factor to seminoma in men and dogs, but in spite of the high incidence of retained testis in the mouse model, we were not able to demonstrate this particular testicular lesion.

Induction of interstitial cell tumors in mouse testis has been reported after prolonged administration of various estrogenic compounds; however, the strain of mouse, the specific agent, and the duration of treatment affect the results. Spontaneously occurring interstitial cell tumors have been reported in hybrid strains of mice as well as in experimentally induced cryptorchid testis of Balb/c mice. These are discussed in review by Mostofi and Bresler[62] and Mostofi and Sesterhenn.[63] Other than the recent findings in the mouse model, the only demonstrated association between malignant growth of the testis and DES exposure has been in a mouse injected with large, repeated doses of the drug by Hooker.[64]

The association of prenatal DES exposure and the development of testicular tumors in men has become a subject of much controversy over the last few years as attention has shifted to identify the exposed male progeny. Some reports, specifically addressing factors for cancer of the testis, list prenatal DES exposure as a risk,[30-34] while other studies show no relationship to hormonal treatment during pregnancy.[65,66] The data in the experimental mouse model support the contention that DES-treated males are at a higher risk for testicular tumors than normal males.

V. EMBRYOLOGICAL CONSIDERATIONS

A role for tissues derived from the embryonic rudiments of female genital organs (Müllerian duct remnants) in the reproductive tract abnormalities of male mice exposed transplacentally to DES was previously proposed by our laboratory.[29] Subsequently, a similar proposition was offered for DES-exposed human males.[9] In the fetus, the definitive sexual structures of the female (oviducts, uterus, cervix, and upper vagina) derived from the Müllerian (paramesonephric) duct while those of the male (epididymis, vas deferens, seminal vesicle, and ejaculatory duct) derived from the Wolffian (mesonephric) duct. Under normal conditions the male accomplished sex differentiation by maintenance and development of the Wolffian duct system with corresponding regression of the Müllerian (female) duct system; the regression of the Müllerian duct (female) in the males exposed to DES during gestation was incomplete with remnants located mainly in the area of the testis (appendix testis) and the seminal colliculus (prostatic utricle). These two anatomical positions were consistent with the epididymal cysts and nodular masses of the coagulating glands and metaplasia of the seminal colliculus observed previously in DES-exposed male mice.[29] More detailed study of the Müllerian remnants in the area of the appendix testis suggests that some lesions of the DES-exposed male genital tract may, in part, be mediated through abnormal differentiation of the female duct system. Retention and hyperplasia of the Müllerian duct have been demonstrated in DES-treated CD-1 male mouse neonates,[15] and Müllerian duct structures were observed in the epididymis adjacent to the vas deferens in 6-month-old male mice treated prenatally with DES.[46] Retention of female structures was reported in newborn male rats given DES prenatally[1] and in DES-exposed human neonates.[47]

When reviewing all the data, there is additional support for the previous hypothesis that, at least for "epididymal cysts", persistent Müllerian duct structures were involved in the pathogenesis of male genital lesions in old animals. Cysts of the epididymis were a common abnormality in the reproductive tracts of both mouse[29] and human males[67] exposed to DES *in utero* (43% and 19%, respectively). Thus, epididymal cysts in humans, like those in mice, may be associated with cysts in tissues derived from Müllerian duct remnants.

The role of tissues derived from the Müllerian duct in the pathogenesis of the prostate/coagulating gland has been addressed in past studies. Previous results associated squamous metaplasia of the coagulating gland and the seminal colliculus with tissues derived from these female rudiments, and demonstrated that the caudal Müllerian duct derivatives, in the area of the prostatic utricle, were well maintained and in some cases had pathological changes after prenatal exposure to DES.[15,29]

Not only were structures derived from Müllerian duct origin affected by prenatal exposure to DES, but structures from mesonephric origin also appear to demonstrate abnormalities. In an early report, mesonephric ducts of the prenatally DES-treated neonate were observed to be hyperplastic and there was evidence of both epithelial and stromal stimulation.[46] Thus, DES also appears to have an effect in the male in the mesonephric structures *in utero*. The relationship of this observation to observed long-term effects of mesonephric-derived tissues such as epididymis and seminal vesicle is unclear. However, the carcinosarcoma of the seminal vesicle[15] and the rete testis adenocarcinoma[21,22] certainly suggest the possibility that these structures are potential targets. Also, it is important to note that cystic mesonephric remnants were seen in 62.5% of prenatally DES-treated females but in only 10% of the corresponding controls.[17] Also a recent report by Haney et al.[16] described paraovarian cysts of mesonephric origin in women exposed prenatally to DES. Moreover, the accumulation of radioactivity is seen in the nuclei of mesenchymal cells in both the mesonephric and Müllerian ducts of 16-day-old male and female fetuses whose mothers were treated with ^{3}H-DES, indicating that both ductal systems are responsive to DES.[68]

VI. CONCLUSION

Although animal studies must be considered thoughtfully if extrapolation to humans is to follow, the prenatal mouse model has provided some interesting comparisons to similarly DES-exposed women such as reduced reproductive capacity, uterine structural malformations, malformed oviduct, and salpingitis isthmica nodosa of the oviduct.[69] Also, in earlier reports on DES-exposed male mice, we suggested cryptorchid testes and epididymal cysts might be a common finding in exposed humans.[29] Therefore, experimental results can be informative and possibly predictive. Hence, continued close surveillance is necessary, especially in view of the young age of the exposed human males and the high incidence of cryptorchidism in the DES-exposed patients.

ACKNOWLEDGMENTS

The authors wish to thank Dr. Bill Bullock, Bowman-Gray School of Medicine, Wake Forest University, Winston-Salem, N.C. for his consultations in pathology. Also the authors are indebted to Ms. Sandra Sandberg for the typing and editorial assistance on this manuscript.

REFERENCES

1. **Greene, R. R., Burrill, M. W., and Ivy, A. C.,** Experimental intersexuality, *Anat. Rec.,* 74, 439, 1939.
2. **Dunn, T. B. and Greene, A. W.,** Cysts of the epididymis, cancer of the cervix, granular cell myoblastoma and other lesions after estrogen injection in newborn mice, *J. Natl. Cancer Inst.,* 31, 425, 1963.
3. **Takasugi, N., Bern, H. A., and De Ome, K. B.,** Persistent vaginal cornification in mice, *Science,* 138, 438, 1962.
4. **Nomura, T. and Kanzaki, T.,** Induction of urogenital anomalies and some tumors in the progeny of mice receiving diethylstilbestrol during pregnancy, *Cancer Res.,* 37, 1099, 1977.
5. **McLachlan, J. A.,** *Estrogens in the Environment,* Elsevier/North Holland, New York, 1980.
6. **Greenwald, P., Nasca, P. C., Burnett, W. S., and Polan, A.,** Prenatal stilbestrol experience of mothers of young cancer patients, *Cancer Brussels,* 31, 568, 1973.
7. **Bibbo, M., Al-Naqeeb, M., Baccarini, I., Gill, W., Newton, M., Sleeper, K., Sonek, M., and Wied, G. L.,** Follow-up study of male and female offspring of DES-treated mothers: a preliminary report, *J. Reprod. Med.,* 15, 29, 1975.
8. **Bibbo, M., Gill, W. B., Azizi, F., Blough, R., Fang, V. S., Rosenfield, R. L., Schumacher, G. F. B., Sleeper, K., Sonek, M. G., and Wied, G. L.,** Follow-up study of male and female offspring of DES-exposed mothers, *Am. J. Obstet. Gynecol.,* 49, 1, 1977.
9. **Gill, W. B., Schumacher, G. F. B., and Bibbo, M.,** Structural and functional abnormalities in the sex organs of male offspring of mothers treated with diethylstilbestrol (DES), *J. Reprod. Med.,* 16, 147, 1976.
10. **Whitehead, E. D. and Leiter, E.,** Genital abnormalities and abnormal semen analyses in male patients exposed to diethylstilbestrol in utero, *J. Urol.,* 125, 47, 1981.
11. **Gill, W. B., Schumacher, G. F. B., Hubby, M. H. et al.,** Male genital tract changes in humans following intrauterine exposure to diethylstilbestrol, in *Developmental Effects of Diethylstilbestrol (DES) in Pregnancy,* Herbst, A. L. and Bern, H. A., Eds., Thieme-Stratton, New York, 1981, 103.
12. **Gill, W. B., Schumacher, G. F. B., and Bibbo, M.,** Genital and semen abnormalities in adult males two and one half decades after in utero exposure to diethylstilbestrol, in *Intrauterine Exposure to Diethylstilbestrol in the Human,* Herbst, A. L., Ed., The American College of Obstetricians and Gynecologists, Chicago, 1978, chap. 7.
13. **Gill, W. B., Schumacher, G. F. B., Bibbo, M., Straus, F. H., II, and Schoenberg, H. W.,** Association of diethylstilbestrol exposure in utero with cryptorchidism, testicular hypoplasia, and semen abnormalities, *J. Urol.,* 122, 36, 1979.
14. **Conley, G. R., Sant, G. R., Ucci, A. A., and Mitcheson, H. D.,** Seminoma and epididymal cysts in a young man with known diethylstilbestrol exposure in utero, *JAMA,* 249, 1325, 1983.
15. **McLachlan, J. A.,** Rodent models for perinatal exposure to diethylstilbestrol and their relation to human disease in the male, in *Developmental Effects of Diethylstilbestrol (DES) in Pregnancy,* Herbst, A. L. and Bern, H. A., Eds., Thieme-Stratton, New York, 1981, 148.
16. **Haney, A. F., Newbold, R. R., Fetter, B. F., and McLachlan, J. A.,** Prenatal diethylstilbestrol exposure in the mouse: effects on ovarian histology and steroidogenesis in vitro, *Biol. Reprod.,* 30, 471, 1984.
17. **McLachlan, J. A., Newbold, R. R., and Bullock, B. C.,** Long-term effects on the female mouse genital tract associated with prenatal exposure to diethylstilbestrol, *Cancer Res.,* 40, 3988, 1980.
18. **Newbold, R. R., Bullock, B. C., and McLachlan, J. A.,** Exposure to diethylstilbestrol during pregnancy permanently alters the ovary and oviduct, *Biol. Reprod.,* 28, 735, 1983.
19. **Byskov, A. G.,** The anatomy and ultrastructure of the rete system in the fetal mouse ovary, *Biol. Reprod.,* 19, 720, 1978.
20. **Byskov, A. G.,** Gonadal sex and germ cell differentiation, in *Mechanisms of Sex Differentiation in Animals and Man,* Austin, C. R. and Edwards, R. G., Eds., Academic Press, New York, 1981, 145.
21. **Newbold, R. R., Bullock, B. C., and McLachlan, J. A.,** Lesions of the rete testis in mice exposed prenatally to diethylstilbestrol, *Cancer Res.,* 45, 5145, 1985.
22. **Newbold, R. R., Bullock, B. C., and McLachlan, J. A.,** Adenocarcinoma of the rete testis, *Am. J. Pathol.,* 125, 625, 1986.
23. **Feek, J. D. and Hunter, W. C.,** Papillary carcinoma arising from rete testis, *Arch. Pathol.,* 40, 399, 1945.
24. **Schoen, S. S. and Rush, B. F., Jr.,** Adenocarcinoma of the rete testes, *J. Urol.,* 82, 356, 1959.
25. **Mostofi, F. K. and Price, E. B., Jr.,** Tumors of the male genital system, in *Atlas of Tumor Pathology,* 2nd Series, Fascicle 8, Armed Forces Institute of Pathology, Washington, D. C., 1973, 127.
26. **Shillitoe, A. J.,** Carcinoma of the rete testes, *J. Pathol. Bacteriol.,* 64, 650, 1952.
27. **Teilum, G.,** *Special Tumors of Ovaries and Testes and Related Extra Gonadal Lesions: Comparative Pathology and Histological Identification,* J. B. Lippincott, Philadelphia, 1971, 407.
28. **Gisser, S. D., Nayak, S., Kaneko, M., and Tchertkoff, V.,** Adenocarcinoma of the rete testis: a review of the literature and presentations of a case with associated asbestosis, *Hum. Pathol.,* 8, 219, 1977.

29. **McLachlan, J. A., Newbold, R. R., and Bullock, B. C.,** Reproductive tract lesions in male mice exposed prenatally to diethylstilbestrol, *Science,* 190, 991, 1975.
30. **Henderson, B. E., Benton, B. D. A., Jing, J., Yu, M. C., and Pike, M. C.,** Risk factors for cancer of the testis in young men, *Int. J. Cancer,* 23, 598, 1979.
31. **Henderson, B. E., Ross, R. K., Pike, M. C., and Casagrande, J. T.,** Endogenous hormones as a major factor in human cancer, *Cancer Res.,* 42, 3232, 1982.
32. **Fink, D. J.,** *DES Task Force Summary Report,* U.S. Government Printing Office, Washington, D.C., 1978, 63.
33. **Loughlin, J. E., Robboy, S. J., and Morrison, A. S.,** Risk factors for cancer of the testis, *N. Engl. J. Med.,* July 10, 112, 1980.
34. **Depue, R. H., Pike, M. C., and Henderson, B. E.,** Estrogen exposure during gestation and risk of testicular cancer, *J. Natl. Cancer Inst.,* 71, 1151, 1983.
35. **Pottern, L. M., Brown, L. M., Hoover, R. N., Javadpour, N., O'Connell, K. J., Stutzman, R. E., and Blattner, W. A.,** Testicular cancer risk among young men: role of cryptorchidism and inguinal hernia, *J. Natl. Cancer Inst.,* 74, 377, 1985.
36. **Yoshitomi, K. and Morii, S.,** Benign and malignant epithelial tumors of the rete testis in mice, *Vet. Pathol.,* 21, 300, 1984.
37. **Gardner, W. U.,** Spontaneous testicular tumors in mice, *Cancer Res.,* 3, 757, 1943.
38. **Huseby, R. A.,** Demonstration of a direct carcinogenic effect of estradiol of Leydig cells of the mouse, *Cancer Res.,* 40, 1006, 1980.
39. **Arai, Y., Mori, T., Suzuki, Y., and Bern, H.,** Long-term effects of perinatal exposure to sex steroids and diethylstilbestrol on the reproductive system of male mammals, in *International Review of Cytology,* Vol. 84, Bourne, G. H. and Danielli, J. F., Eds., Academic Press, New York, 1983, 235.
40. **Boylan, E. S.,** Morphological and functional consequences of prenatal exposure to diethylstilbestrol in the rat, *Biol. Reprod.,* 19, 854, 1978.
41. **Jones, L. A.,** Long term effects of neonatal administration of estrogen and progesterone, alone or in combination, on male Balb/c and Balb/cf C3H mice, *Proc. Soc. Exp. Biol. Med.,* 165, 117, 1980.
42. **Takasugi, N.,** Testicular damages in neonatally estrogenized adult mice, *Endocrinol. Jpn.,* 17(2), 277, 1970.
43. **Jost, A.,** Embryonic sexual differentiation (morphology, physiology, abnormalities), in *Hermaphroditism, Genital Anomalies and Related Endocrine Disorders,* Jones, H. W. and Scott, W. W., Eds., Williams & Wilkins, Baltimore, 1971, 16.
44. **Josso, N., Picard, J. Y., and Tran, D.,** The antiMüllerian hormone, *Recent Prog. Horm. Res.,* 33, 117, 1977.
45. **Wolf, E.,** L'action due diethylstilbestrol sur les organes genitaux de l'embryon de poulet, *C. R. Acad. Sci Ser.,Paris* 208, 1532, 1939.
46. **Newbold, R. R. and McLachlan, J. A.,** Effects of DES on developing fetal reproductive tracts, *J. Toxicol. Environ. Health,* 4, 491, 1978.
47. **Driscol, S. G. and Taylor, S. H.,** Effects of prenatal maternal estrogen on the male urogenital system, *Obstet. Gynecol.,* 56, 537, 1980.
48. **Newbold, R. R., Carter, D. B., Harris, S. E., and McLachlan, J. A.,** Molecular differentiation of the mouse genital tract: altered protein synthesis following prenatal exposure to diethylstilbestrol, *Biol. Reprod.,* 30, 459, 1984.
49. **Newbold, R. R., Suzuki, Y., and McLachlan, J. A.,** Müllerian duct maintenance in heterotypic organ culture after in vivo exposure to diethylstilbestrol, *Endocrinology,* 115, 1863, 1984.
50. **Thorborg, J. V.,** On the influence of oestrogenic hormones on the male accessory genital system, *Acta Endocrinol. Copenhagen Suppl.,* 2, 1, 1948.
51. **Mori, T.,** Effects of early postnatal injections of estrogen on endocrine organs and sex accessories in male C3H/MS mice, *J. Fac. Sci. Univ. Tokyo Sect. 4,* 11, 243, 1967.
52. **Arai, Y.,** Nature of metaplasia in rat coagulating glands induced by neonatal treatment with estrogen, *Endocrinology,* 86, 918, 1970.
53. **Arai, Y., Suzuki, Y., and Nishizuka, Y.,** Hyperplastic and metaplastic lesions in the reproductive tract of male rats induced by neonatal treatment with diethylstilbestrol, *Virchows Arch. A,* 376, 21, 1977.
54. **Arai, Y., Chen, C. Y., and Nishizuka, Y.,** Cancer development in male reproductive tract in rats given diethylstilbestrol at neonatal age, *Gann,* 69, 861, 1978.
55. **Rustia, M.,** Role of hormone imbalance in transplacental carcinogenesis induced in Syrian golden hamsters by sex hormones, *J. Natl. Cancer Inst. Monogr.,* 51, 77, 1979.
56. **Vorherr, H., Messer, R. H., Vorherr, U. F., Jordan, S. W., and Kornfeld, M.,** Teratogenesis and carcinogenesis in rat offspring after transplacental and transmammary exposure to diethylstilbestrol, *Biochem. Pharmacol.,* 28, 1865, 1979.
57. **Arai, Y.,** Metaplasia in male rat reproductive accessory glands induced by neonatal estrogen treatment, *Experimentia,* 24, 180, 1984.

58. **Newbold, R. R., Bullock, B. C., and McLachlan, J. A.,** Progressive proliferative changes in the oviduct following developmental exposure to diethylstilbestrol, *Teratogen. Carcinogen Mutagen.*, 5, 473, 1985.
59. **Newbold, R. R. and McLachlan, J. A.,** Effects of perinatal exposure to diethylstilbestrol on oviductal development, in *The Fallopian Tube: Basic Studies and Clinical Contributions,* Siegler, A. M., Ed., Futura Publishing Company, New York, 1986, chap. 14.
60. **Newbold, R. R., Bullock, B. C., and McLachlan, J. A.,** Lesions in Müllerian remnants of male mice exposed prenatally to diethylstilbestrol, *Tert. Ca. Mut.*, 7, 377, 1987.
61. **von Bomhard, D., Pukkavesa, C., and Haenichen, T.,** The ultrastructure of testicular tumors in the dog, *J. Comp. Pathol.*, 88, 49, 1978.
62. **Mostofi, F. K. and Bresler, V. M.,** Tumors of the testes, in *IARC (Int. Agency Res. Cancer) Sci. Publ.*, 23, 325, 1979.
63. **Mostofi, F. K. and Sesterhenn, I.,** Neoplasms of the male reproductive system, in *The Mouse in Biomedical Research,* Foster, H. L., Small, J. D., and Fox, J. G., Eds., Academic Press, New York, 1982, chap. 23.
64. **Hooker, C. W., Gardner, W. U., and Pfeiffer, C. A.,** Testicular tumors in mice receiving estrogens, *JAMA,* 115, 443, 1940.
65. **Beard, C. M., Melton, L. J., III, O'Fallon, M., Noller, K. L., and Benson, R. C.,** Cryptorchidism and maternal estrogen exposure, *Am. J. Epidemiol.*, 120, 707, 1984.
66. **Leary, F. J., Resseguie, L. J., Kurland, L. T., O'Brien, P. C., Emslander, R. F., and Noller, K. L.,** Males exposed in utero to diethylstilbestrol, *JAMA,* 252, 2984, 1984.
67. **Gill, W. B., Schumacher, G. F. B., and Bibbo, M.,** Pathological semen and anatomical abnormalities of the genital tract in human male subjects exposed to diethylstilbestrol in utero, *J. Urol.*, 117, 477, 1977.
68. **Stumpf, W. E., Narbaitz, R., and Sar, M.,** Estrogen receptors in the fetal mouse, *J. Steroid Biochem.*, 12, 55, 1980.
69. **Newbold, R. R. and McLachlan, J. A.,** Diethylstilbestrol associated defects in murine genital tract development, in *Estrogens in the Environment,* McLachlan, J. A., Ed., Elsevier, New York, 1985, 288.

Chapter 8

ESTROGENS AND IMMUNITY: LONG-TERM CONSEQUENCES OF NEONATAL IMPRINTING OF THE IMMUNE SYSTEM BY DIETHYLSTILBESTROL

T. Kalland and R. Holmdahl

TABLE OF CONTENTS

I. Introduction 112

II. Estrogens and Immunity 112
   A. Estrogen Influence on Physiological Immunity 112
      1. Sexual Dimorphism of the Immune Response 112
      2. Influence of Castration and Treatment with Estrogen of Adult Animals 113
      3. Effects on "Nonspecific" Immune Functions 113
      4. Effects on T Cell-Mediated Immunity 113
      5. Effects on B Cell-Mediated Immunity 114
   B. Estrogen Influence on Autoimmunity 115
      1. Estrogen Influences on B Cell-Mediated Autoimmunity; The Lupus Disease 115
      2. Estrogen Influences on T Cell-Mediated Autoimmunity; The Rheumatoid Disease 115
   C. Summary of Estrogen Effects on Immunity in Adult Animals 116

III. Imprinting Effects on the Immune System by Neonatal Exposure to Diethylstilbestrol 117
   A. Pathotoxicology of Lymphoid Organs 117
   B. T Cell-Mediated Immunity 117
   C. B Cell-Mediated Immunity 118
   D. Natural Killer Cell Activity and Tumor Susceptibility 118
   E. Summary of Imprinting Effects on the Immune System after Neonatal Exposure to Diethylstilbestrol 120

IV. Imprinting of the Immune System: Further Perspectives 120

References 121

I. INTRODUCTION

It is well documented that perinatal treatment of experimental animals with substances that have sex steroid activity interferes with the normal development of a number of organs and thus may have long-lasting and serious biological consequences.[1] The brain, pituitary gland, mammary gland, and genital tract are targets for the nonsteroidal estrogen diethylstilbestrol (DES) in the perinatal period. More recently, the immune system has been included among the organ systems being affected by DES.[2-4] The immune system of the mouse undergoes a series of developmental changes during the perinatal period which may make it particularly vulnerable to damage.[5] Similar critical developmental stages have been identified for other organ systems, such as sex differentiation of the central nervous system or hepatic enzymes which are permanently altered by neonatal sex steroid exposure, the phenomenon of imprinting.[6,7]

Estrogen effects on the immune system of adult experimental animals are extensively documented.[8-10] In contrast to estrogen effects after neonatal exposure, these effects are reversible. However, disturbances of endocrine regulation as a consequence of perinatal estrogen treatment may lead to secondary effects on an already impaired immune system. To provide a background for the discussion of the possible participation of the immune system in the developmental effects of perinatal estrogen treatment, a review of more recent findings on the modulatory effects of estrogen on physiological immunity as well as autoimmunity is given.

II. ESTROGENS AND IMMUNITY

A. Estrogen Influence on Physiological Immunity

1. Sexual Dimorphism of the Immune Response

It has long been known that males and females respond differently to immunologic stimuli. The most common finding is an enhanced B cell function in females compared with males. The T cell function has been reported to be either enhanced or depressed in females depending on the experimental system. Eidinger and Garrett[11] found that female mice developed a better antibody response to both T-dependent and T-independent antigens compared with males. The mitogenic response of both T and B cells[12] as well as more complex biological reactions such as transplantation reactions[13] are enhanced in females compared with males. We have found that the T cell-dependent antibody response towards type II collagen in mice is relatively depressed in females.[14] Human females have higher IgM levels compared with males, while lymphocytes from females display a reduced cytotoxic function compared with lymphocytes from males.[15]

It is possible that parts of these observed differences could be explained by the actions of sex hormones. However, since both X chromosomes[16-18] and Y chromosomes[19,20] are believed to harbor genes that can accelerate B cell responses, they may also influence the observed sexual differences in the immune responses.

Attempts to correlate hormone levels and immune response during the estrous cycle show that the responsiveness of murine spleen cells to both B cell and T cell mitogens are highest at proestrus and metaestrus as compared with estrus and diestrus.[12,21]

Pregnancy-associated changes in immune reactivity have been reported repeatedly. Immunosuppression, at least locally in the uterus, is necessary for the prevention of a maternal-fetal rejection response and appears to affect primarily cell-mediated reactions.[22] Pregnancy does prevent skin transplant rejections in rat uteri, and the same effect was obtained with a combined estrogen-progesterone treatment of virgin females.[22] The response to the contact allergen picryl chloride was diminished in pregnant compared with pseudopregnant mice. Estrogen treatment of pseudopregnant mice in this system induced the same degree of suppression.[23] Furthermore, T cell-dependent antibody responses have been shown to be depressed in pregnancy.[24] The number of B cells is essentially unchanged, while the number of T cells is temporarily decreased during murine pregnancy.[25] Likewise, the T-helper (CD4) cell subset is specifically decreased during human pregnancy.[26]

It is possible that the high levels of estrogens during pregnancy contribute to the observed generalized suppression of T cell immunity, although other candidates, such as progestins, are likely to be potent immunosuppressors, especially locally in the fetoplacental unit.[27] However, estrogens probably play an important role and the pregnant state may be a physiological explanation for the immunomodulatory effects of estrogen.

2. Influence of Castration and Treatment with Estrogen of Adult Animals

To directly investigate the in vivo role of estrogens on the immune system, animals have to be manipulated either by gonadectomy and/or administration of estrogens. Both methods have drawbacks. Castration drastically reduces physiological levels of estrogen but also of other hormones, e.g., progesterone. In addition, it induces a trauma with an undefined stress hormone response that may severely affect immune functions. Estrogen treatment, on the other hand, has usually been performed with implanted capsules which induce serum levels that frequently exceed physiological estrogen levels. In addition, estrogen treatment does not mimic the normal cycling physiology.[28,29] However, we think with these considerations that when the large number of manipulative experiments are taken together they form a general picture of estrogen-mediated effects on the normal immune system.

3. Effects on ''Nonspecific'' Immune Functions

It is clearly established that estrogens are potent stimulators of the reticuloendothelial system[30,31] and peritoneal macrophages.[32] Two effects on macrophages have been postulated: immediate activation of various functions including phagocytic capacity and a later phase characterized by stimulated proliferation.[32] Recently, spleen macrophages were reported to produce enhanced amounts of Interleukin-1 at physiological levels of estradiol in vitro.[33] Other properties of fully activated macrophages such as secretion of certain proteases or tumor cytotoxicity were not enhanced after estrogen treatment.[32] This suggests that estrogen directly or indirectly stimulates distinct subsets or stages of macrophage differentiation. However, the macrophage-stimulating capacities of different estrogens and estrogen-analogs do not correlate with estrogenicity.[34]

4. Effects on T Cell-Mediated Immunity

A number of reports show that castration of females facilitates T cell-mediated immune reactions. These include transplant rejection,[13] T cell proliferation in vitro,[35] and induction of thymic hyperplasia.[36] Furthermore, T cell-dependent immunity is depressed by treatment with estrogens: transplant rejection,[37] delayed type hypersensitivity,[38,39] and T cell proliferation in vitro.[34,40] In spite of the clear evidence that different T cell-mediated effector functions are depressed by an estrogenic influence, neither the mechanism of action nor the cellular targets have been clearly defined.

Several investigators have suggested that the estrogen effect on T cell-mediated immunity in adult animals is mediated via the thymus.[35,41] The thymus has for a long time been known to be sensitive to estrogen.[42] Doses exceeding 0.8 μg estradiol or 0.02 μg DES per mouse induce significant thymus atrophy and depress T cell activation as tested with phytohemagglutinin (PHA) and mixed lymphocyte cultures in vitro.[34] Different groups[35,41] have shown that depression of lymphocyte cultures in vitro could be induced with addition of serum from estradiol-treated euthymic rats but not with serum from thymectomized controls. In the mouse, a specific thymosin has been reported to be induced by estradiol treatment, providing additional evidence of a relationship between the thymus and estrogen.[28] Previous studies have indicated that the cytosolic fraction from thymic epithelial cells, but not from thymocytes, contains estrogen receptors.[36] It has therefore been suggested that the thymic reticular cells are responsible for thymic involution and the secretion of T cell suppressor factors.[36,43] However, the restriction of estrogen receptors in the thymus to the reticular cells is far from established since it has recently become clear that both occupied and unoccupied forms of the receptor are located in the nuclear compartment, and only trace amounts in the cytosol.[44] In fact, it has recently been proposed that blastoid lymphocytes, within or outside the thymus, express estrogen receptors and are sensitive for estrogen treatment.[45] Estradiol treatment of thymectomized and shamectomized mice induces an equally severe depression of various T cell functions.[34] Estrogens strongly suppress the proliferative response of T and B lymphocytes to mitogens in vitro, although at very high concentrations.[40,46] These latter results indicate that a direct effect of estrogen on a T cell blastoid stage cannot be excluded as an explanation of estrogenic suppression of T cell functions in vivo.

Another post-thymic mechanism that has been proposed is an estrogen-mediated activation of suppressor macrophages that can depress T cell activation.[40,47] Plastic-adherent spleen cells (macrophages) derived from DES-treated mice, but not from control mice, induced suppression when cocultured with T cells.[47] Unfortunately we have not found any reports on how antigen-presenting cell functions, normally displayed by subsets of macrophages, are affected by estrogen. Such data would be of importance in relation to the estrogen effects of T-helper cell functions.

5. Effects on B Cell-Mediated Immunity

B cell responses have been reported either to be stimulatory or suppressed by the action of estrogen.[12,47-50] In general, the outcome of the estrogenic modulation of the B cell response tends to depend on the dose of estrogen; high doses suppress, while low doses enhance the response. Thus, treatment of mice with high doses of (2 to 200 μg) DES suppressed the in vitro spleen cell response with the B cell mitogen lipopolysaccharide (LPS), while treatment with low doses of (0.02 to 2 μg) DES enhanced the response.[33,51] When the influence of low doses of estradiol on the antibody responses to T cell-dependent (sheep red blood cells [SRBC]) and T cell-independent (TNP-Ficoll, TNP-LPS) antigens were compared, it was found that T cell-independent antigens were most sensitive for enhancement.[49] Only the IgM response, but not the IgG, was stimulated in the response to SRBC. These data suggest that with low-dose estrogen treatment, the estrogen-mediated depression of T helper cell function influences the T cell-dependent antibody response. It is likely that estrogen in low (physiological) concentration stimulates B cells as such — an effect that is most evident for T cell-independent B cell activation.

An alternative explanation for the observed enhancement of B cell responses by estrogens is that T-suppressor cells are inactivated. Paavonen et al.[53] showed that treatment of in vitro mitogen-stimulated cultures of human peripheral blood leukocytes with physiological doses of estradiol (780 to 2600 pmol/ℓ) increased the number of IgM plasma cells. They also showed that addition of an Fc-gamma-bearing T cell population from control cultures normalized the enhanced B cell differentiation effect. Another possible T-suppressor cell, the Ly2 (the human counterpart is CD8)-positive cell, has also been suggested to be the target for estrogen. Treatment of mice with low doses of estradiol (1 μg on alternate days for 2 to 4 weeks) reduces the number of Ly2-positive cells in the thymus as well as in peripheral lymphoid organs.[54] Furthermore, estrogen receptors have been demonstrated in the CD8-positive subset of human peripheral T cells.[55] Although the suppressor function of CD8/Ly2-positive cells has not directly been demonstrated in the context of estrogen enhancement of B cell immunity, there are number of reports that CD8/Ly2-positive cells under other experimental conditions can function as effector-suppressor cells.[56] Other possible mechanisms for the enhancement of B cell immunity with low doses of estrogens are not better documented. Besides a poorly defined direct modulation of the B cells (see above), it is possible that stimulated macrophages could secrete lymphokines or complement factors that promote B cell activation.[57] It has also been suggested that in vivo stimulation of the reticuloendothelial system may redistribute the antigen to more immunogenic locations and thereby enhance B cell activation.[50]

B. Estrogen Influence on Autoimmunity

1. Estrogen Influences on B Cell-Mediated Autoimmunity; The Lupus Disease

Systemic lupus erythematosus (SLE) is a severe disease with generalized autoimmune manifestations. Usually there is an advanced B cell hyperreactivity in which the circulating immunoglobulin levels are increased 3 to 10 times compared with normal. A wide range of autoantibodies, such an anti-DNA, anti-nuclear factor, and rheumatoid factors, are concomitantly elevated. Circulating immune complexes precipitate in the capillaries and the T cell immunity is depressed.

The importance of sex has long been recognized for susceptibility to SLE. Females, especially females in childbearing years, are much more susceptible than men, with an incidence ratio of 9:1.[58] Both pregnancy[59] and consumption of estrogen-containing contraceptive pills[60] increase the risk for development of disease, and the severity is often observed to vary with the menstrual cycle.[61] Castration decreases the risk for development of lupus disease.[62] These findings suggest that estrogen exaggerates the development of lupus disease in humans.

Murine experimental models for SLE have been carefully characterized. One such model is the F1 hybrid between the NZB and NZW strains. The NZB/W mouse develops a murine lupus with B cell hyperreactivity, T cell hyporeactivity, and with systemic autoimmune manifestations similar to what is seen in humans.[63] Female mice are more susceptible to the disease than males.[63] Thus, the murine lupus apparently provides a good model for studies on hormonal modulation of autoimmunity. Roubinian et al.[64] showed that a main accelerating factor for the disease in females was the influence of estrogen. Prepubertal castration of NZB/W mice combined with estradiol-containing silastic capsule implants caused a greatly enhanced disease.

Estrogen treatment raised the levels of various autoantibodies.[64,65] Both spontaneous and immunization-induced autoantibodies reactive with DNA were elevated by treatment with estrogen.[65] The hormone modulation of the autoantibody response most likely accounts for the observed sex difference since sex chromosome-linked effects have been shown not to play a major role.[18,66]

2. Estrogen Influences on T Cell-Mediated Autoimmunity; The Rheumatoid Disease

While lupus autoimmunity is well characterized regarding the influence of sex steroids, other forms of autoimmunity are not. This discrepancy may explain why lupus autoimmunity has been the general model in the explanation of how estrogen modulates autoimmune diseases. Hence, it has been difficult to explain the differences between, for instance, SLE and rheumatoid arthritis (RA), in responsiveness to pregnancy and contraceptive pill consumption, in spite of a similar sex distribution.[67]

RA is a disease of unknown origin which may be autoimmune and which most likely is dependent on T cell-mediated immunity.[68,69] Rheumatoid symptoms decrease during pregnancy[70] and during the post-ovulatory, estrogen-producing phase of the estrous cycle.[71] Furthermore, estrogen-containing contraceptive pill consumption or menopausal treatment have been reported by several investigators to reduce the incidence of RA.[72-74] In summary, the estrogen influence seems to protect from rheumatoid disease and can therefore not explain the increased female susceptibility. Other factors, such as the influence of X chromosomes, must be taken into consideration. Nevertheless, RA and SLE are regulated differently by estrogen. In order to understand the mechanisms operating, we have to turn to experimental arthritis models.

One such model is collagen arthritis, which can be induced in several species such as mice, rats, guinea pigs, and apes by a single immunization with cartilage type II collagen.[75,76] The disease is dependent on a functional thymus,[77] and autoreactive T-helper cells can mediate the disease.[78] In addition, T cell-dependent antibodies reactive with autologous type II collagen seem to be of importance in the pathogenesis.[79] The disease in mice is more readily induced in males than females.[80,81] Furthermore, pregnancy delays the onset of arthritis.[82,83] These findings also suggest that collagen arthritis is influenced by female sex hormones.

In fact estrogen is a potent modulator of collagen arthritic disease. Castration of female DBA/1 mice increases the disease susceptibility to the same level as males.[80] Treatment of castrated DBA/1 females with low doses of estradiol (0.2 μg twice a week) suppressed both the incidence and severity of the disease.[84] In collagen arthritis, there are several lines of evidence that estradiol treatment suppresses T cell activation. Treatment of mice with low doses of estradiol suppressed activation of collagen-reactive T-helper cells. Furthermore, a generation of the T cell-dependent IgG anticollagen antibodies was suppressed by estradiol treatment but the levels of IgM antibodies were enhanced. Hirahara et al.[82] could show similar phenomena by pregnancy in mice immunized with type II collagen. Although T cell functions are suppressed by estradiol treatment of mice immunized with type II collagen, this effect is not mediated by modulation of thymic endothelial cells. Adult thymectomy did not abrogate the estradiol-induced suppression of arthritis of T cell function.[85] It is therefore possible that other estrogen-mediated mechanisms may operate to suppress T cell activation. As discussed previously, an effect on the activation of T-helper cells, either directly on the T-helper cell or on the antigen-presenting cell, cannot be excluded.

C. Summary of Estrogen Effects on Immunity in Adult Animals

Both in physiological and pathological immunity, estrogen seems to depress antigen-specific activation of T lymphocytes. B lymphocytes and macrophage-like cells, on the other hand, can be stimulated by estrogen. This dualistic phenomenon is clearly demonstrated by the variable influence of estrogen on autoimmune disease. Thus, by treatment with estrogen, the lupus autoimmunity is exaggerated with an enhanced autoantibody formation, and T cell-dependent autoimmunity is depressed with the suppression of T-helper cell functions.

Although estrogens seem to have variable effects on different parts of the immune system, it is possible that these could depend on a common mechanism that is not yet defined. Estrogens stimulate cells in the reticuloendothelial system and certain subsets of macrophages. Stimulated reticuloendothelial cells may secrete thymosin which depletes progenitors of T-helper or T-suppressor cells in the thymus. These alterations of different T cell subsets are one explanation for an enhanced B cell response and a decreased T-helper cell activation. It is also possible that suppressor macrophages are induced that can abrogate the antigen-mediated stimulation of T-helper cells.

It seems reasonable to look for a physiological explanation for the various effects of estrogen in the maintenance of pregnancy, i.e., to protect both the mother and the fetus. Because of this, the female should have a different profile of her immune system. This would also result in a different pattern of susceptibility to pathogens. Thus, estrogen treatment has been shown both to protect from[86] and reduce resistance[87,88] to various viral and bacterial pathogens. It is also reasonable that the different profile of the female immune system could result in a different susceptibility to various autoimmune diseases.

III. IMPRINTING EFFECTS ON THE IMMUNE SYSTEM BY NEONATAL EXPOSURE TO DIETHYLSTILBESTROL

A. Pathotoxicology of Lymphoid Organs

Treatment of neonatal mice with high doses of estrogen results in the development of a wasting syndrome with similarities to that seen following neonatal thymectomy: severe growth retardation and high mortality.[89] The wasting syndrome apparently is secondary to acute thymic necrosis following milligram doses of estrogens in neonatal as well as adult mice.[90] No signs of a wasting syndrome were seen with the standard dose (5 μg DES daily for the first 5 days of life) used in studies leading to long-term immune deviations without acute effects on mortality. However, moderate to severe degeneration of cortical lymphocytes was observed clearly exceeding that normally found in the thymic cortex in the neonatal period.[91] The thymus apparently is a preferential target for estrogen-mediated toxicity among tissues not normally included among estrogen-responsive organs. While several studies have indicated that the reticuloepithelial cells of the thymus contain sex steroid receptors, [36,60,92-94] evidence of estrogen-binding components in the neonatal thymus has only been obtained using a fluorescent estradiol-albumin conjugate as a marker.[43] The effect on the neonatal thymus was reversible, and thymus weight and histology returned to normal within 6 weeks with some evidence of compensatory weight increase. Adult animals neonatally exposed to DES revealed no alterations in body weight, or weight or histology of lymphoid organs.[89,95] The relevance of the acute toxicity on the thymus for the development of the DES-associated long-term effects on the immune system may be related to disturbance of T cell differentiation at a critical period of life, as is discussed in the following.

B. T Cell-Mediated Immunity

The persistent modulation of the immune response after neonatal manipulation, imprinting, was demonstrated for the first time in experiments analyzing the delayed-hypersensitivity response to the contact allergen oxazolone.[96] Female mice, 6- to 9-months old, neonatally exposed to DES showed a significantly reduced ability to mount a delayed-hypersensitivity response, indicating a persistent disturbance in cellular immunity. This observation was later verified by Luster et al.[97] by measuring the delayed-hypersensitivity response to the recall antigen PPD. The delayed-hypersensitivity response is dependent on specific stimulation of T-helper/inducer lymphocytes (phenotype Lyt 1^+,L3T4$^+$,Lyt2$^-$) capable of releasing biologically active mediators, which in turn recruit nonspecific inflammatory cells. Further evidence for alterations in T lymphocyte responses was obtained from comparison of the response of spleen lymphocytes from neonatally DES-treated female mice to the T cell mitogens Con A and PHA, which was clearly depressed even in the oldest animals studied at 17 months of age.[97-99]

Female mice of the Balb/cfC3H strain develop a natural immune reactivity to mammary tumor virus-associated antigens, which is detected as antigen-specific cytotoxic T cell activity. Blair[3] showed that neonatally DES-treated mice did not develop cytotoxic T cells capable of eliminating mammary tumor virus-infected cells. Thus DES treatment permanently alters the expression of immune reactivity to the antigens of persisting virus that is also introduced in the neonatal period. There is a correlation between this diminution of immune reactivity and subsequent mammary tumor development in the neonatally treated females.

The ability to provoke a graft-vs.-host reaction is impaired in lymphoid cells from DES-exposed mice; this also is indicative of a T-helper cell defect.[99] To directly examine the possible effect of neonatal treatment with DES on T-helper cells, their relative frequencies as well as functional capacity in adult mice were examined.

The number of cells expressing the T cell marker Thy1 was reduced from 40% in control animals to about 25% in those neonatally exposed to DES.[100] Ly antigens are a series of surface markers expressed on T cells in distinct combinations corresponding, to some extent, to functional subclasses. In general, T cells with helper function have the surface phenotype $Ly1^{+}2^{-}$. Recently, all T cells have been found to express Ly1, although cells with helper function have a quantitative higher expression. Phenotypic characterization of mouse T-helper lymphocytes should be performed with the L3T4 marker,[101] which so far has not been used in the study of DES effects on the immune system. Cytotoxic T lymphocytes and cells with suppressor activity are contained in the $Ly2^{+}$ T cell population, while cells of the $Ly123^{+}$ class have been implicated as precursors of both Ly1 and Ly2 cells, as suppressors in some systems, and as amplifier and intermediary cells.[102] The T lymphocyte population of neonatally DES-exposed mice was disproportionate with respect to subclasses of cells expressing different Ly antigens. The proportion of cells expressing high amounts of Ly1 cells was reduced with a concomitant increase in proportion of cells with the Ly123 phenotype.[100] The disproportion increased with increasing age, indicating an underlying developmental defect aggravated by time. The immune response is the net result of interactions between various cells involving complex regulatory circuits, and any conclusion on function based only on phenotype is fraught with misinterpretation. T-helper cell function was therefore studied in an in vitro response to the T cell-dependent antigen SRBC.[52] The impairment in the antibody response was then shown to reside in the T cell and not the B cell population of DES-treated mice. This finding was not unexpected since the frequency of T-helper cells in spleen cells from DES-treated mice was reduced. When the same number of enriched T-helper cells from control and DES-treated mice were added as helper cells in the culture system, there was no difference in helper activity between T cells from control or neonatally DES-treated mice. Neither the number nor the activity of in vivo primed Ly2 suppressor cells was altered in DES-treated mice. The ability of macrophages from DES-treated mice to function as antigen-presenting cells was also comparable to that of macrophages from control mice.[52] In conclusion, suggestive evidence indicates that the reduced primary antibody response in DES-exposed mice is a consequence of a decreased population of T-helper cells. The central role of the T-helper cell in most aspects of T-mediated immunity makes it conceivable that this defect may be underlying many of the immune deviations induced by DES.

C. B Cell-Mediated Immunity

Studies on the proliferative response to the B cell mitogen LPS of lymphocytes from neonatally DES-treated mice indicated that even the B cell system was affected.[21,98] The number of B cells in the spleen and peripheral blood, however, was normal.[100] The number of antibody-producing cells after a single immunization with the T-dependent antigen SRBC or the T-independent antigen LPS was depressed in perinatally DES-exposed mice.[21,52,97] The number of direct plaque-forming cells (PFC) to SRBC was reduced 40% and the number of PFC to LPS reduced 30%. No effect on either the proliferative response to Con A or the antibody response to SRBC was found after adrenalectomy or ovariectomy in control or neonatally DES-treated animals.[21,103] However, the response to LPS was reduced in control animals as a consequence of ovariectomy or adrenalectomy, abrogating the difference between control and DES-exposed animals. The difference was restored by additional estradiol treatment. These data indicate that T-independent B cell responses are secondary to endocrine alterations, while T-dependent responses are secondary to an ovary- and adrenal-independent defect in the T-helper cell population as discussed previously. Thus DES-induced alterations of the two major arms of adaptive immunity are apparently related to qualitatively different mechanisms.

D. Natural Killer Cell Activity and Tumor Susceptibility

Natural killer (NK) cells are lymphocytes spontaneously cytotoxic to a variety of tumor cells, virally infected cells, and some microorganisms.[105] They have been implicated in the defense against tumors, in particular in the resistance to metastasis. NK cells also recognize some immature normal cells of the hematopoietic system and may play a regulatory role in hematopoiesis.[104] NK activity of neonatally DES-exposed female mice from different mouse strains is reduced.[103] No evidence of humoral or cellular suppressor factors to NK cells induced by DES could be detected, nor did ovariectomy or adrenalectomy affect the depressed NK activity of DES-treated mice.[103,105] Moreover, NK activity was reduced in all lymphoid compartments tested and could not be restored by treatment with interferon or interferon inducers.[106] Examination of the number of NK cells in a single cell assay revealed that the actual number of NK cells was decreased, while the individual NK cell was able to lyse tumor targets in an apparently normal fashion.[106] NK cells are derived from stem cells in the bone marrow and are continuously dependent on an intact bone marrow function. In a recent study, we were able to show that the DES-induced impairment of NK cells could be related to a defect in the generation of mature NK cells from bone marrow progenitors.[106] Adoptive transfer of bone marrow from control mice to syngeneic lethally irradiated DES recipients restored NK activity to normal levels. Transplantation of bone marrow from neonatally DES-treated mice to control animals resulted in subnormal levels of NK activity compared with mice transplanted with normal bone marrow. The DES-induced impaired NK activity clearly is related to persistent defects at the stem cell level and is not related to alterations in the internal milieu of the animals. We have recently established a bone marrow culture system where NK cells can be generated from immature progenitors in the presence of Interleukin-2.[107] In this culture system, 2×10^{-7} M DES inhibited the generation of NK cells from bone marrow cells of 5-day-old, but not from adult, mice.[121] At concentrations of DES $>10^{-6}$, NK cell formation was inhibited even in adult bone marrow cultures. This latter concentration of DES also inhibited murine and human NK cells in vitro.[108,109] These in vitro data further indicate that the remarkable sensitivity of the neonatal immune system to DES, as compared to that of adult mice, might be related to direct effects of DES or its metabolites on lymphoid progenitors at a particular state of their differentiation.

The in vivo relevance of depressed NK activity as a defense against tumor development was studied by transplantation of NK-sensitive as well as NK-resistant syngeneic tumors. The elimination of intravenously injected radiolabeled tumor cells from various organs has been shown to be closely correlated with in vitro NK activity in the same animals in a variety of experimental conditions.[110] The AKR lymphomas I-51 and I-522 have a grossly similar growth pattern in NK-deficient mice but differ in NK sensitivity.[111] Elimination of the NK-sensitive I-522 lymphoma was strongly reduced in neonatally DES-treated AKR mice, while no difference was found for the NK-resistant I-51 tumor. Similarly, when the same tumors were inoculated subcutaneously, a significantly higher incidence of I-522 lymphomas developed in neonatally DES-treated mice than in control mice, with no difference for I-51. The susceptibility to primary carcinogenesis was investigated by induction of sarcomas with different doses of 3-methylcholanthrene.[111] The yield of local sarcomas as a function of time after intramuscular injection of 3-methylcholanthrene was significantly higher in neonatally DES-treated mice than in controls when low doses of the carcinogen were injected. When higher doses of methylcholanthrene were used, no difference was seen. This latter observation may be accounted for by the immunosuppressive effect of methylcholanthrene per se[110] which also affects NK activity.[113,114] Taken together, these results indicate that neonatally DES-exposed female mice have an impaired NK cell function resulting in an increased susceptibility to transplanted as well as primary carcinogen-induced tumors. Moreover, Blair[3] has shown that neonatally DES-treated mice of susceptible strains also have an increased incidence of mammary tumor virus-induced tumors correlating with the lack of virus-specific cytotoxic T lymphocytes. The incidence of DMBA-induced mammary tumors in the rat was significantly increased following prenatal exposure to DES.[115] Moreover, transplacental treatment of hamsters with DES increased the susceptibility to chemical carcinogens in adult life.[116] It should be noted that further immunosuppression of neonatally DES-treated mice did not increase the incidence of tumors.[117] The possible role of the immune system in the spontaneous development of tumors of the genital tract in perinatally DES-exposed mice has not been studied in detail. However, transplantation of the cervicovaginal anlage from DES-treated to control thymectomized mice did not increase the frequency of cervicovaginal tumors.[117] T cells therefore apparently play a limited role in the control of these tumors.

E. Summary of Imprinting Effects on the Immune System after Neonatal Exposure to Diethylstilbestrol

Exposure of the developing immune system to estrogens causes persistent effects on immune reactivity, in contrast to the reversible modulation seen in adult mice. Estrogens apparently affect the immune system during a critical ontogenetic period, leading to disturbances in differentiation of mature lymphoid cells. A variety of immune parameters are affected, which can be referred to disturbances in the development of T-helper and NK cells. Accumulating evidence based on sharing of surface markers, genetic studies, and reciprocal regulation of precursors indicates that NK and T cells share an early common progenitor.[118,119] DES may therefore interfere with the development of the postulated NK/T lymphocyte progenitors, leading to permanent impairment of their function. Certain of the long-term effects of DES on B cell-related immune functions were shown to be secondary to endocrinological alterations. It should therefore be emphasized that perinatal DES exposure affects the immune system not only directly by interfering with hematopoietic progenitors, but also indirectly via hormonal alterations. Hormones have been shown to modulate the immune response of adult animals and to also have a strong impact on the expression of autoimmune disease, as was discussed. The study of the biological consequences of perinatal DES exposure has mostly concentrated on implications for tumor development. We regard it highly probable, however, that disturbances in immune regulation will lead to autoimmune manifestations and enhance susceptibility to infections. The murine model has been extremely useful in predicting the pathological changes of the cervicovaginal region of young women exposed to DES *in utero*.[120] Several well-characterized murine models (Section II.B) are also available for the study of possible autoimmune manifestations related to DES-induced immune alterations and would constitute an important direction for future research.

IV. IMPRINTING OF THE IMMUNE SYSTEM: FURTHER PERSPECTIVES

Classic teratology has concentrated on morphological alterations induced by developmental insults. The long-term effects of DES on the endocrine and immune systems demonstrate that more subtle although permanent alterations not evident as gross organ malformations may be induced during development. Such functional teratogenesis may pass unattended due to the wide biological variation in the response of individuals in biological assays. With the increasing awareness of the importance of the immune system in the defense against infections and tumors, the possible role of perinatal exposure to agents with imprinting activity on the immune system, as demonstrated for DES, should be carefully considered. One of the major unresolved problems in immunology today is a proper understanding of the tolerance phenomenon, i.e., the lack of reactivity towards self. Tolerance induction is mainly a characteristic of the immature immune system and can be regarded as imprinting by self components. Tolerance is the central issue of immune deviations such as allergy and autoimmune disease. The demonstration that exogenous factors can have strong imprinting effects on the immature immune system emphasizes that similar effects should not be excluded in the search for an etiology of immune disturbances manifested later in life.

REFERENCES

1. **Herbst, A. L. and Bern, H. A., Eds.,** *Developmental Effects of Diethylstilbestrol (DES) in Pregnancy,* Thieme-Stratton, New York, 1981.
2. **Kalland, T.,** Effects of Neonatal Exposure to Diethylstilbestrol on the Immune System in Female Mice, Ph.D. thesis, University of Bergen, Bergen, Norway, 1980.
3. **Blair, P. B.,** Immunologic consequences of early exposure of experimental rodents to diethylstilbestrol and steroid hormones, in *Developmental Effects of Diethylstilbestrol (DES) in Pregnancy,* Herbst, A. L. and Bern, H. A., Eds., Thieme-Stratton, New York, 1981, 167.
4. **Kalland, T.,** Long term effects on the immune system of an early life exposure to diethylstilbestrol, in *Environmental Factors in Human Growth and Development,* Hunt, V. R., Smith, M. K., and Worth, D., Eds., Cold Spring Harbor Laboratory, Cold Spring Harbor, N.Y., 1982, 217.
5. **Murgita, R. A. and Wigzell, H. A.,** Regulation of immune functions in the fetus and newborn, *Prog. Allergy,* 29, 54, 1983.
6. **McEven, B. S.,** Interactions between hormones and nerve tissue, *Sci. Am.,* 235, 48, 1976.
7. **Skett, P. and Gustafsson, J. Å.,** Imprinting of enzyme systems of xenobiotic and steroid metabolism, in *Reviews in Biochemical Toxicology,* Hodgson, E., Bend, J., and Philpot, R., Eds., Elsevier/North-Holland, Amsterdam, 1979, 27.
8. **Dougherty, T. F.,** Effects of hormones on lymphatic tissue, *Physiol. Rev.,* 32, 379, 1952.
9. **Grossman, C. J.,** Regulation of the immune system by sex steroids, *Endocr. Rev.,* 5, 435, 1984.
10. **Forsberg, J.-G.,** Short-term and long-term effects of estrogen on lymphoid tissues and lymphoid cells with some remarks on the significance for carcinogenesis, *Arch. Toxicol.,* 55, 79, 1984.
11. **Eidinger, D. and Garrett, T. J.,** Studies of the regulatory effects of the sex hormones on antibody formation and stem cell differentiation, *J. Exp. Med.,* 136, 1098, 1972.
12. **Krzych, U., Strausser, H. R., Bressler, J. P., and Goldstein, A. L.,** Effects of sex hormones on some T and B cell functions, evidenced by differential immune expression between male and female mice and cyclic pattern of immune responsiveness during the estrous cycle in female mice, *Am. J. Reprod. Immunol.,* 1, 73, 1981.
13. **Graff, R. J., Lappe, M. A., and Snell, G. D.,** The influence of the gonads and adrenal glands on the immune response to skin grafts, *Transplantation,* 7, 105, 1969.
14. **Holmdahl, R., Jansson, L., Larsson, E., Rubin, K., and Klareskog, L.,** Homologous type II collagen induces chronic and progressive arthritis in mice, *Arthritis Rheum.,* 29, 106, 1986.
15. **Butterworth, M., McClellan, B., and Allansmith, M.,** Influence of sex on immunoglobulin levels, *Nature (London),* 214, 1224, 1967.
16. **Wortis, H. H., Burkly, L., Hughes, D., Roschelle, S., and Waneck, G.,** Lack of mature B-cells in nude mice with X-linked immune deficiency, *J. Exp. Med.,* 154, 903, 1982.
17. **Gill, T. J. and Kunz, H. W.,** Genetic and cellular factors in the immune response. II. Evidence for the polygenic control of the antibody response from further breeding studies and from pedigree analysis, *J. Immunol.,* 106, 980, 1971.
18. **Raveche, E. S., Tjio, J. H., Boegel, W., and Steinberg, A. D.,** Studies of the effects of sex hormones on autosomal and X-linked genetic control of induced and spontaneous antibody production, *Arthritis Rheum.,* 22, 1177, 1979.
19. **Murphy, E. D. and Roths, J. B.,** A Y chromosome associated factor in strain BXSB producing accelerated autoimmunity and lymphoproliferation, *Arthritis Rheum.,* 22, 1188, 1979.
20. **Steinberg, R. T., Miller, M. L., and Steinberg, A. D.,** Effect of the BXSB Y chromosome accelerating gene on autoantibody production, *Clin. Immunol. Immunopathol.,* 35, 67, 1985.
21. **Kalland, T.,** Ovarian influence on mitogen responsiveness of lymphocytes from mice neonatally exposed to diethylstilbestrol, *J. Toxicol. Environ. Health,* 6, 67, 1980.
22. **Watnick, A. S. and Russo, R. A.,** Survival of skin homografts in uteri of pregnant and progesterone-estrogen treated rats, *Proc. Soc. Exp. Biol. Med.,* 128, 1, 1968.
23. **Carter, J.,** The effect of progesterone, oestradiol and HCG on cell-mediated immunity in pregnant mice, *J. Reprod. Fertil.,* 46, 211, 1976.
24. **Baines, M. G. and Pross, H. F.,** Impairment of the thymus-dependent humoral immune response by syngeneic or allogeneic pregnancy, *J. Reprod. Immunol.,* 4, 337, 1982.
25. **Lala, P. K., Chatterjee-Hasrouni, S., Kearns, M., Montgomery, B., and Colavincenzo, V.,** Immunobiology of the feto-maternal interface, *Immunol. Rev.,* 75, 87, 1983.
26. **Vanderbeeken, Y., Vlieghe, P., Delespesse, G., and Duchateau, J.,** Characterization of immunoregulatory T cells during pregnancy by monoclonal antibodies, *Clin. Exp. Immunol.,* 48, 118, 1982.
27. **Stites, D. P. and Siiteri, P. K.,** Steroids as immunosuppressants in pregnancy, *Immunol. Rev.,* 75, 117, 1983.
28. **Allen, L. S., McClure, J. E., Goldstein, A. L., Barkley, M. S., and Michael, S. D.,** Estrogen and thymic hormone interactions in the female mouse, *J. Reprod. Immunol.,* 6, 25, 1984.

29. **Ryan, K. D. and Schwartz, N. B.,** Changes in serum hormone levels associated with male-induced ovulation in group-housed adult female mice, *Endocrinology*, 106, 959, 1980.
30. **Nicol, T., Bilbey, D. L. J., Charles, L. M., Cordingley, J. L., and Vernon-Roberts, B.,** Oestrogen: the natural stimulant of body defence, *J. Endocrinol.*, 30, 277, 1964.
31. **Steven, W. M. and Snook, T.,** The stimulatory effects of diethylstilbestrol and diethylstilbestrol diphosphate on the reticuloendothelial cells of the rat spleen, *Am. J. Anat.*, 144, 339, 1975.
32. **Boorman, G. A., Luster, M. I., Dean, J. H., and Wilson, R. E.,** The effect of adult exposure to diethylstilbestrol in the mouse on macrophage function and numbers, *J. Reticuloendothel. Soc.*, 28, 547, 1980.
33. **Flynn, A., Finke, J. H., and Hilfiker, M. L.,** Estrogen stimulated production of IL-1 from human placental derived macrophages, *Immunobiology*, 163, 279, 1982.
34. **Luster, M. I., Hayes, H. T., Korach, K., Tucker, A. N., Dean, J. H., Greenlee, W. F., and Boorman, G. A.,** Estrogen immunosuppression is regulated through estrogenic responses in the thymus, *J. Immunol.*, 133, 110, 1984.
35. **Grossman, C. J., Sholiton, L. J., and Roselle, G. A.,** Estradiol regulation of thymic lymphocyte function in the rat: mediation by serum thymic factors, *J. Steroid Biochem.*, 16, 683, 1982.
36. **Grossman, C. J., Sholiton, L. J., and Nathan, P.,** Rat thymic estrogen receptor. I. Preparation, location and physicochemical properties, *J. Steroid Biochem.*, 11, 1233, 1979.
37. **Waltman, S. R., Burde, R. M., and Berrios, J.,** Prevention of corneal homograft rejection by estrogens, *Transplantation*, 11, 194, 1971.
38. **Kappas, A., Jones, H. E. H., and Roitt, I. M.,** Effects of steroid sex hormones on immunological phenomena, *Nature (London)*, 198, 902, 1963.
39. **Haukaas, S. A. and Kalland, T.,** Effects of diethylstilbestrol and estramustine phosphate (Estracyt) on delayed hypersensitivity response to oxazolone in male mice, *Prostate*, 3, 149, 1982.
40. **Haukaas, S. A. and Kalland, T.,** Effects of diethylstilbestrol and estramustine phosphate (Estracyt) on lymphoid cell populations and mitogen responsiveness in male mice, *J. Urol.*, 128, 862, 1982.
41. **Stimson, W. H. and Hunter, I. C.,** Oestrogen-induced immunoregulation mediated through the thymus, *J. Clin. Lab. Immunol.*, 4, 27, 1980.
42. **Golding, G. T. and Ramirez, F. T.,** Ovarian and placental hormone effects in normal, immature albino rats, *Endocrinology*, 12, 804, 1928.
43. **Kalland, T. and Forsberg, J.-G.,** Occurrence of estrogen binding components in mouse thymus, *Immunol. Lett.*, 1, 293, 1980.
44. **Gorski, J., Welshons, W. V., Sakai, D., Hansen, J., Walent, J., Kassis, J., Shull, J., Stack, G., and Campen, C.,** Evolution of a model of estrogen action, *Recent Prog. Horm. Res.*, 42, 297, 1986.
45. **Gulino, A., Screpanti, I., Torrisi, M. R., and Frati, L.,** Estrogen receptors and estrogen sensitivity of fetal thymocytes are restricted to blast lymphoid cells, *Endocrinology*, 117, 47, 1985.
46. **Ablin, R. J., Bruns, G. R., Guinan, P. D., and Bush, I. M.,** The effects of estrogen on the incorporation of 3H-thymidine by PHA stimulated peripheral blood lymphocytes, *J. Immunol.*, 113, 705, 1974.
47. **Luster, M. I., Boorman, G. A., Dean, J. H., Luebke, R. W., and Lawson, L. D.,** The effect of adult exposure to diethylstilbestrol in the mouse: alterations in immunological functions, *J. Reticuloendothel. Soc.*, 28, 561, 1980.
48. **Stern, K. and Davidsohn, I.,** Effect of estrogen and cortisone on immune hemoantibodies in mice of different inbred strains, *J. Immunol.*, 74, 479, 1955.
49. **Brick, J. E., Wilson, D. A., and Walker, S. E.,** Hormonal modulation of responses to thymus-independent and thymus-dependent antigens in autoimmune NZB/W mice, *J. Immunol.*, 134, 3693, 1985.
50. **Sljivic, V. S., Clark, D. W., and Warr, G. W.,** Effects of oestrogen and pregnancy on the distribution of sheep erythrocytes and the antibody response in mice, *Clin. Exp. Immunol.*, 20, 179, 1975.
51. **Haukaas, S. A.,** Suppression of antibody response of male mice exposed to diethylstilbestrol or estramustine phosphate (Estracyt), *Prostate*, 4, 375, 1983.
52. **Kalland, T.,** Alterations of antibody response in female mice after neonatal exposure to diethylstilbestrol, *J. Immunol.*, 124, 194, 1980.
53. **Paavonen, T., Andersson, L. C., and Adlercreutz, H.,** Sex hormone regulation of in vitro immune response. Estradiol enhances human B cell maturation via inhibition of suppressor T cells in Pokeweed mitogen stimulated cultures, *J. Exp. Med.*, 154, 1935, 1981.
54. **Ansar Ahmed, S., Dauphinee, M. J., and Talal, N.,** Effects of short term administration of sex hormones on normal and autoimmune mice, *J. Immunol.*, 134, 204, 1985.
55. **Cohen, J. H. M., Danel, L., Cordier, G., Saez, S., and Revillard, J.-P.,** Sex steroid receptors in peripheral T cells: absence of androgen receptors and restriction of estrogen receptors to OKT8-positive cells, *J. Immunol.*, 131, 2767, 1983.
56. **Dorf, M. E. and Benacerraf, B.,** Suppressor cells and immunoregulation, *Annu. Rev. Immunol.*, 2, 127, 1984.

57. **Melchers, F. and Andersson, J.,** Factors controlling the B-cell cycle, *Annu. Rev. Immunol.*, 4, 13, 1986.
58. **Inman, R.,** Immunologic sex differences and the female predominance in systemic lupus erythematosus, *Arthritis Rheum.*, 21, 849, 1978.
59. **Mund, A., Swison, J., and Rothfield, N.,** Effect of pregnancy on the course of SLE, *JAMA*, 183, 917, 1963.
60. **Jungers, P., Dougados, M., Pelissier, C., Kuttenn, F., Tron, F., Lesavre, P., and Bach, J.-F.,** Influence of oral contraceptive therapy on the activity of systemic lupus erythematosus, *Arthritis Rheum.*, 25, 618, 1982.
61. **Rose, E. and Pillsbury, D. M.,** Lupus erythematosus (erythematoides) and ovarian function: observations on a possible relationship with a report of 6 cases, *Ann. Intern. Med.*, 21, 1022, 1944.
62. **Yocum, M. W., Grossman, J., Waterhouse, C., Abraham, G. N., May, A. G., and Condemi, J. J.,** Monozygotic twins discordant for systemic lupus erythematosus. Comparison of immune response, autoantibodies, viral antibody titers, gamma globulin, and light chain metabolism, *Arthritis Rheum.*, 18, 193, 1975.
63. **Theofilopoulos, A. N. and Dixon, F. J.,** Murine models of systemic lupus erythematosus, *Adv. Immunol.*, 37, 269, 1985.
64. **Roubinian, J. R., Talal, N., Greenspan, J. S., Goodman, J. R., and Siiteri, P. K.,** Effect of castration and sex hormone treatment on survival, anti-nucleic acid antibodies, and glomerulnephritis in NZB/NZW F1 mice, *J. Exp. Med.*, 147, 1568, 1978.
65. **Steinberg, A. D., Melez, K. A., Raveche, E. S., Reeves, J. P., Boegel, W. A., Smathers, P. A., Taurog, J. A., Weinlein, L., and Duvic, M.,** Approach to the study of the role of sex hormones in autoimmunity, *Arthritis Rheum.*, 22, 1170, 1979.
66. **Raveche, E. S., Klassen, L. W., and Steinberg, A. D.,** Sex differences in formation of anti-T-cell antibodies, *Nature (London)*, 263, 415, 1976.
67. **Lahita, R. G.,** Sex steroids and the rheumatic diseases, *Arthritis Rheum.*, 28, 121, 1985.
68. **Janossy, G., Panai, G., Duke, O., Bofill, M., Poulter, L. W., and Goldstein, G.,** Rheumatoid arthritis: a disease of T-lymphocyte/macrophage immunoregulation, *Lancet*, 2, 839, 1981.
69. **Klareskog, L., Forsum, U., Scheynius, A., Kabelitz, D., and Wigzell, H.,** Evidence in support of a self-perpetuating HLA-DR dependent delayed type cell reaction in rheumatoid arthritis, *Proc. Natl. Acad. Sci. U.S.A.*, 79, 3632, 1982.
70. **Oka, M. and Vainio, U.,** Effect of pregnancy on the prognosis and serology of rheumatoid arthritis, *Acta Rheumatol. Scand.*, 12, 47, 1966.
71. **Latman, N.,** Relation of menstrual cycle phase to symptoms of rheumatoid arthritis, *Am. J. Med.*, 74, 957, 1983.
72. **Vandenbroucke, J. P., Valkenburg, H. A., Boersma, J. W., Cats, A., Festen, J. J. M., Huber-Bruning, O., and Rasker, J. J.,** Oral contraceptives and rheumatoid arthritis: further evidence for a preventive effect, *Lancet*, 2, 839, 1982.
73. **Vandenbrocke, J. P., Witteman, J. C. M., Valkenburg, H. A., Boersma, J. W., Cats, A., Festen, J. J. M., Hartman, A. P., Huber-Bruning, O., Rasker, J. J., and Weber, J.,** Noncontraceptive hormones and rheumatoid arthritis in perimenopausal and postmenopausal women, *JAMA*, 255, 1299, 1986.
74. **Allebeck, P., Ahlbom, A., Ljungström, K., and Allander, E.,** Do oral contraceptives reduce the incidence of rheumatoid arthritis?, *Scand. J. Rheumatol.*, 13, 140, 1984.
75. **Trentham, D. E., Townes, A. S., and Kang, A. H.,** Autoimmunity to type II collagen: an experimental model of arthritis, *J. Exp. Med.*, 146, 857, 1977.
76. **Courtenay, J. S., Dallman, M. J., Dayan, A. D., Martin, A., and Mosedale, B.,** Immunization against heterologous type II collagen induces arthritis in mice, *Nature (London)*, 283, 666, 1980.
77. **Klareskog, L., Holmdahl, R., Larsson, E., and Wigzell, H.,** Role of T-lymphocytes in collagen II induced arthritis in rats, *Clin. Exp. Immunol.*, 51, 117, 1983.
78. **Holmdahl, R., Klareskog, L., Rubin, K., Larsson, E., and Wigzell, H.,** T-lymphocytes in collagen II induced arthritis in mice: characterization of arthritogenic collagen II specific T-lymphocyte lines and clones that can transfer arthritis, *Scand. J. Immunol.*, 22, 295, 1985.
79. **Stuart, J. M. and Dixon, F. J.,** Serum transfer of collagen induced arthritis in mice, *J. Exp. Med.*, 158, 378, 1983.
80. **Holmdahl, R., Jansson, L., and Andersson, M.,** Female sex hormones suppress development of collagen-induced arthritis in mice, *Arthritis Rheum.*, 29, 1501, 1986.
81. **Holmdahl, R., Jansson, L., Gullberg, D., Forsberg, P.-O., Rubin, K., and Klareskog, L.,** Incidence of arthritis and autoreactivity of anti-collagen antibodies after immunization of DBA/1 mice with heterologous and autologous collagen II, *Clin. Exp. Immunol.*, 62, 639, 1985.
82. **Hirahara, F., Woolley, P. H., Luthra, H. S., Coulam, C. B., Griffiths, M. M., and David, C. S.,** Collagen-induced arthritis and pregnancy in mice: the effects of pregnancy on collagen-induced arthritis and the high incidence of infertility in arthritic female mice, *Am. J. Reprod. Immunol. Microbiol.*, 11, 44, 1986.

83. **Holmdahl, R.,** Experimental Collagen Induced Arthritis. Pathogenetic Mechanisms and Relevance for Rheumatoid Arthritis, Dissertation 2/85, Faculty of Medicine, Uppsala University, Sweden, 1985.
84. **Holmdahl, R., Jansson, L., Meyerson, B., and Klareskog, L.,** Oestrogen induced suppression of collagen arthritis. I. Long term oestradiol treatment of DBA/1 mice reduces severity and incidence of arthritis and decreases the anti type II collagen immune response, *Clin. Exp. Immunol.,* 70, 372, 1987.
85. **Holmdahl, R. and Jansson, L.,** Estrogen induced suppression of collagen arthritis. III. Adult thymectomy does not affect the course of arthritis or the estrogen mediated suppression of T cell immunity, submitted.
86. **von Haam, E. and Rosenfeld, I.,** The effect of the various sex hormones upon experimental pneumococcus infections in mice, *J. Infect. Dis.,* 70, 243, 1942.
87. **Friedman, S. B., Grota, L. J., and Glasgow, L. A.,** Differential susceptibility of male and female mice to encephalomyocarditis virus: effects of castration, adrenalectomy and the administration of sex hormones, *Infect. Immun.,* 5, 637, 1972.
88. **Toivanen, P.,** Enhancement of staphylococcal infection in mice by estrogens, I. *Ann. Med. Exp. Fenn.,* 45, 138, 1967.
89. **Reilly, R. W., Thompson, J. S., Bielski, R. K., and Severson, C. D.,** Estradiol-induced wasting syndrome in neonatal mice, *J. Immunol.,* 98, 321, 1967.
90. **Thompson, J. S., Severson, C. D., and Reilly, R. W.,** Autoradiographic study of the effect of estradiol and irradiation on nucleic acid metabolism of the thymus and lymph node of mice, *Radiat. Res.,* 40, 46, 1969.
91. **Kalland, T., Fossberg, T. M., and Forsberg, J.-G.,** Effect of estrogen and corticosterone on the lymphoid system in neonatal mice, *Exp. Mol. Pathol.,* 28, 76, 1978.
92. **Brodie, J. Y., Hunter, I. C., Stimson, W. H., and Green, B.,** Specific estradiol-binding in cytosols from thymus glands from normal and hormone treated male rats, *Thymus,* 1, 337, 1980.
93. **Seiki, K., Imanishi, Y., and Haruku, Y.,** Accumulation and binding of 3-H-estradiol by lymphoid tissues of castrated mice, *Endocrinol. Jpn.,* 25, 289, 1978.
94. **Thompson, E. A.,** The effects of estradiol upon the thymus of the sexually immature female mouse, *J. Steroid Biochem.,* 14, 167, 1981.
95. **Ways, S. C. and Bern, H. A.,** Longterm effects of neonatal treatment with cortisol and estrogen in female Balb/c mouse, *Proc. Soc. Exp. Biol. Med.,* 160, 94, 1979.
96. **Kalland, T. and Forsberg, J.-G.,** Delayed hypersensitivity response to oxazolone in neonatally estrogenized mice, *Cancer Lett.,* 4, 141, 1978.
97. **Luster, M. I., Boorman, G., Dean, J. H., Luebke, R. W., and Lawson, L. D.,** Effects of in utero exposure to diethylstilbestrol on the immune response in mice, *Toxicol. Appl. Pharmacol.,* 47, 287, 1979.
98. **Kalland, T., Strand, Ö., and Forsberg, J.-G.,** Long term effects of neonatal estrogen treatment on mitogen responsiveness of mouse spleen lymphocytes, *J. Natl. Cancer Inst.,* 63, 413, 1979.
99. **Ways, S. C., Blair, P. B., Bern, H. A., and Staskawitcz, M. O.,** Immune responsiveness of adult mice exposed neonatally to diethylstilbestrol, steroid hormones, or vitamin A, *J. Environ. Pathol. Toxicol.,* 3, 207, 1980.
100. **Kalland, T.,** Decreased and disproportionate T cell population in adult mice after neonatal exposure to diethylstilbestrol, *Cell. Immunol.,* 51, 55, 1980.
101. **Sarmiento, M., Dialynas, D. P., Lanchi, D. W., Lorber, M. R., Loken, M. I., and Fitch, F. W.,** *Immunol. Rev.,* 68, 135, 1982.
102. **Cantor, H. and Boyse, E. A.,** Regulation of the immune response by T-cell subclasses, *Contemp. Top. Immunobiol.,* 7, 47, 1977.
103. **Akslen, L. A. and Kalland, T.,** Adrenal gland influence on immune disturbances induced by neonatal exposure to diethylstilbestrol, *Cell. Mol. Biol.,* 28, 587, 1982.
104. **Roder, J. C., Karre, K., and Kiessling, R.,** Natural killer cells, *Prog. Allergy,* 28, 66, 1981.
105. **Kalland, T.,** Reduced natural killer cell activity in mice after neonatal exposure to diethylstilbestrol, *J. Immunol.,* 124, 1297, 1980.
106. **Kalland, T.,** Exposure of neonatal female mice to diethylstilbestrol persistently impairs NK activity through reduction of effector cells at the bone marrow level, *Immunopharmacology,* 7, 127, 1984.
107. **Kalland, T.,** Generation of natural killer cells from bone marrow precursors in vitro, *Immunology,* 57, 493, 1986.
108. **Kalland, T. and Haukaas, S. A.,** Effects of diethylstilbestrol and estramustine phosphate (Estracyt) on natural killer cell activity and tumor susceptibility in male mice, *Prostate,* 5, 649, 1984.
109. **Kalland, T. and Campbell, T.,** Effects of diethylstilbestrol on human natural killer cells in vitro, *Immunopharmacology,* 8, 19, 1984.
110. **Riccardi, C., Puccetti, C., Santoni, A., and Herberman, R. B.,** Rapid in vivo assay of mouse natural killer cell activity, *J. Natl. Cancer Inst.,* 63, 1041, 1979.
111. **Kalland, T. and Forsberg, J.-G.,** Natural killer cell activity and tumor susceptibility in female mice treated neonatally with diethylstilbestrol, *Cancer Res.,* 41, 5134, 1981.

112. **Stjernswärd, J.,** Immunosuppression by carcinogens, *Antibiot. Chemother. (Basel),* 15, 213, 1969.
113. **Kalland, T. and Forsberg, J.-G.,** 3-methylcholanthrene: transient inhibition of the lytic step of mouse natural killer cells, *J. Natl. Cancer Inst.*, 71, 385, 1983.
114. **Kalland, T.,** The effect of 3-methylcholanthrene on mouse natural killer cells in vitro, *Int. J. Immunopharmacol.*, 6, 299, 1984.
115. **Boylan, E. S. and Colhoon, R. E.,** Mammary tumorigenesis in the rat following prenatal exposure to diethylstilbestrol and postnatal treatment with 7,12-dimethylbenz(a)anthracene, *J. Toxicol. Environ. Health,* 5, 1059, 1979.
116. **Rustia, M. and Shubik, P.,** Effect of transplacental exposure to diethylstilbestrol on carcinogenic susceptibility during postnatal life, *Cancer Res.*, 39, 4636, 1979.
117. **Ways, S. C., Bern, H. A., and Blair, P. B.,** Effect of immunosuppression on neonatally diethylstilbestrol-induced genital tract lesions and tumor development in female mice, *J. Natl. Cancer Inst.*, 73, 863, 1984.
118. **Hackett, J., Bennett. M., and Kumar, V.,** Origin and differentiation of natural killer cells. I. Characteristics of transplantable NK cell precursors, *J. Immunol.*, 134, 3731, 1985.
119. **Kalland, T.,** Interleukin 3 is a major negative regulator of the generation of natural killer cells from bone marrow precursors, *J. Immunol.*, 137, 2268, 1986.
120. **Forsberg, J.-G.,** Physiological mechanisms of diethylstilbestrol organotropic carcinogenesis, *Arch. Toxicol.*, 2, 263, 1979.
121. **Kalland, T.,** unpublished data.

Chapter 9

PSYCHOLOGICAL AND ANATOMICAL CONSEQUENCES OF PRENATAL EXPOSURE TO ANDROGENS IN FEMALE RHESUS*

Robert W. Goy, Hideo Uno, and Samuel A. Sholl

TABLE OF CONTENTS

I. Introduction 128

II. Behavioral Consequences of Prenatal Androgens 128

III. Effects on Juvenile Behavior 128

IV. Effects on Adult Sexual Behavior and Menarche 129

V. Principles of Hormonal Action 133

VI. Hormonal Effects on Reproductive Tract 133

VII. Timing of Anatomical Changes 133

VIII. Hormonal Effects on Adult Anatomy 134

IX. Types of Urogenital Organs 135
- A. Type A 135
- B. Type B 135
- C. Type C 135

X. Microscopic Study of the Prostatic Glands 139

XI. Summary and Discussion 139

References 141

* Supported by Grants RR00167 and HD18865 from the National Institutes of Health, U.S. Public Health Service. Publication Number 26-032 of the Wisconsin Regional Primate Research Center.

I. INTRODUCTION

During the past 15 years, a number of reports have been published dealing with the psychological characteristics of genotypic female rhesus monkeys experimentally virilized prior to birth by transplacental exposure to androgens injected into the pregnant mother.[1-4] These past reports dealt with the expression of social behavior of these pseudohermaphrodites while they were living in enduring social groups of familiar peers. For the most part, the reports dealt with development of behavior during the first year of postnatal life, a period that includes the transition from a state of infancy and dependence on the mother to the state of juvenile semi-independence from the mother. In interpreting the results of these past studies, it is important to bear in mind that the animals were exposed to androgens only prior to birth. Therefore, the behavior they displayed during juvenile development from 3 to 12 months of postnatal age was not directly influenced by the concurrent administration of androgens. Moreover, their behavior was not influenced by endogenous gonadal hormones since puberty does not occur until about 30 months of age in rhesus monkeys.

II. BEHAVIORAL CONSEQUENCES OF PRENATAL ANDROGENS

The psychological findings from these past studies of juvenile female pseudohermaphrodites are briefly summarized here. The female offspring from mothers injected with testosterone propionate (10 to 15 mg/day) from the 40th to at least the 90th day of gestation displayed patterns of behavior that more closely resembled those of the genetic male than the genetic female. The pronounced tendency of the prenatally androgenized females to express behaviors typical of a masculine gender role was evident in their relationships with their mothers as well as with their peers. With their mothers, female pseudohermaphrodites from this treatment regimen showed frequencies of mounting that did not differ statistically from those of normal males. Normal females rarely mounted their mothers. With their juvenile peers, these pseudohermaphrodites displayed frequencies of mounting behavior and levels of rough play that were intermediate between those of normal males and females. In marked contrast to comparable studies in laboratory rodents, prenatal treatments with dihydrotestosterone propionate were nearly as effective in establishing or inducing masculine gender role behavior as treatments with testosterone propionate.[5] Similarity of these two hormones in altering mounting of peers over the first 39 months of life is shown in Figure 1.

In more recent work, behavioral consequences of varying the timing and duration of prenatal exposure to testosterone propionate have been assessed. The parameters of treatment are illustrated in Figure 2. Treatments were designed to test for psychological virilization associated with short durations of exposure (15 days) beginning at 26, 40, 50, 60, or 100 days of gestation. In addition, for treatments beginning at gestational day 40, the duration was systematically varied to last for 15, 25, 35, or 50 or more successive days. Finally, following the discovery that treatments begun on day 40 and lasting for at least 25 days produced reliable psychological masculinization, a group was added to the study for which treatment began at day 115 of gestation and lasted for 25 days (until day 140).

III. EFFECTS ON JUVENILE BEHAVIOR

Results obtained from studies of the juvenile social behavior from all of the treatments described in Figure 2 and also for normal males and females are illustrated in Figures 3 to 5. Mounting of mothers was most pronounced in females androgenized prenatally for at least 50 days beginning on day 40. Moreover, there was a graded increase in the frequency of display of this masculine behavior directly related to duration of treatment (Figure 3).

FOOT-CLASP MOUNTING OF PEERS

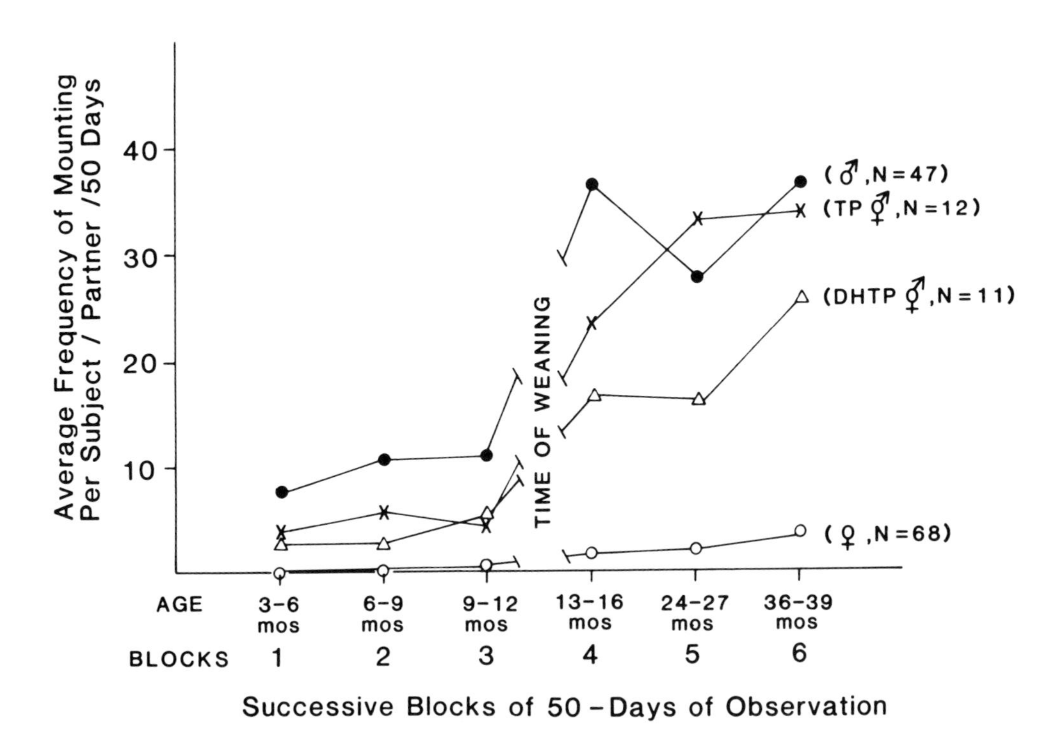

FIGURE 1. Average frequency of mounts displayed to peers by male, female, and prenatally androgenized female rhesus during 50-day blocks of observation carried out from 3 to 39 months of age. Females androgenized prenatally were obtained by injecting their mothers during pregnancy (day 40 through 94 of the 168 day gestation) with either testosterone propionate (TP) or dihydrotestosterone propionate.

Readers' attention is directed to the particular finding that treatments initiated on day 100 or later in gestation had no reliable effect on this kind of juvenile mounting behavior.

Masculine social behavior with peers was affected differently from the masculine behavior directed toward the mother. Treatments as short as 15 days, regardless of the time of initiation, caused elevations in the frequency of performance of rough play with peers (Figure 4). There is also suggestive evidence that treatments started late in gestation were more effective in augmenting rough play than those started early.

The influence of the various prenatal treatments on mounting peers was complex. Although there was evidence for a graded effect related to duration, treatments begun as late as 115 days of gestation were as effective as treatments initiated on day 40. The effectiveness of these late treatments is important because they demonstrate conclusively that for nonhuman primates, as for laboratory rodents, psychological masculinization can be induced independently of genital virilization. None of the females exposed to androgen from day 100 or day 115 showed any genital virilization, although at birth there was a mild clitoral hypertrophy. This clitoral hypertrophy, noticeable at birth, appeared to be transitory since it was unremarkable in all subjects by 3 months of age.

IV. EFFECTS ON ADULT SEXUAL BEHAVIOR AND MENARCHE

Some of the females (a randomly selected subsample) androgenized from day 40 through day 94 or later have been studied as young adults for their tendency to display adult male

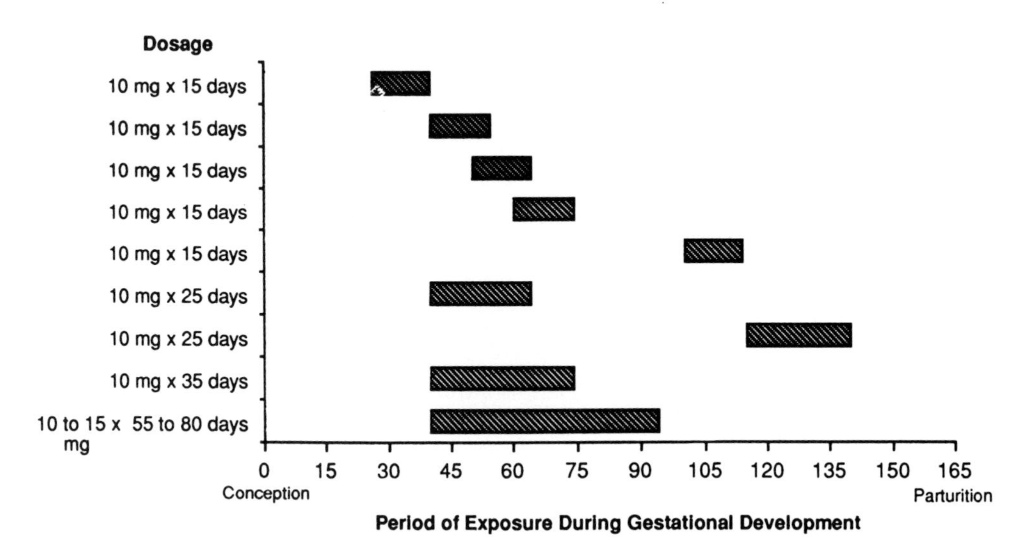

FIGURE 2. Summary of treatments of testosterone propionate given to pregnant females. Bars in the figure represent the duration of each treatment. For most treatments dosage was constant at 10 mg/day.

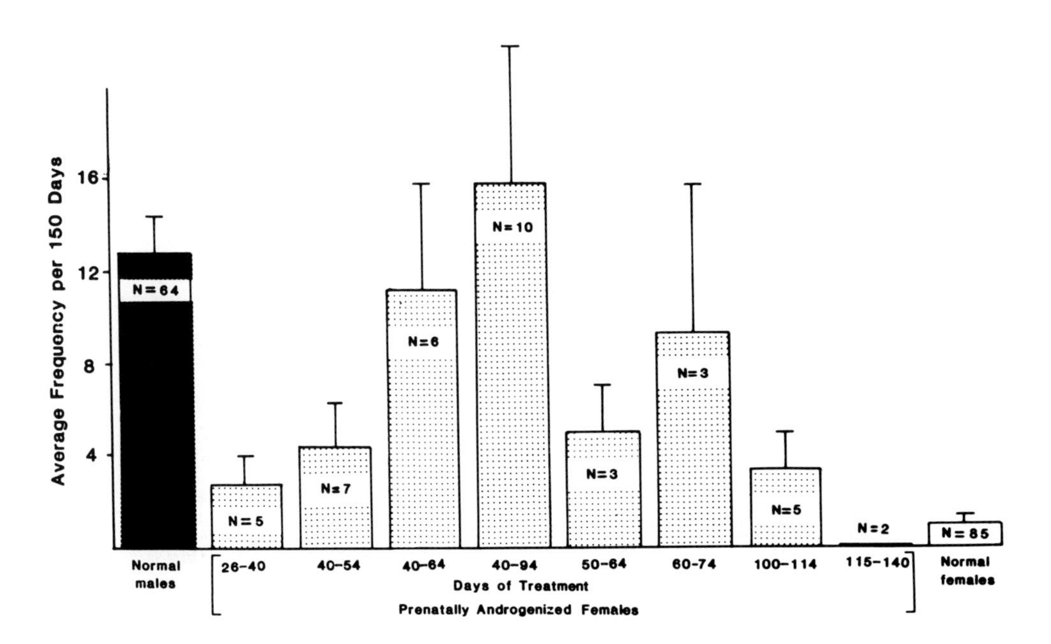

FIGURE 3. The average frequency of display of mounts to mothers by normal male and female rhesus and females exposed prior to birth to testosterone propionate during varying gestational intervals. Measurements of behavior were made for 150 consecutive days when subjects were 3 to 12 months of age.

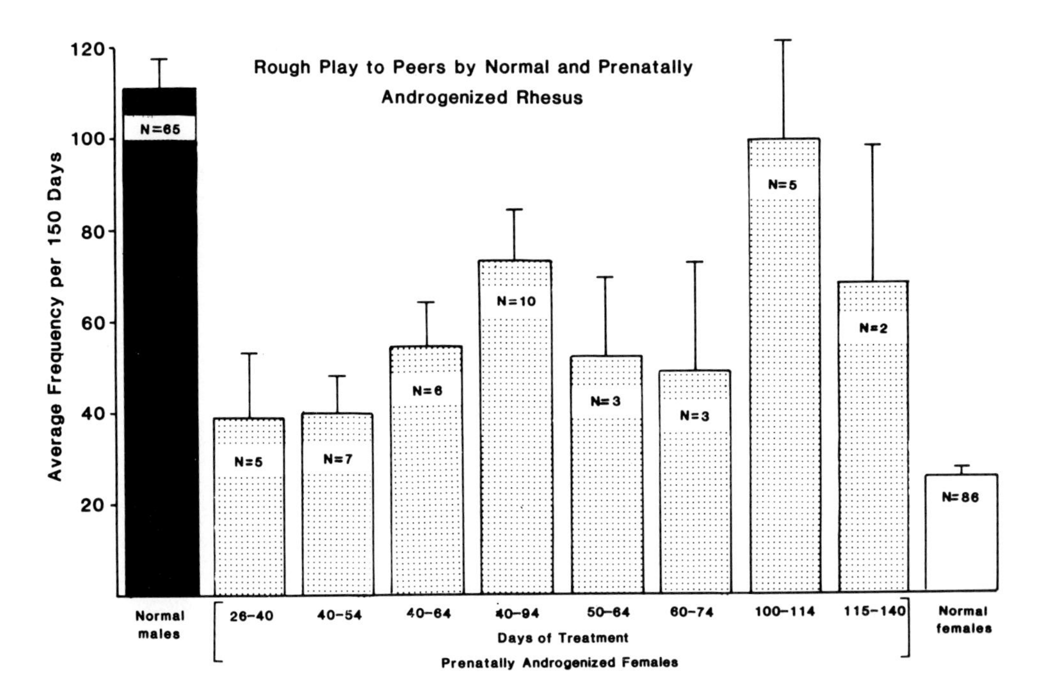

FIGURE 4. The average frequency of display of rough play with peers by normal male and female rhesus and females exposed prior to birth to testosterone propionate during varying gestational intervals. Measurements of behavior were made for 150 consecutive days when subjects were 3 to 12 months of age.

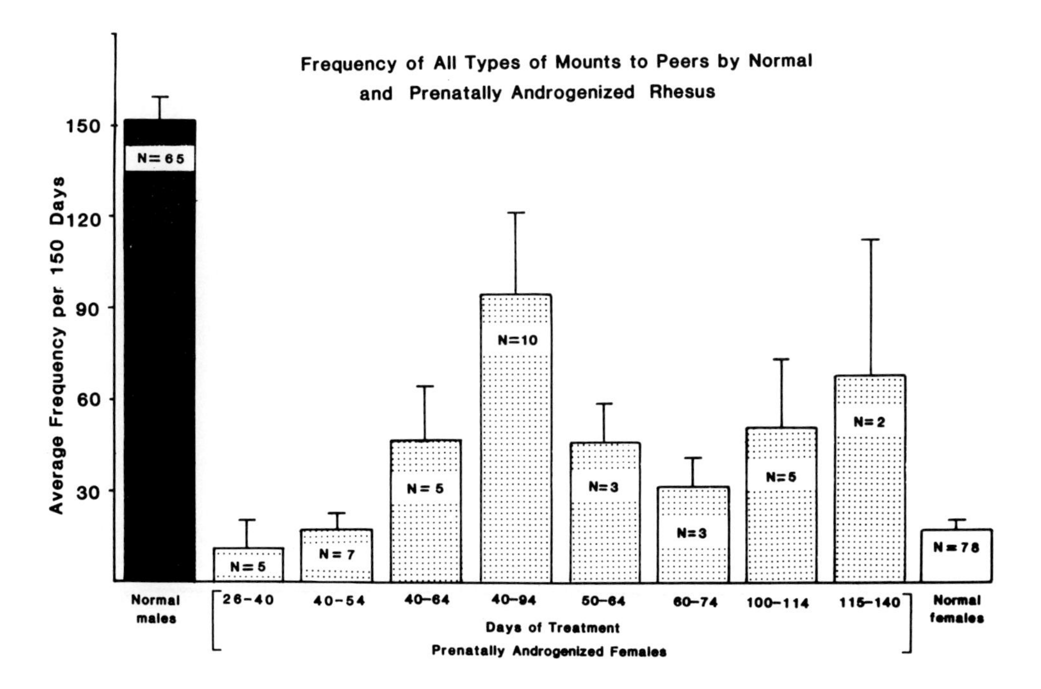

FIGURE 5. The average frequency of display of mounts to peers by normal male and female rhesus and females exposed prior to birth to testosterone propionate for varying gestational intervals. Measurements of behavior were made for 150 consecutive days when subjects were 3 to 12 months of age.

Table 1
AGE AT MENARCHE IN NORMAL AND PRENATALLY ANDROGENIZED RHESUS

Prenatal treatment			Age at menarche	
Hormone	Starting (day of gestation)	Duration (days)	N	Mean ± S.E.
Dihydrotestosterone	40	55	9	1187.4 ± 46.2
Testosterone	40	15	7	978.7 ± 30.7
	40	25	7	1048.7 ± 44.7
	40	55	9	1088.9 ± 49.8
	115	25	7	908.7 ± 36.8
None (Controls)			41	894.5 ± 22.4

sexual behavior.[6] Following ovariectomy, the pseudohermaphrodites were injected daily with testosterone propionate (2 mg/kg body weight) and given standardized tests with receptive normal female partners. Their behavior in these tests was compared with that shown by identically tested castrated males and ovariectomized normal females also injected daily with testosterone propionate. In these tests, pseudohermaphroditic females were indistinguishable from the males, and both groups differed reliably from the normal females in the expression of masculine sexual behavior. Hence, one effect of the prenatal androgen can be described as enhancing the responsiveness of the tissues mediating adult male sexual behavior to hormonal stimulation in adulthood. The pseudohermaphrodites used in these studies of adult masculine sexual behavior were produced by prenatal exposure to either testosterone propionate or dihydrotestosterone propionate. Regardless of the type of androgen used prior to birth, both types of pseudohermaphrodite were equally responsive to stimulation with testosterone propionate in adulthood. This finding also contrasts markedly with the results from similar studies of rodents in which the perinatal administration of dihydrotestosterone alone to genetic females has been shown to have no effect on increasing responsiveness of the substrate for male sexual behavior to testosterone in adulthood.

Androgens given for at least 50 days, beginning on day 40 of gestation, to female fetal rhesus have also been shown to modify their adult female sexual behavior. The effect is different from that commonly demonstrated in rodents for which the perinatal or prenatal androgens have been shown to suppress the responses indicative of female sexual receptivity. So far, no evidence supporting suppression of receptive behavior has been found for rhesus pseudohermaphroditic females. However, certain responses that are part of the normal female sexual behavior system were suppressed by the prenatal androgens. In particular, those responses indicative of female proceptivity were much less frequently displayed in sexual situations by pseudohermaphrodites than by normal females despite equality of ovarian hormone stimulation.[7,8]

Primates appear to differ from rodent models in another important way in terms of the long-term effects of prenatal androgenization. In rodents, a reproducible consequence of androgens administered during a critical period is the establishment of an anovulatory syndrome following or concurrent with puberty. This effect has not been found in rhesus with any of the androgen treatments given. Moreover, anovulatory syndrome is not associated with adrenogenital syndrome in humans. Human females with adrenogenital syndrome may, however, show delayed menarche,[9] and rhesus females exposed prenatally to androgens show a parallel phenomenon (Table 1). It is interesting to point out that results in Table 1 suggest that 25 days of treatment beginning on day 40 is effective in delaying menarche, but the same duration of treatment beginning on day 115 produces no statistically significant delay compared with control values for age at menarche.

V. PRINCIPLES OF HORMONAL ACTION

Certain principles describing the manner in which androgens present at early stages of development alter the behavior of genetic females can be gleaned from the data on laboratory primates and other mammals. To begin with, these hormones act either primarily or exclusively to influence systems of behavior that are normally expressed dimorphically within the species under consideration. Stated more concretely, the androgens only act to augment the behavior patterns that a normal male shows more often than a normal female. A second principle is that androgens act to suppress, inhibit, or otherwise reduce the display of behavior patterns normally characteristic of the female. This second principle has been very well documented in studies of female sexual receptivity in rodents[10] and even in dogs[11] and sheep,[12] but so far no convincing evidence for the suppression of female receptive behavior has been found in primates.

A third principle adduced from the abundant data is that androgens primarily affect the quantitative expression of a behavioral trait, and they do not seem to determine whether a behavioral response can or cannot be displayed. Fourth, the androgens acting at early developmental stages, especially during those times conceptualized as "critical periods", produce permanent and irreversible alterations in the predispositions and behavioral tendencies. In this respect, androgens acting during critical periods differ dramatically from the actions of androgen in adult animals where the alterations in behavior are reversible at will by alternately supplying and withdrawing androgenic stimulation and/or support. Finally, it should be emphasized that androgens present during critical stages can induce changes in behavioral systems that do not require any concurrent stimulation with androgen at the time the responses are displayed. This fifth descriptive principle is well illustrated by the development of juvenile male gender role behavior in rhesus monkeys, and its generality is evidenced by the thorough studies of the ontogeny of micturitional behavior in pseudohermaphroditic dogs.[13]

VI. HORMONAL EFFECTS ON REPRODUCTIVE TRACT

The principles describing androgenic influences on the differentiation of tissues mediating diverse behavioral patterns are strikingly similar to those describing androgenic influences on the differentiation and development of reproductive tract structures. Among mammals, a variety of chemical substances ranging from synthetic progestagens[14] to natural androgens[15] have been shown to virilize anatomical features of genotypic female fetuses. The experimental or clinical administration of androgens to pregnant females during critical periods has been shown to masculinize Wolffian ducts and external genitalia of female offspring in mice,[16] hamsters,[17] rats,[18] guinea pigs,[10,19] dogs,[20] sheep,[21] marmosets,[22] rhesus monkeys,[15] and humans.[14] Morphogenic potency of these compounds depends upon the specific chemical as well as the time in pregnancy and the corresponding developmental stage of the embryo or fetus. Assuming adequate dosage, the anatomical virilization that results when timing varies can be either mild and transient or nearly complete and permanent. Similarly, when timing is optimal, different chemical substances can be shown to vary in their virilizing potency. In accomplishing these effects on the female fetus, the substances mimic and duplicate the effects of the naturally produced and secreted androgens from the developing fetal testis. That is to say, the effects of androgens during these early developmental stages are not genotype specific, and effects are not discriminably different in the XX and the XY fetus when all conditions are optimized in both sexes.

VII. TIMING OF ANATOMICAL CHANGES

The time in early development when androgen-mimetic or androgenic substances produce their effects differs according to the species and to the target system or target organ. Thus, changes in Wolffian derivatives can be initiated at an earlier time than changes in urogenital sinus derivatives or changes in the genital tubercle. In general, organs of the reproductive tract manifest their susceptibility to androgenic influences in a cephalocaudal temporal gradient. Moreover, there is a strong indication that diverse organs of the reproductive tract have requirements for a specific chemical substance at the cellular level so that the virilizing change can be brought about. For example, Wolffian derivatives require testosterone and urogenital sinus derivatives require dihydrotestosterone, a 5α-reduced metabolite of testosterone.[23,24] A logical inference based on our current knowledge of target organ specificity is that substances incapable of mimicking the 5α-reduced metabolites of testosterone would not result in the production of a prostate or the remodeling of the genital tubercle into a penis that incorporates a male urethra,[25,26] although the same substance might cause retention and development of epididymis, vas deferens, and seminal vesicles.

The differences among species with respect to timing are related primarily to differences in developmental rate or schedule. There is no known exception to the general rule that primordia responsive to androgens appear later in development than the differentiation of the primitive gonad, however some species appear to require only brief exposure to the virilizing substances whereas the duration of exposure required is relatively long in others. For example, complete, or nearly complete, differentiation of male reproductive tract structures can be produced experimentally in rats by exposure of female fetuses to appropriate androgens from day 14 postcoitus to about day 5 of postnatal age (approximately 13 days of exposure). A comparable degree of completeness cannot be effected in rhesus monkeys, however, unless exposure is initiated at about 40 days postcoitus and maintained until day 90 of gestation, about 50 days of exposure in all.[27] The obvious difference that some species can be virilized by exposure after birth and some by exposure prior to birth is clearly related to the timing of birth relative to development of the androgen-responsive primordia. Regardless of species differences in timing, however, in all mammalian species the structures of the male reproductive tract (excluding the testis) depend upon actions of androgenic substances during critical periods for their retention, development, and differentiation.

VIII. HORMONAL EFFECTS ON ADULT ANATOMY

The androgenic differentiation of Wolffian duct derivatives in genetic female rhesus has been well demonstrated in the earlier studies from van Wagenen's laboratory.[15] Not included in this early work, however, was any estimate of the responsiveness of these structures to androgens administered during adult stages. We have recently had the opportunity to administer androgens to adult pseudohermaphroditic female rhesus for varying periods of time shortly before their sacrifice for unrelated reasons. During autopsy, we were able to harvest the reproductive tracts of these androgen-stimulated pseudohermaphrodites and compare relevant structures with those in unstimulated pseudohermaphrodites. The opportunistic nature of these evaluations precluded the possibility of establishing a balanced research design at the present time, but the results suggest that derivatives that are responsive to androgens in adulthood cannot be fully established if prenatal exposure to androgen extends only from the 50th through the 64th day of gestational age.

Seven animals were treated with prenatal injections of testosterone for varying gestation periods, and three different types of the persistent urogenital sinus and virilized reproductive organs are listed in Table 2. Five animals were treated with testosterone (150 mg total over 15 days), 40 to 54 days (2) or 50 to 64 days (3) during the gestational period. Two other animals were treated with testosterone (750 mg total over 50 days) from the 39th through the 88th gestational day (females 1619 and 1664 in Table 2).

One animal from the low-dose and two animals from the high-dose group were injected

Table 2
TYPES OF GENITAL ANATOMY ASSOCIATED WITH COMBINATIONS OF PRENATAL AND POSTNATAL TREATMENTS WITH TESTOSTERONE PROPIONATE

	Prenatal period		Postnatal period				Age/body wt at autopsy	
ID	Gestation day	Dose/day (mg)	Age (years)	Period (days)	Dose/day (mg)	Anatomy of genital organs	Age (years)	Body wt (kg)
AF84	40—54	10		None		Type A	8.6	6.2
79143	40—54	10		None		Type B	6.5	5.6
78011	50—64	10		None		Type B	7.11	5.1
79078	50—64	10		None		Type B	6.7	5.0
78012	50—64	10	8	15	16	Type B	8.10	8.1
1619	39—48	25	18	27	5	Type C	18.6	8.9
	49—68	15		20	7.5			
	69—88	10		56	10			
1664	39—48	25	18	7	5	Type C	18.11	4.8
	49—68	15						
	69—88	10						

with testosterone, in varying doses and durations, at 8 years 9 months, 18 years 5 months, and 18 years 11 months of age.

IX. TYPES OF UROGENITAL ORGANS

A. Type A

Development of the external genital organs, the labia majora and minora, and the vestibule are entirely missing. A small hole of the urogenital sinus opens in the midgenital fossa below the pubic symphysis. The clitoris, a small nodule, presents above the hole and is unexposed to the skin surface (Figure 6). Internally, the major abnormality represents the persistence of the urogenital sinus which forms a narrow tubular extension between the vaginal vault and the external orifice of the urogenital sinus. The urethra opens into the urogenital sinus near the external orifice (Figure 7, Type A). The body and cervix of the uterus and vagina were generally well developed. This mild form of genital tract virilization was found in female AF74 (Table 2).

B. Type B

Externally, the glans penis and prepuce protrude from the genital fossa below the pubic symphysis (Figure 8). There is no bulging of the scrotal sac on the anal side of the penis. The distal portion of the penile body, with an average length of about 3 cm, protrudes beyond the body wall. The persistent urogenital sinus further extends and forms the penile urethra (Figure 7, Type B, and Figure 9). The uterine body, cervix, and vaginal vault are well developed, and the distal end of the vagina communicates with the proximal portion of the penile urethra. The distal vagina does not have an external orifice.

Marked atrophy of the uterine body and cervix associated with flattening of the wall of the vagina was noted in a similar case ovariectomized 3 months prior to postnatal treatment with testosterone (Figure 10). These changes in uterus and vagina are believed due to ovariectomy prior to treatment with testosterone in adulthood, and they are not attributed to any action of testosterone per se.

C. Type C

This type of anomaly involves the development of the masculine type of the Wolffian

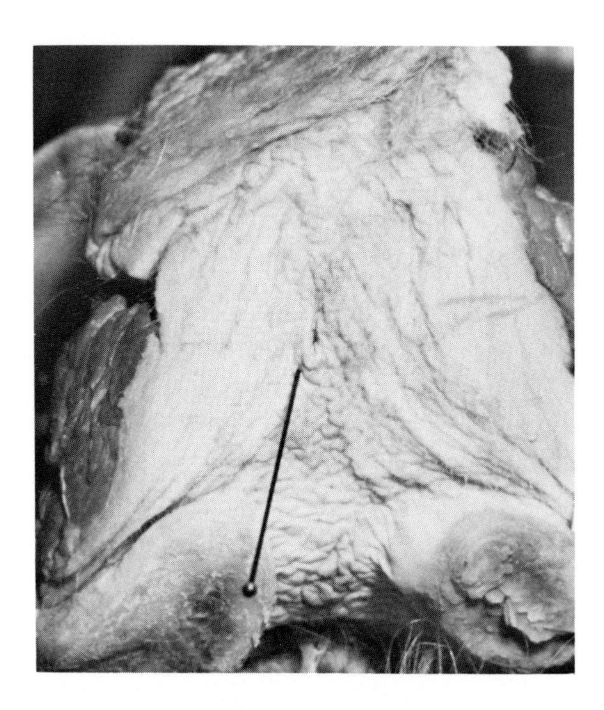

FIGURE 6. External genitalia of female rhesus exposed to androgen prenatally from the 40th through the 54th day of gestation. Type A, see text for description. The pin insertion indicates a residual external opening of the urogenital sinus.

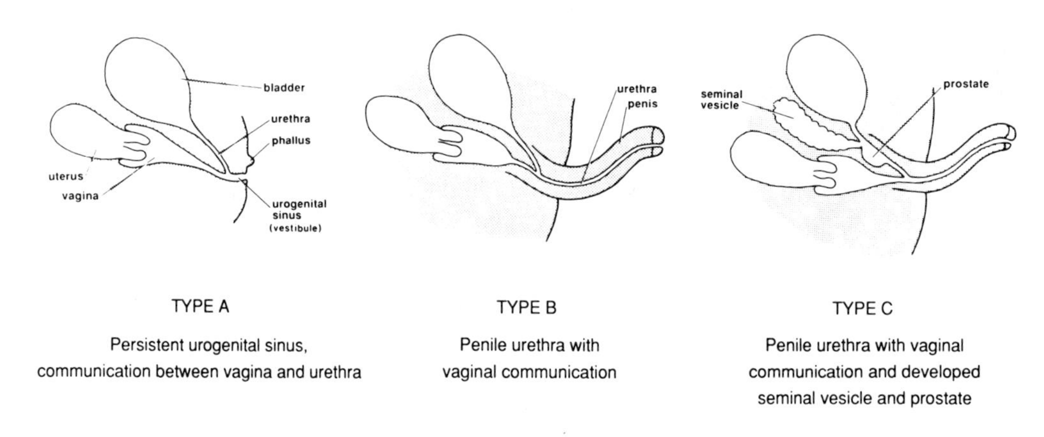

FIGURE 7. Types of genital tract masculinization of female rhesus resulting from different parameters of treatment with testosterone propionate prior to birth. Type A: mild virilization associated with treatment given from day 40 through 54 of gestation. This association was not consistent, and other females given the same treatment showed a greater degree of virilization. Type B: more frequently associated with treatments from day 50 through 64. Type C: the most extensive virilization of the cases studied, which was associated with treatments lasting from day 40 through 88 or 89 shown in Table 2.

duct resulting in a seminal vesicle and presence of the prostate, both of which show strong functional activation from the testosterone injections during adulthood (Figure 6, Type C, and Figure 11). The persistent urogenital sinus and development of the penile urethra are similar to Type B. The multilocular lobes of the seminal vesicle protrude bilaterally into the excavatio vesicouterina. The prostatic tissue is seen in the proximal end of the seminal

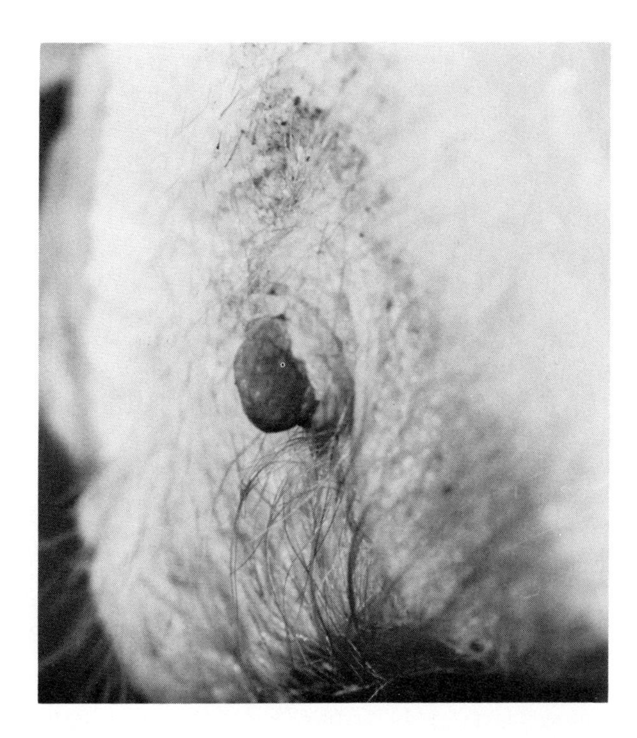

FIGURE 8. External genitalia (Type B) of female 79078 (Table 2) treated from day 50 through 64 of gestational development. Glans penis protrudes through the prepuce, a penile urethra is present, and the external vaginal orifice is obliterated. A scrotal sac is not well developed. No postnatal androgen was given.

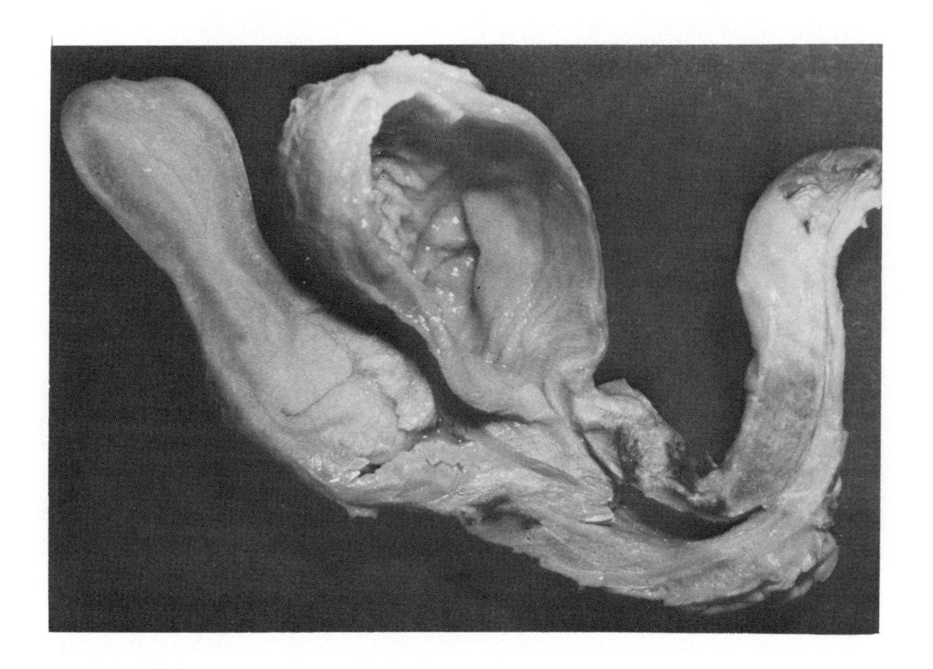

FIGURE 9. Reproductive tract (Type B) of same female depicted in Figure 8.

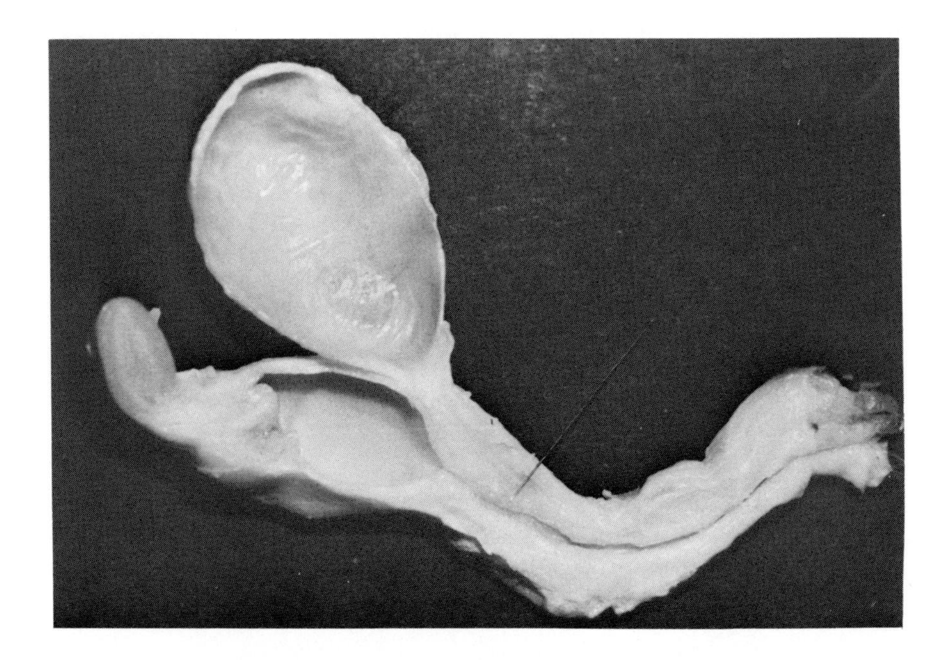

FIGURE 10. Reproductive tract (Type B) of female 78012 (Table 2) treated from day 50 through 64 of gestational development. This female was ovariectomized 3 months prior to postnatal treatment with androgen.

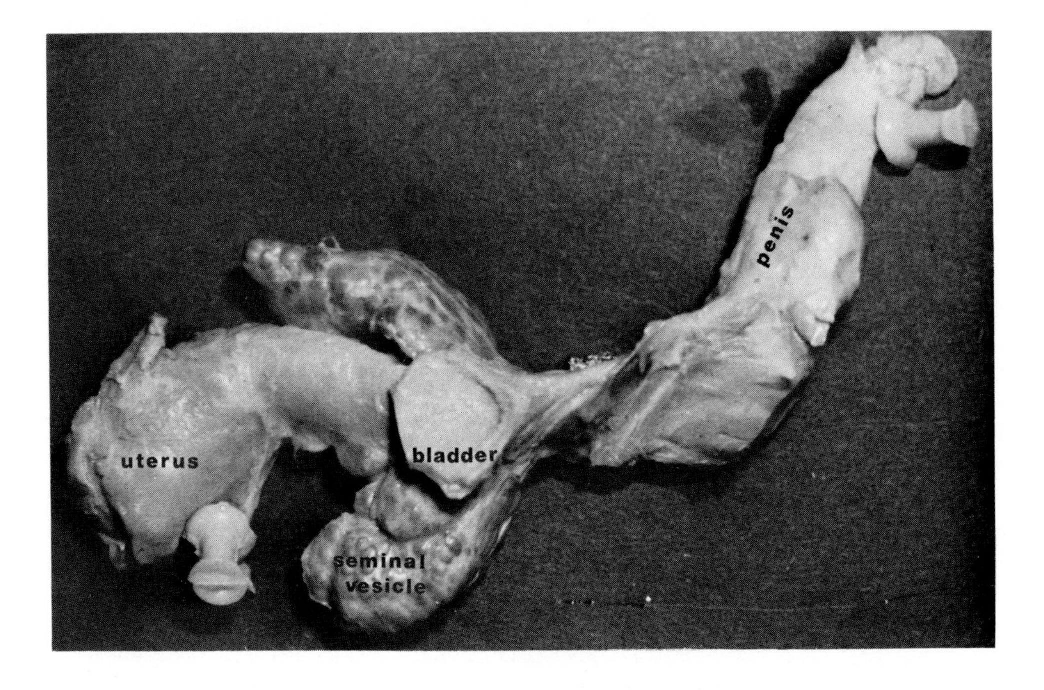

FIGURE 11. Reproductive tract (Type C) of female 1664 (Table 2) showing seminal vesicles that have been very extensively stimulated with testosterone treatment in adulthood. The prostate is not visible in this photograph.

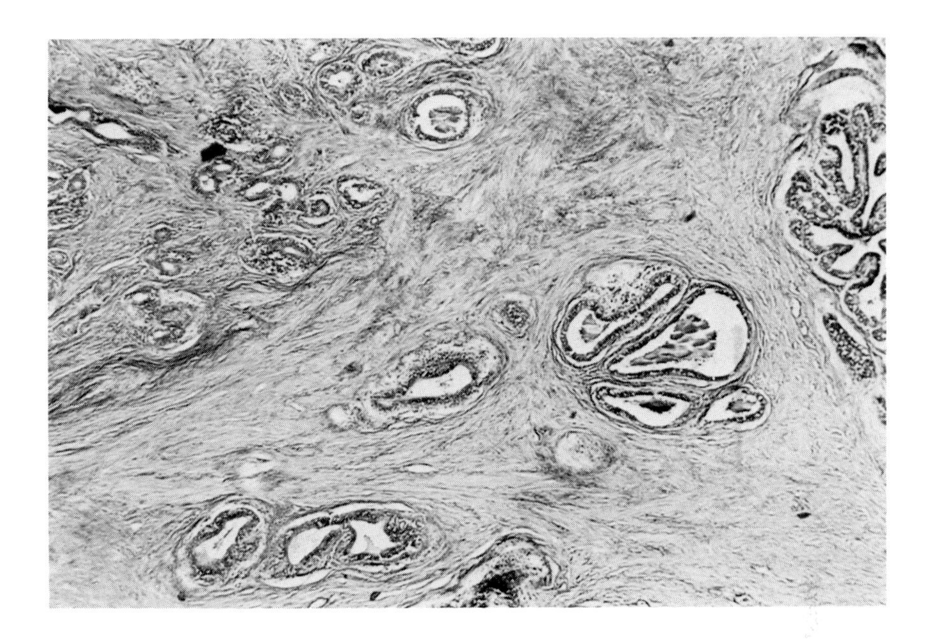

FIGURE 12. Photomicrograph of prostate from female 1619 stimulated with high doses of testosterone in adulthood for a period of 105 consecutive days (see Table 2).

vesicle and is tightly attached with the posterior wall of the urethra near the bladder opening. The uterus and deeper portions of the vagina develop well in these two cases of Type C anomaly. This type of anomaly was seen in two cases (females 1619 and 1664, Table 2) which were given large doses (750 mg) of testosterone for a long period during gestation (39 to 88 gestation days).

X. MICROSCOPIC STUDY OF THE PROSTATIC GLANDS

There was no presence of microscopically detectable prostatic tissues in either Type A or B anomalies. In Type C anomaly, the prostatic glandular tissues representing the presence of glandular acinus and fibromuscular tissue in the interstitium were found in the posterior wall of the urethra between the bladder and communication site of the vagina.

The prostatic gland in case 1619 that was given the postnatal treatment with large doses of testosterone for a long time in adulthood showed well-developed glandular acini with hyperplasia of the glandular epithelium (Figure 12) compared with an inert appearance of the gland in case 1664 given a low dose of testosterone for a short period (Figure 13).

As can be seen from the results of this anatomical survey, androgens prior to birth in rhesus females do not suppress the development of female characteristics derived from the Müllerian anlagen (uterus, cervix, and deep portions of the vagina). These typically female structures develop normally in all mammals unless acted upon by a Müllerian-inhibiting factor from the fetal testis. Lower and more distal portions of the vagina, however, can be inhibited by the presence of androgen prenatally. The urogenital sinus anlagen that normally give rise to the vaginal vestibule are completely converted by the actions of prenatal androgens into the penile urethra which, in pseudohermaphroditic females of Types B and C, communicates directly with the deep portion of the vagina.

XI. SUMMARY AND DISCUSSION

This survey of anatomical and psychological virilization of female rhesus monkeys brought

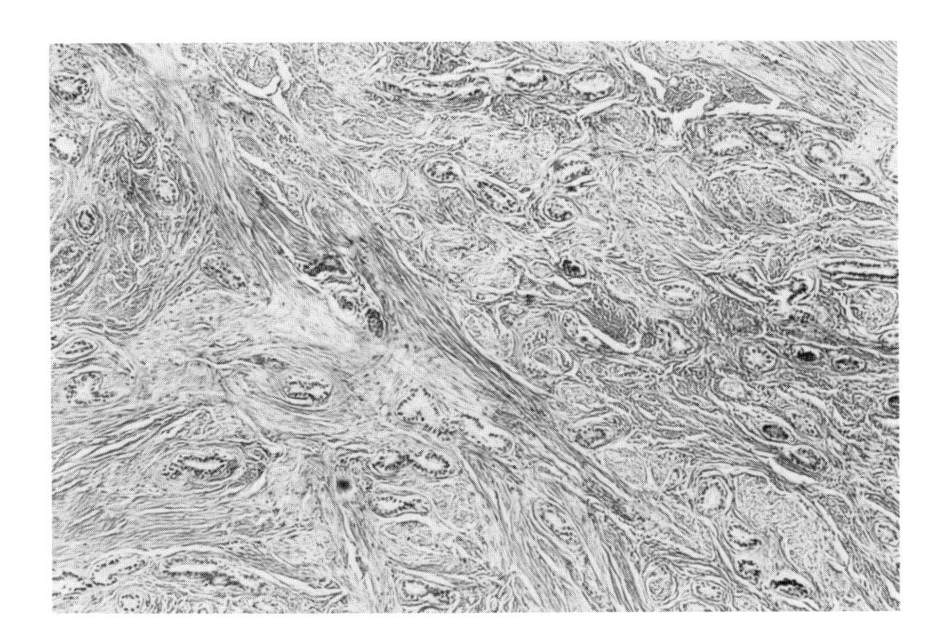

FIGURE 13. Photomicrograph of prostate from female 1664 stimulated with testosterone in adulthood for only 7 consecutive days prior to sacrifice (see Table 2).

about by transplacental delivery of androgens indicates the continuities and divergences between primates and nonprimate mammals. Continuities are evident in the anatomical modifications of the developing reproductive tract. In fact, the actions of androgen on the sexually neutral embryo appear to correspond completely throughout all mammalian species studied. Divergences occur, however, in the actions of androgens on those tissues, presumably the central nervous system, that mediate the expression of behavior and neuroendocrine coordination. For both of these integrative functions, the rhesus monkey is nearly equally sensitive to both testosterone and dihydrotestosterone, a nonaromatizable androgen. In contrast, among nonprimate species studied, the estrogenic metabolites of testosterone have been shown to be essential for psychological masculinization and for masculinization of the hypothalamic-pituitary-gonadal axis.[28] The reasons for this mammalian evolutionary divergence in the dependence upon estrogens for brain masculinization are not understood since there is clear evidence for the presence of both androgen and estrogen receptors in fetal primate brain[29,30] as well as evidence for phylogenetic continuity of brain metabolism of steroids.[31] Nevertheless, the ultimate contrast of this difference in estrogen dependence is made evident by comparing the behavior of humans and rats afflicted with the genetic disorder resulting in testicular feminizing syndrome. In both species, afflicted individuals have a deficiency of androgen receptors and a normal complement of estrogen receptors. Both produce normal or supranormal amounts of testosterone, and both aromatize testosterone to its estrogenic metabolites. Behaviorally, however, afflicted rats show mounting behavior (masculine traits) and suppressed female sexual response.[32] Afflicted humans, however, show no masculine psychological characteristics and normal female psychosexual traits.[33]

While endogenous estrogens, principally those derived from androgen aromatization, do not appear to masculinize humans, there is less certainty about influences of exogenous estrogens administered to the mother during pregnancy. Assuming that diethylstilbestrol (DES) does not differ in some critical unknown way in its actions from the actions of the naturally occurring estrogens, then recent research suggests that exogenous estrogens given prior to birth may have masculinizing effects on human females. This is suggested by data on cognitive processes,[34] and also by very recent studies of psychosexual orientation[35] of women exposed to DES in large doses prior to birth. Not to be overlooked, however, is the demonstrable fact that these large doses of DES administered to pregnant females do not impair ovarian function or delay menarche in their female offspring.[36,37]

It seems more than likely that the lingering questions about species differences in hormonal requirements for the prenatal modification of masculine and feminine traits in behavior and neuroendocrine function will not be resolved without additional research. What seems certain at this time, however, is that in some species the brain mechanisms mediating behavior and neuroendocrine functions do not have hormonal requirements identical to those of the reproductive tract structures.

REFERENCES

1. **Goy, R. W. and Resko, J. A.,** Gonadal hormones and behavior of normal and pseudohermaphroditic nonhuman female primates, *Recent Prog. Horm. Res.*, 28, 707, 1972.
2. **Goy, R. W.,** Development of play and mounting behavior in female rhesus virilized prenatally with esters of testosterone or dihydrotestosterone, in *Recent Advances in Primatology,* Vol. 1, Chivers, D. J. and Herbert, J., Eds., Academic Press, London, 1978, 449.
3. **Goy, R. W. and Robinson, J. A.,** Prenatal exposure of rhesus monkeys to potent androgens: morphological, behavioral, and physiological consequences. *Banbury Report II. Environmental Factors in Human Growth and Development,* Cold Spring Harbor Laboratory, Cold Spring Harbor, N.Y., 1982, 355.
4. **Goy, R. W. and Kemnitz, J. W.,** Early, persistent, and delayed effects of virilizing substances delivered transplacentally to female rhesus fetuses, in *Application of Behavioral Pharmacology in Toxicology,* Zbinden, G., Cuomo, V., Racagni, G., and Weiss, B., Eds., Raven Press, New York, 1983, 303.
5. **Goy, R. W.,** Differentiation of male social traits in female rhesus macaques by prenatal treatment with androgens: variation in type of androgen, duration, and timing of treatment, in *Fetal Endocrinology,* Novy, M. J. and Resko, J. A., Eds., Academic Press, New York, 1981, 319.
6. **Pomerantz, S. M., Goy, R. W., and Roy, M. M.,** Expression of male-typical behavior in adult female pseudohermaphroditic rhesus: comparisons with normal males and neonatally gonadectomized males and females, *Horm. Behav.*, 20, 483, 1986.
7. **Thornton, J. and Goy, R. W.,** Female-typical sexual behavior of rhesus and defeminization by androgens given prenatally, *Horm. Behav.*, 20, 129, 1986.
8. **Pomerantz, S. M., Roy, M. M., Thornton, J. E., and Goy, R. W.,** Expression of adult female patterns of behavior by male, female, and pseudohermaphroditic female rhesus, *Biol. Reprod.*, 33, 878, 1985.
9. **Jones, H. W., Jr. and Verkauf, B. S.,** Congenital adrenal hyperplasia: age at menarche and related events in puberty, *Am. J. Obstet. Gynecol.*, 109, 292, 1971.
10. **Goy, R. W., Bridson, W. E., and Young, W. C.,** The period of maximal susceptibility of the prenatal female guinea pig to masculinizing actions of testosterone propionate, *J. Comp. Physiol. Psychol.*, 57, 166, 1964.
11. **Beach, F. A.,** Hormonal modification of sexually dimorphic behavior, *Psychoneuroendocrinology,* 1, 3, 1975.
12. **Clarke, I. J., Scaramuzzi, R. J., and Short, R. V.,** Sexual differentiation of the brain: endocrine and behavioural responses of androgenized ewes to oestrogen, *J. Endocrinol.*, 71, 175, 1976.
13. **Beach, F. A.,** Effects of gonadal hormones on urinary behavior of dogs, *Physiol. Behav.*, 12, 1005, 1974.
14. **Wilkins, L., Jones, H. W., Holman, G. H., and Stempfel, J.,** Masculinization of the female foetus associated with administration of oral and intramuscular progestogens during gestation: non-adrenal female pseudohermaphroditism, *J. Clin. Endocrinol. Metab.*, 18, 559, 1959.
15. **van Wagenen, G. and Hamilton, J. B.,** The experimental production of pseudohermaphroditism in the monkey, in *Essays in Biology,* University of California Press, Berkeley, 1943, 581.
16. **Turner, D. C., Haffen, R., and Struett, H.,** Some effects of testosterone on sexual differentiation of female albino mice, *Proc. Soc. Exp. Biol. Med.*, 42, 107, 1939.
17. **Bruner, J. A. and Witschi, E.,** Testosterone induced modifications of sex development in female hamsters, *Am. J. Anat.*, 79, 293, 1946.
18. **Greene, R. R., Burrill, M. W., and Ivy, A. C.,** Experimental intersexuality. The effect of antenatal androgens on sexual development of female rats, *Am. J. Anat.*, 65, 415, 1939.

19. **Dantchakoff, V.,** Realisation du sexe a volonte par inductions hormonales. II. Inversions et deviations de l'histogenese sexuelle chez l'embryon de mammifere genetiquement femelle, *Bull. Biol. Fr. Belg.*, 71, 269, 1937.
20. **Beach, F. A. and Kuehn, R. E.,** Coital behavior in dogs. X. Effects of androgenic stimulation during development on masculine mating response in females, *Horm. Behav.*, 3, 347, 1970.
21. **Short, R. V.,** Sexual differentiation of the brain of the sheep, *INSERM*, 32, 121, 1977.
22. **Abbott, D. H. and Hearn, J. P.,** The effects of neonatal exposure to testosterone on the development of behaviour in female marmoset monkeys, in *Sex, Hormones, and Behaviour*, Ciba Foundation Symp. 62 (new series), Excerpta Medica, New York, 1979, 299.
23. **Schultz, F. M. and Wilson, J. D.,** Virilization of the Wolffian duct in the rat fetus by various androgens, *Endocrinology*, 94, 979, 1974.
24. **Goldstein, J. L. and Wilson, J. D.,** Genetic and hormonal control of male sexual differentiation, *J. Cell. Physiol.*, 85, 365, 1975.
25. **Imperato-McGinley, J., Guerrero, L., Gautier, T., and Peterson, R. E.,** Steroid 5α-reductase deficiency in man: an inherited form of male pseudohermaphroditism, *Science*, 186, 1213, 1974.
26. **Imperato-McGinley, J., Peterson, R. E., and Gautier, T.,** Male pseudohermaphroditism secondary to 5α-reductase deficiency: a review, in *Fetal Endocrinology*, Novy, M. J. and Resko, J. A., Eds., Academic Press, New York, 1981, 359.
27. **Wells, L. J. and van Wagenen, G.,** Androgen-induced female pseudohermaphroditism in the monkey *(Macaca mulatta):* anatomy of the reproductive organs, *Carnegie Inst. Washington Publ.* 603, Contrib. Embryol. Vol. XXXV, 93, 1954.
28. **Goy, R. W. and McEwen, B. S.,** *Sexual Differentiation of the Brain*, MIT Press, Cambridge, Mass., 1980.
29. **Pomerantz, S. M., Fox, T. O., Sholl, S. A., Vito, C. C., and Goy, R. W.,** Androgen and estrogen receptors in fetal rhesus monkey brain and anterior pituitary, *Endocrinology*, 116, 83, 1985.
30. **Sholl, S. A. and Pomerantz, S. M.,** Androgen receptors in the cerebral cortex of fetal female rhesus monkeys, *Endocrinology*, 119, 1625, 1986.
31. **Sholl, S. A., Goy, R. W., and Uno, H.,** Differences in brain uptake and metabolism of testosterone in gonadectomized, adrenalectomized male and female rhesus monkeys, *Endocrinology*, 111, 806, 1982.
32. **Olsen, K. L.,** Genetic determinants of sexual differentiation, in *Hormones and Behaviour in Higher Vertebrates*, Balthazart, J., Prove, E., and Gilles, R., Eds., Springer-Verlag, Heidelberg, 1983, 138.
33. **Masica, D. N., Money, J., and Ehrhardt, A. A.,** Fetal feminization and female gender identity in the testicular feminizing syndrome of androgen insensitivity, *Arch. Sex. Behav.*, 1, 131, 1971.
34. **Hines, M.,** Prenatal gonadal hormones and sex differences in human behavior, *Psychol. Bull.*, 92, 56, 1982.
35. **Erhardt, A. A., Meyer-Bahlburg, H. F., Rosen, L. R., Feldman, J. F., Veridiavo, N. P., Zimmerman, I., and McEwen, B. S.,** Sexual orientation after prenatal exposure to exogenous estrogen, *Arch. Sex. Behav.*, 14, 57, 1985.
36. **Barnes, A. B.,** Menstrual history of young women exposed in utero to diethylstilbestrol, *Fertil. Steril.*, 32, 148, 1979.
37. **Meyer-Bahlburg, H. F., Ehrhardt, A. A., Rosen, L. R., Feldman, J. F., Veridiano, N. P., Zimmerman, I., and McEwen, B. S.,** Psychosexual milestones in women prenatally exposed to diethylstilbestrol, *Horm. Behav.*, 18, 359, 1984.

Chapter 10

EFFECTS ON FEMALE OFFSPRING AND MOTHERS AFTER EXPOSURE TO DIETHYLSTILBESTROL

J. Rotmensch, K. Frey, and A. L. Herbst

TABLE OF CONTENTS

I. Introduction 144

II. Epidemiology 144

III. DES-Associated Adenocarcinoma 145

IV. Treatment of CCA 146

V. Survival 147

VI. DES-Associated Changes in the Female Lower Genital Tract 149

VII. Management of the DES-Exposed Female 152

VIII. Results of Colposcopic Examination 154

IX. Reproductive and Gynecological Surgical Experience 154

X. DES and Breast Carcinoma 155

XI. Histogenesis 156

References 158

I. INTRODUCTION

In 1938, diethylstilbestrol (DES) (Figure 1) was synthesized and then marketed as an inexpensive, orally active, nonsteroidal estrogen.[1] In the late 1940s studies reported beneficial effects of DES administration in high-risk pregnancies to prevent complications such as abortions, premature delivery, preeclampsia, and intrauterine death.[2,3] In 1948, Smith reported a reduced incidence of pregnancy wastage in 632 patients treated during gestation with DES.[3] It was believed that DES stimulated placental estrogen and progesterone production and thus improved the fetal environment.

The efficacy of DES in pregnancy was first questioned in 1953 when Ferguson reported that DES did not reduce the rate of pregnancy complications in a controlled study.[4] Dieckmann, in a double-blind study, also reported no improvement in fetal outcome with the use of DES.[5] Subsequently, the use of DES in pregnancy declined, but the drug was used, albeit less frequently, until 1971.

In 1970, Herbst and Scully reported seven cases of vaginal adenocarcinoma in young women between the ages of 15 and 22 in the Boston area.[6] Subsequently, Herbst et al. reported that 7 of 8 mothers of patients with vaginal clear cell adenocarcinoma (CCA) had been treated with DES during the first trimester of pregnancy.[7] Greenwald then reported an additional five cases of vaginal carcinoma associated with DES in women 15 to 19 years of age.[8] Prior to these reports, vaginal carcinoma was rarely diagnosed in the young female and these initial cases exceeded the number of clear cell vaginal carcinomas reported in the world literature. Other reports followed, confirming the association between prenatal use of DES and the development of vaginal and cervical CCA in the offspring. In 1971, the Food and Drug Administration in the U.S. proscribed the use of DES in pregnancy.[9]

II. EPIDEMIOLOGY

In 1971, a registry was established to obtain information regarding the epidemiological, clinical, and pathological aspects of CCA diagnosed in women born after 1940. The size of the DES-exposed population has been difficult to establish due to the lack of precise information regarding the frequency of drug use for pregnancy support. After the initial favorable report, DES usage reached its peak in the early 1950s, at which time some centers prescribed DES for approximately 5 to 7% of all pregnancies.[10] In the 1960s, it has been estimated that <1%, or between 100,000 and 160,000 of all liveborn female infants, were exposed to DES.[11] It is likely that approximately two million pregnant women in the U.S. were treated with DES.

As of 1986, over 530 cases of vaginal or cervical CCA had been accessioned into the registry. Of these cases, approximately 65% have a history of ingestion of DES or an equivalent nonsteroidal estrogen, 25% had no known maternal hormone ingestion, and 10% received an unidentified medication for a high-risk pregnancy. The actual risk of an exposed woman developing CCA is extremely small. Melnick et al. recently estimated the risk of CCA among DES-exposed females born between 1948 and 1965 to be 1 case per 1000 women through age 34. The age of patients diagnosed with CCA has ranged from 9 to 34 years, with 91% of DES-exposed cases diagnosed between the ages of 15 and 28 (Figure 2). The median age at diagnosis was 19 years.[12] These figures are consistent with an earlier study which estimated the risk to be 0.14 to 1.4 cases per 1000 exposed, with an identical median age of diagnosis of 19 years.[13]

In a recent case-control study by Herbst et al., 156 registry cases with confirmed DES exposure were compared to 1848 exposed women of similar age without cancer with respect to possible risk factors related to the development of CCA. The relative risk for the development of CCA was found to be significantly higher for those daughters whose mothers

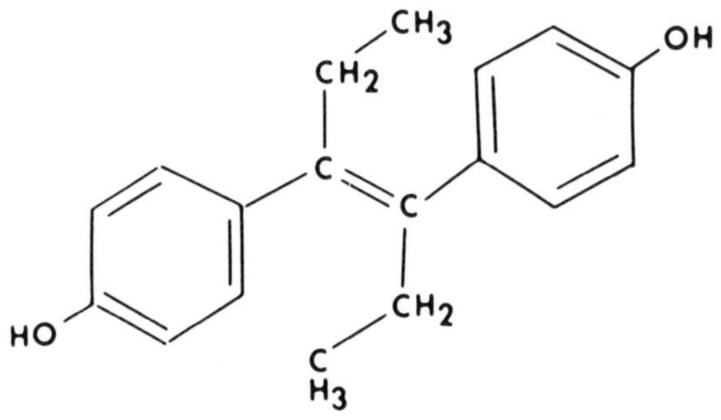

FIGURE 1. Structure of diethylstilbestrol.

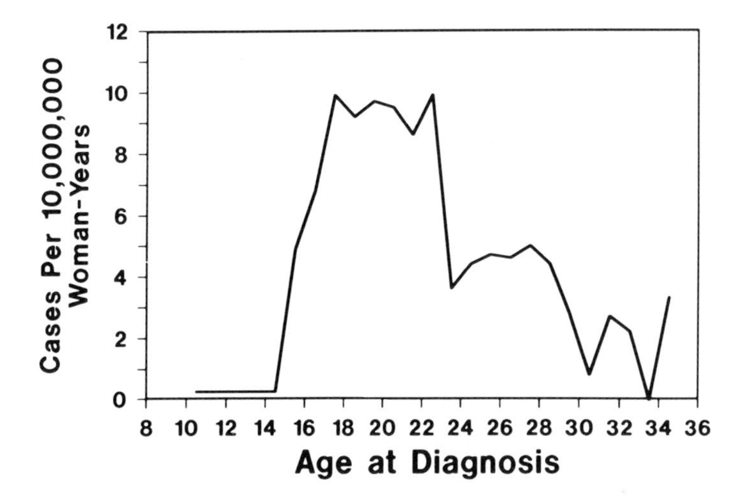

FIGURE 2. Incidence of CCA among white females born in the U.S. (From Melnick, S., Cole, P., Anderson, D., et al., *N. Engl. J. Med.*, in press. With permission.)

began DES before the 12th week of pregnancy. Additional factors found to be significant were a mother's history of prior spontaneous abortion and a fall season of birth (winter conception). The time of exposure to DES during pregnancy remained a significant risk factor even after the maternal history of prior spontaneous abortion was adjusted.[14]

III. DES-ASSOCIATED ADENOCARCINOMA

Approximately 60% of the CCA recorded in the registry are vaginal according to criteria established by international staging classification (FIGO), with the remainder classified as cervical. The predominant site of development of the vaginal lesions has been the anterior third of the vagina, whereas the cervical lesions usually develop on the ectocervix. Grossly, these tumors have ranged in size from 0.2 to 10 cm and have varied in appearance (Figure 3): polypoid, ulcerated, and nodular formations have all been noted. The fact that some tumors have not been visible due to their subepithelial location emphasizes the importance of careful palpation during pelvic examination for detection of these cancers.

Histologically, clear cell carcinomas have exhibited three predominant patterns (Figure 4). The tubulocystic type, characterized by tubule and cyst formation without papillae, is the most common. The solid pattern, consisting of solid nests or masses of neoplastic

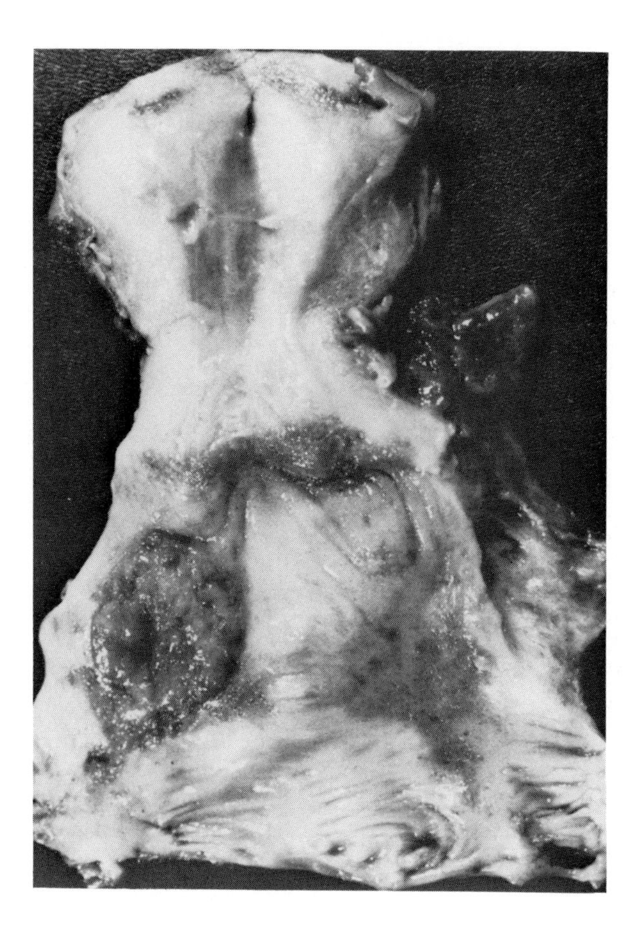

FIGURE 3. Gross appearance of a CCA on the anterior vaginal wall after the uterus and vagina have been opened. Adenosis is present on the posterior wall and edge of tumor. (From Herbst, A. L. and Scully, R. E., *Cancer*, 25, 745, 1970. With permission.)

epithelial cells, is the second most common, while the least common is the papillary pattern, which resembles the tubulocystic type but has papillae which protrude into the tubule and cyst walls. At times, the carcinoma has an endometrioid pattern resembling an endometrial carcinoma.

IV. TREATMENT OF CCA

The majority of CCA patients have been diagnosed with early stage disease. The FIGO classification for cervical and vaginal cancers is shown in Table 1. Stage I disease has been diagnosed in 59% and Stage II in 33% of patients. The treatment regimens have been similar to those implemented for squamous cell carcinoma of the cervix or vagina, taking into account the young age of these patients. For Stage I or early Stage II disease, either radical surgery or radiation therapy have been used effectively. For Stage II and more advanced vaginal or cervical carcinomas, radiation therapy has been the primary modality of treatment. In more advanced disease, treatment has been decided based on tumor location, extent of disease, and the size of the lesion. Surgery, such as anterior, posterior, or total pelvic exenteration, or pelvic radiation has been used as the primary therapy for more advanced tumors localized to central pelvic structures.

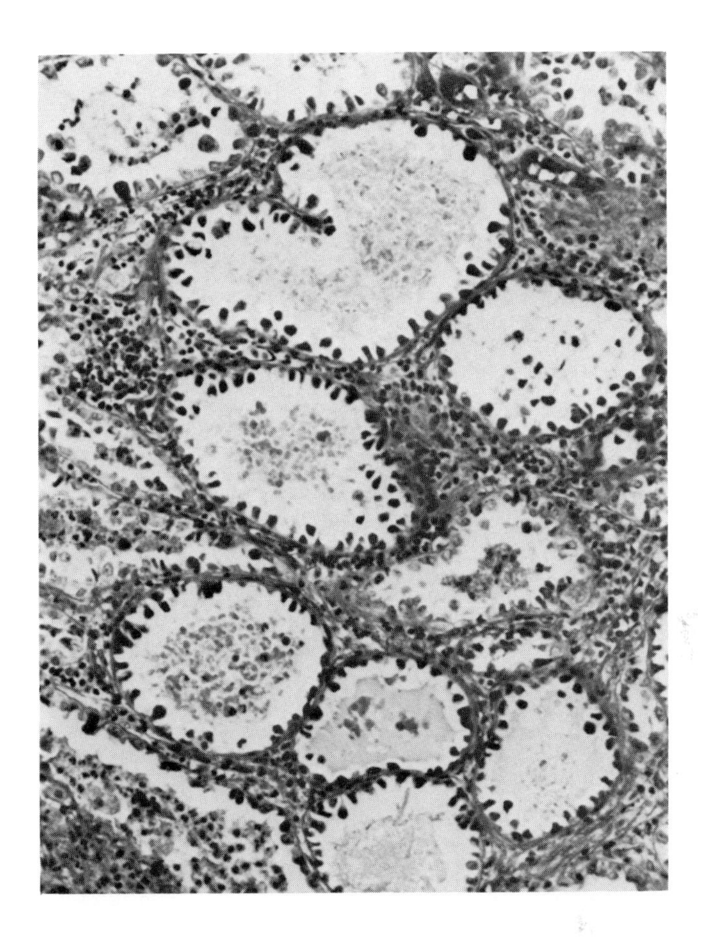

FIGURE 4. CCAs: (A) tubulocystic (H.E.; magnification × 180) showing hobnails extruding into lumen; (B) solid pattern (H.E., magnification × 300); (C) papillary pattern (H.E., magnification × 50). (From Scully, R. E., Robboy, S. J., and Herbst, A. L., *Ann. Clin. Lab. Sci.*, 4, 222, 1974, Institute for Clinical Science, Copyright 1974; Scully, R. E., Robboy, S. J., and Welch, W. R., Proc. Symp. DES, *Obstet. Gynecol.*, p. 17, 1978. Reprinted with permission from The American College of Obstetricians and Gynecologists.)

Local excision with and without pelvic lymphadenectomy has been reported for early lesions. However, local vaginal excision alone is not advocated due to the high recurrence rate noted in patients so treated. Local radiation therapy combined with surgical excision appears to be an acceptable form of therapy for selected early vaginal CCA, particularly for small vaginal tumors not located near the cervix. A pelvic lymphadenectomy is usually performed to be certain that metastatic disease has not developed. A few patients so treated have subsequently been able to bear children.[15]

V. SURVIVAL

Patient survival is related to tumor stage, lesion size, depth of invasion, and predominant tumor histology as well as the age of the patient at the time of diagnosis. Overall 5-year survival rates have been reported as 80%. For early stage disease, survival rates have been similar for both cervical and vaginal carcinomas. Stage I disease has a 5-year survival rate

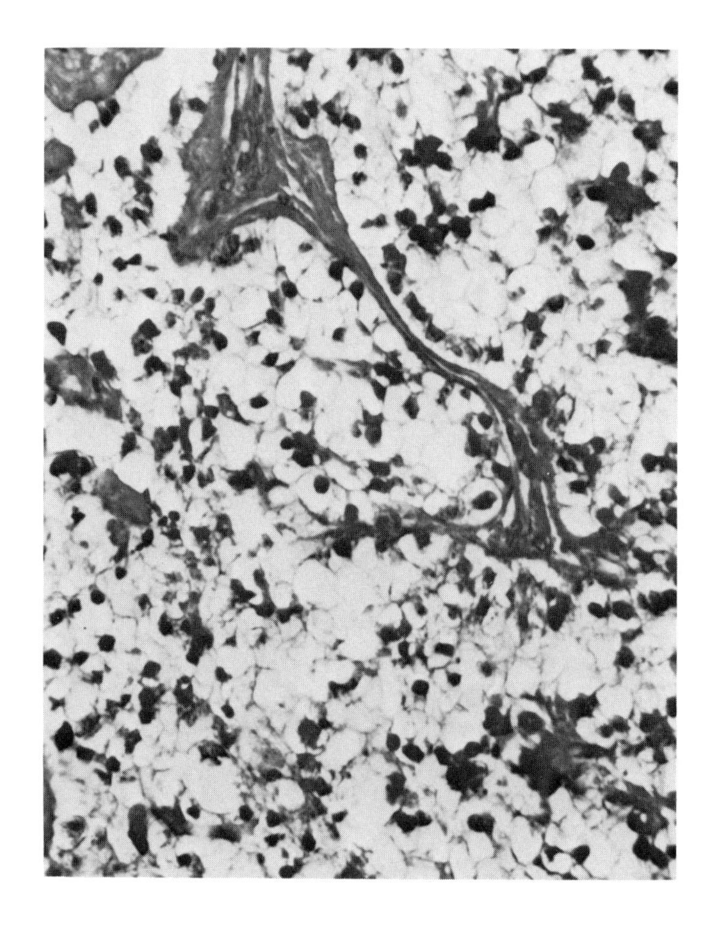

FIGURE 4B.

of approximately 90% for both vaginal and cervical tumors. The 5-year survival rate for Stage II vaginal and IIa cervical disease has been approximately 80%, whereas the survival for Stage IIb carcinoma of the cervix has been less, i.e., approximately 60%. Survival in advanced disease has been poor. Stage III disease, regardless of primary site, has had a survival rate of approximately 35% and no 5-year survivals have been recorded for Stage IV cases.[16]

Patients with CCA over age 19 years have a better prognosis than younger patients. This fact is related to the observation that the older patients more commonly have a predominantly tubulocystic histologic pattern and this pattern has been found to have a more favorable prognosis than other histologic types.[16]

Tumor size and depth of invasion are both correlated with the frequency of lymphatic involvement and patient prognosis. Herbst et al. reported that for all Stage I and cervical Stage IIa cases when the tumor volume was >6 cm^2 the rate of nodal involvement was 24%. In contrast, 12% of those lesions from 1.1 to 6.0 cm^2 in size had lymph node metastasis. In regard to depth of invasion, only 11% of tumors with 6 mm or less of penetration were noted to have lymphatic involvement, while 24% of those with invasion deeper than 6 mm had lymph node involvement.[16] The stage of disease is also related to the frequency of metastasis. Approximately 17% of vaginal Stage I and cervical Stage I and IIa carcinomas have metastasized to pelvic lymph nodes, while about half of the vaginal Stage II and cervical Stage IIb lesions have had pelvic lymph node metastasis.[17]

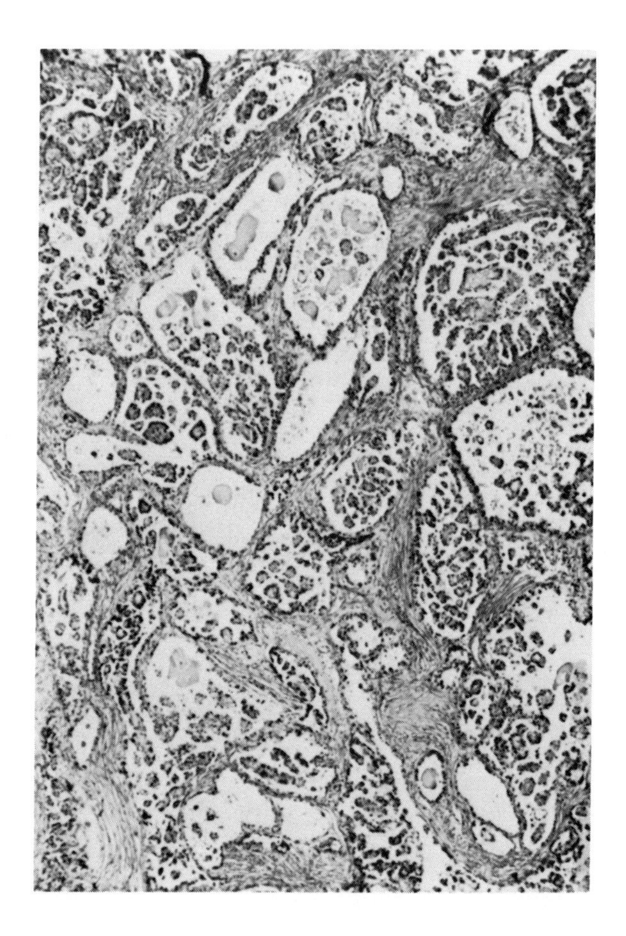

FIGURE 4C.

The overall rate of recurrence has been 21%, with the majority of recurrences occurring within 3 years of diagnosis, but recurrences have been reported as long as 9 years after initial diagnosis. The lungs and supraclavicular areas as well as the pelvis are the most common sites of recurrence. Treatment, depending on the site of recurrence, has consisted mainly of radical surgery or radiation or a combination of one of these with chemotherapy for systemic disease. No consistently effective chemotherapy has been identified, but regressions have been noted with multiple-agent protocols consisting of *cis*-platinum and 5-FU as well as adriamycin and alkylating agents. Most of those patients with persistent or recurrent disease have died.[17,18]

VI. DES-ASSOCIATED CHANGES IN THE FEMALE LOWER GENITAL TRACT

A large proportion of DES-exposed females have ectopic glandular or mucinous epithelium on the vagina (adenosis) or cervix (cervical ectropion). The presence of adenosis in DES-exposed females has been reported to vary from 35 to 90%, depending on risk factors such as age, maternal dosage, length of exposure, and the definition of adenosis used.[19-21] Part of the wide variation in published incidence rates can be attributed to varying criteria used to diagnose adenosis. Some series have included cases of cervical columnar epithelium as well as vaginal adenosis, while others required the presence of squamous metaplasia. However, the term adenosis should be restricted to columnar or mucinous epithelium occurring in the vagina (Figure 5).

Table 1
FIGO STAGING OF VAGINAL AND CERVICAL CARCINOMAS

Stage	Carcinoma of the Vagina
I	Limited to vaginal wall
II	Invades the subvaginal tissues but has not extended to the pelvic wall
III	Reaches the pelvic wall
IV	Extends into the mucosa of the bladder or rectum or outside the true pelvis
	If the tumor involves the external os of the cervix, it is classified as cervical and if the vulva is involved it is classified as vulvar.

Stage	Carcinoma of the Cervix
I	Confined to cervix
IIa	Involves upper two thirds of the vagina but has not infiltrated the parametrium
IIb	Involves upper two thirds of the vagina with parametrial infiltration but has not reached the pelvic side wall
III	Reaches the pelvic side wall and/or lower one third of the vagina and/or hydronephrosis or nonfunction of kidney due to tumor
IV	Extends into the mucosa of bladder or rectum or outside the true pelvis

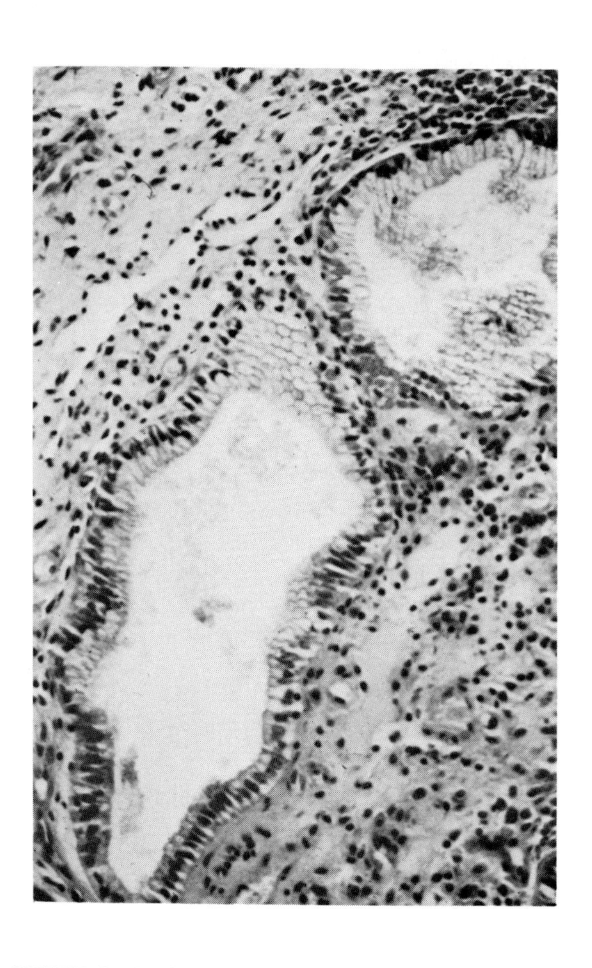

FIGURE 5. Vaginal adenosis. The glands are lined with endocervical epithelium. (Magnification × 200.) (From Herbst, A. L. and Scully, R. E., *Cancer,* 25, 745, 1970. With permission.)

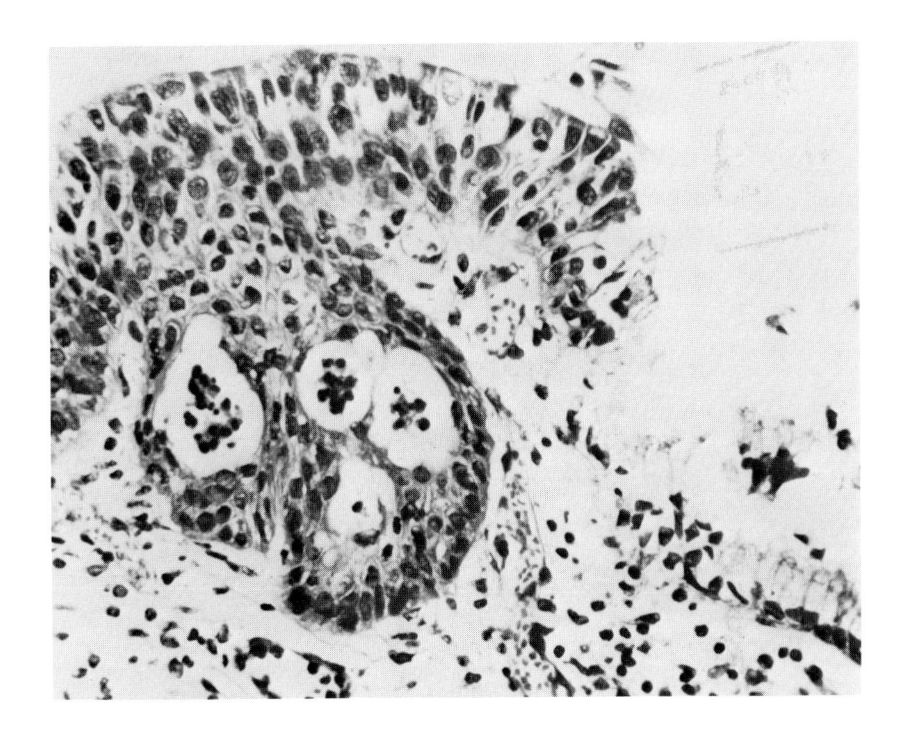

FIGURE 6. Squamous metaplasia of surface epithelium and underlying glands. (Magnification × 310.) (From Herbst, A. L., Scully, R. E., and Robboy, S. J., *Hosp. Pract.*, 10, 51, 1975. With permission.)

The highest rate of adenosis has been in those patients who were exposed early in pregnancy. It appears that the larger the ingested dose and the earlier in pregnancy DES was initiated, the greater the risk for adenosis.[20,21] The frequency of adenosis is also related to age. Women over age 26 are found to have decreased frequency of adenosis due in part to the columnar epithelium of adenosis healing by squamous metaplasia. Squamous metaplasia, which may represent a healing process or replacement of the glandular epithelium, is frequently found in women exposed to DES (Figure 6). Noller et al. found that in approximately 30% of 452 exposed women the epithelial changes had regressed or disappeared within a 3-year followup period.[22]

Two cell types which have been identified in the columnar epithelium which characterizes adenosis are mucinous cells resembling the endocervical epithelium and mucin-free cells which are somewhat similar to fallopian tube epithelium and proliferative endometrium.[17] Often, atypical tuboendometrial cells have been found adjacent to areas of CCA.[23] However, the risk of adenosis developing into a carcinoma is small. Currently, only ten cases of CCA are known to have developed under observation in DES-exposed females followed for vaginal adenosis or cervical ectropion.[24]

Structural changes of the female upper and lower genital tract have been seen in patients with *in utero* DES exposure (Figures 7 and 8). In 20 to 50% of DES-exposed females, transverse ridges, cervical collars, hoods, coxcombs, hypoplastic cervixes, and pseudopolyps have been described.[25-27] Kaufman et al. reported on 267 exposed women who underwent hysterosalpingograms, 69% of which demonstrated some abnormality.[26] The most frequently observed abnormalities were T-shaped uterus, small uterine cavity, and constriction rings. These upper genital tract abnormalities have not been confined to the uterine cavity; fallopian tube anomalies have also been described. At laparoscopy for infertility evaluation, DES-

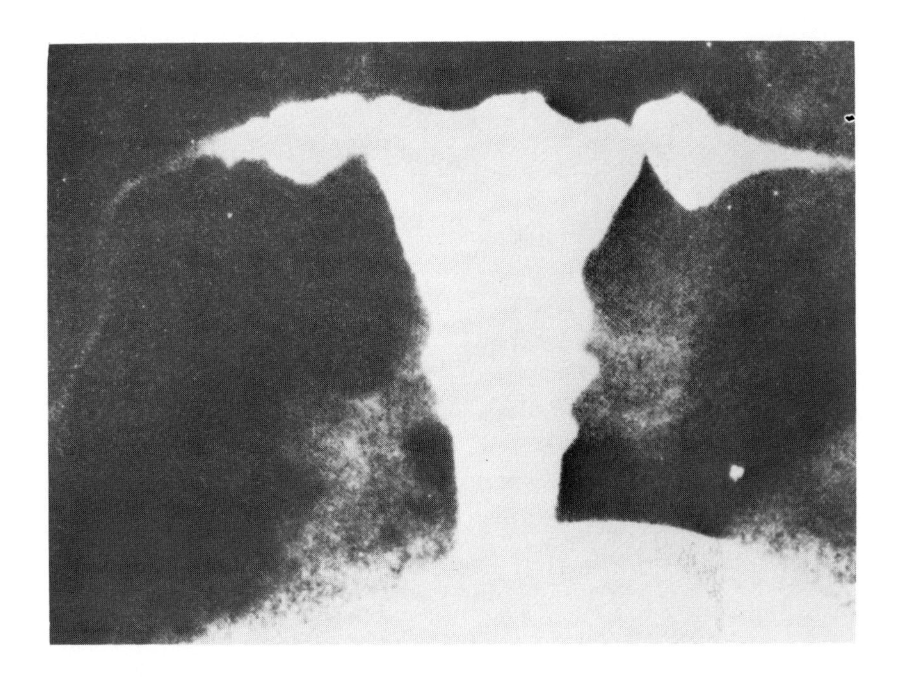

FIGURE 7. Hysterosalpingogram showing proximal portion of horn-like extension from uterine cavity. Uterine cavity is somewhat irregular. (From Kaufman, R. H., Adam, E., Binder, G. L. et al., *Am. J. Obstet. Gynecol.*, 137, 299, 1980. With permission.)

exposed females have been reported to have foreshortened tubes with pinpoint ostia and withered fimbriae, but these changes have not been commonly observed in DES-exposed women.[28] Over time, some of the cervico-vaginal abnormalities have also regressed or disappeared.[29,30] It has been reported that in 53% of patients with vaginal ridges there has been a decrease or disappearance of the abnormalities[29] (Table 2).

VII. MANAGEMENT OF THE DES-EXPOSED FEMALE

Any women suspected of having been exposed to DES should be carefully screened with an effort made to document maternal medication history. The screening examination consists of careful inspection, palpation of the entire vagina and cervix, cytological sampling, colposcopy, and biopsy of suspicious areas.

Inspection and palpation have been important in the examination of patients with adenosis to identify structural abnormalities and nodular or ulcerated areas in the vagina that are suspicious for carcinoma. Since the majority of lesions has occurred on the anterior vaginal wall, rotation of the speculum to adequately visualize the entire vaginal wall is necessary. Areas of adenosis have appeared as bright red and granular or as polypoid, fungating, or ulcerated lesions.

Cervical and vaginal cytology should be obtained separately. Cytology has on occasion detected a CCA prior to clinical evidence and is useful for detecting neoplastic changes of presquamous epithelium (see following).[31] To improve upon this lack of sensitivity, quadrants of the vagina and cervix should be separately sampled.

Colposcopy is often useful to adequately evaluate the DES-exposed patient. Colposcopic examination has made it possible to identify abnormal areas for biopsy and followup in subsequent examinations. During the examination, the entire transformation zone and the

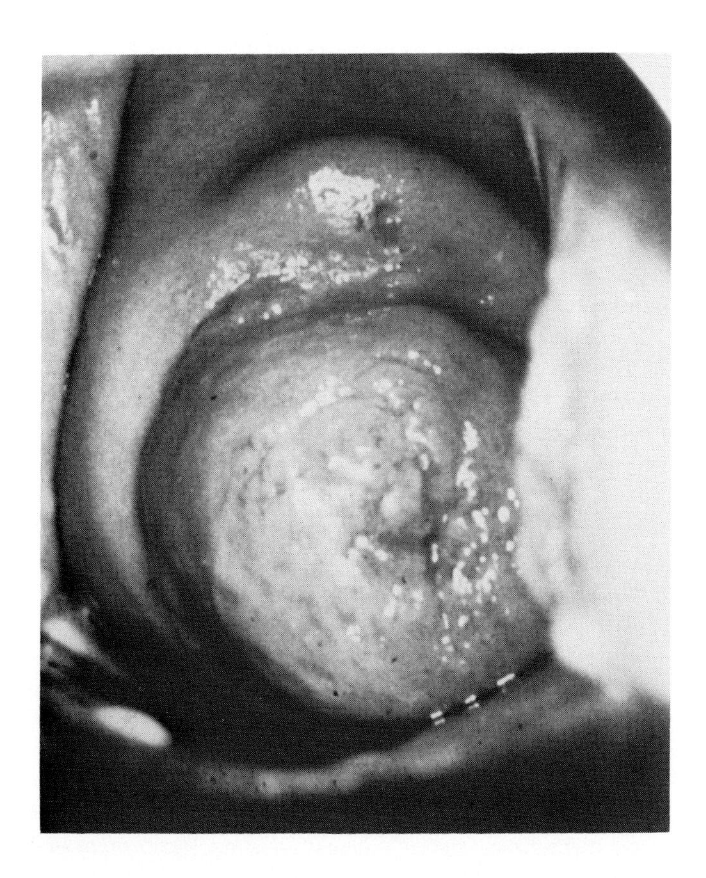

FIGURE 8. Concentric ridge in the cervix (left). A circular fold gives the appearance of a hood covering the cervix. (From Robboy, S. J., Scully, R. E., Herbst, A. L. et al., *J. Reprod. Med.,* 15, 13, 1975. With permission.)

Table 2
CHANGES IN LOWER GENITAL TRACT ABNORMALITIES IN EXPOSED FEMALES

	No change	Decreased area or extent	Disappearance
Cervical ectropion n = 121	30 (25%)	53 (44%)	38 (31%)
Cervicovaginal ridge n = 123	58 (47%)	30 (24%)	35 (29%)

Adapted from Antonioli, D. A., Burke, L., Friedman, E. A., et al., *Am. J. Obstet. Gynecol.,* 137, 847, 1980.

extent of the lesion should be noted. Care must be taken to visualize all vaginal folds and structural abnormalities. Vaginal adenosis often appears as grapelike reddish areas. Colposcopic changes in DES-exposed females can be difficult to interpret without histological confirmation. Areas of punctation and mosaicism may represent metaplasia. Colposcopic findings suspicious for neoplasia and all abnormal lesions should be biopsied.

After examination by colposcopy, the vagina and cervix may be stained with half-strength

Lugol's solution to delineate normal glycogen-containing tissues from nonglycogen areas. This technique has provided useful information to help evaluate changes in the size and location of abnormal areas.

VIII. RESULTS OF COLPOSCOPIC EXAMINATION

Most areas of adenosis and ectropion have undergone progressive metaplasia and are eventually replaced by normal glycogenated mucosa. Both cervical adenosis and structural abnormalities may eventually disappear in the DES-exposed female. In the DES adenosis (DESAD) study, there has been a decrease in the frequency of vaginal-epithelial changes proportional to the patient's age.

The relationship between *in utero* exposure to DES and dysplasia is as yet unclear.[32-35] However, active metaplasia of the glandular epithelium has often been misinterpreted on colposcopy. Followup studies have shown that abnormal colposcopic findings have often been caused by active metaplasia. Welch reported that 80% of biopsies from mosaic and punctated areas from 171 DES-exposed females showed active metaplasia alone.[32] In 1978, Robboy and co-workers reported only a 2.1% incidence rate of dysplasia in exposed women, with the majority of cases being mild dysplasia.[33] However, in a 1984 DESAD followup study, Robboy et al. reported a significantly higher incidence of dysplasia and carcinoma *in situ* (15.7 vs. 7.9 cases per 1000 person-years of followup) among exposed women vs. controls in a matched cohort.[34] In sum, the possible association between dysplasia and DES exposure *in utero* remains unclear, but DES patients require regular followup and screening for squamous cell neoplasia with therapy undertaken when an authentic premalignant lesion is found.

Future techniques may be used to assist in the identification of premalignant epithelium to differentiate it from metaplasia. One technique is DNA nuclear content analysis for cellular ploidy. Normal cells have a diploid value, metaplastic cells and occasionally mildly dysplastic cells have a tetraploid value, while biopsies with an aneuploid value are neoplastic.

If a DES-exposed woman has a negative cytologic evaluation and no evidence of vaginal epithelial changes, an annual examination is probably adequate. Women with large areas of vaginal epithelial changes or those on whom a complete evaluation is difficult should be examined more frequently, usually twice yearly. Generally the areas of adenosis and ectopia become smaller over a period of years as a result of the replacement of the columnar epithelium by squamous metaplasia.

IX. REPRODUCTIVE AND GYNECOLOGICAL SURGICAL EXPERIENCE

Efforts to determine the effect of DES exposure on ovarian and menstrual functions, fertility, and future pregnancy outcome have yielded inconclusive results. With regard to ovarian and menstrual functions, Bibbo et al. reported that 18% of exposed females compared to 10% of controls had irregular cycles.[36] In a followup study of the same population, Herbst et al. found a 10% incidence of menstrual irregularities in 226 DES-exposed women compared to 4% in 203 unexposed patients.[37] However, some investigators have found no difference between DES-exposed and unexposed populations. Barnes reported no difference in age of menarche or frequency of menstrual irregularities between controls and DES-exposed patients.[38]

The effect of DES exposure, if any, on fertility has also not been conclusively determined. Bibbo et al. reported in a matched cohort study that a significantly lower proportion of DES-exposed patients had achieved pregnancy (18%) than those who had received a placebo (33%).[36] Herbst et al. confirmed this difference in the same cohort when they reported that among women not always protected by contraceptives, significantly more unexposed women

Table 3
FIRST PREGNANCY OUTCOME

Evaluable pregnancy outcome	DES-exposed n = 150	Unexposed n = 181
Evaluable outcomes	114	128
Term birth	59 (52%)	106 (83%)
Premature[a]	23 (20%)	8 (6%)
Spontaneous abortion		
week 14 to 26	5 } (21%)	2 } (11%)
week ≤13	19 }	12 }
Ectopic pregnancy	8 (7%)	0

[a] Births were scored as premature if the duration of pregnancy was 26 weeks or more and birth weight was less than 2500 g. All but one premature baby in each group survived.

From Herbst, A. L., Hubby, M. M., Azizi, F., et al., *Am. J. Obstet. Gynecol.*, 141, 1019, 1981. With permission.

had become pregnant than exposed women (92 vs. 75%).[39] Other investigators have reported no difference in the rate of infertility among DES-exposed females vs. controls.[40,41]

Studies have indicated an increased risk of pregnancy wastage in DES-exposed females. The major categories are ectopic pregnancy, premature labor, and midtrimester loss (Table 3).[39-42] Barnes et al. reported a greater risk of unfavorable pregnancy outcome among 618 DES-exposed females when compared with matched controls.[40] It has been suggested that the incidence of cervical incompetence is increased due to the high prevalence of cervical anomalies. Herbst and associates have reported that nonviable births were more common among DES-exposed patients with cervical ridges.[37] In a non-case-control study, Goldstein has also described DES patients with cervical malformations and hypoplasia who have developed incompetent cervixes during the second trimester of pregnancy.[43] Herbst et al. reported a 7% rate of ectopic pregnancies among 150 exposed women and none among 181 controls.[39]

Although exposure to DES *in utero* is associated with unfavorable pregnancy outcome, over 80% of DES-exposed women who achieved pregnancy have delivered at least one live-born infant.[37,40] Due to the increased risk during pregnancy, careful prenatal care is mandatory and the patients should be managed as having a high-risk pregnancy.

X. DES AND BREAST CARCINOMA

Attention has also been focused on the possibility that mothers who took DES during pregnancy are at greater risk for breast carcinoma. No clear determination can be made based on the current evidence. In two recent large cohort studies, an elevated risk of breast carcinoma has been found among women exposed to DES while pregnant. Hadjimichael et al. reported a relative risk of 1.37% for breast carcinoma among those exposed in a study of 3315 mothers.[44] Greenberg and co-workers have reported that the incidence of breast cancer per 100,000 women-years was 134 in exposed females compared to 93 in those unexposed, yielding a crude relative risk of 1.4%.[45] However, an increased risk was not detected in several earlier, smaller studies. Brian and co-workers at the Mayo Clinic studied 408 women given DES who were enrolled in the DESAD project and found no increase in the risk of breast cancer compared with the general population.[46] Reporting in 1981, Hubby et al. described a nonsignificant difference of 34 cases of breast cancer among 655 exposed mothers (5.21%) vs. 28 cases in 645 unexposed women (4.3%).[47] The two groups were known to be similar with respect to reproductive and menstrual history, age, and history of hormonal therapy. The possible relationship of DES exposure to breast carcinoma in females who took DES during pregnancy is an active area of research.

XI. HISTOGENESIS

An understanding of the histogenesis of the human lower genital tract may help in the interpretation of clinical changes seen due to the transplacental passage of DES early in pregnancy. CCA associated with DES is believed to be Müllerian in origin rather then mesonephric (Figure 9). The Müllerian system arises from the paramesonephric duct during embryogenesis. At 6 weeks, a fetus develops a Müllerian cleft in the pronephric duct. A tip of the mesothelial cells then splits off and develops the Müllerian ducts. At 8 weeks of development, the Müllerian ducts proceed caudally and meet at the urogenital sinus. During the 12th through 16th weeks of fetal development, the two Müllerian ducts fuse and begin to proceed cranially to the ligamentum vaginale (round ligament) forming the precursors of the uterus and upper vagina. The midline fusion and canalization of the Müllerian ducts are completed by the 16th week. A squamous thickening and vaginal plate forms at the urogenital sinus which ultimately folds in to form the lower vagina. When the plate opens into the fused Müllerian ducts, the canalization has been completed.[48-52]

The DES-associated changes of the female genital tract are suggestive of a disorder in the union and fusion of the Müllerian duct. Anatomical changes such as the hypoplastic and T-shaped uterine cavity may result from the imperfect fusion of the Müllerian ducts. The lower tract changes suggest an imperfect fusion of the descending Müllerian ducts and the ingrowing vaginal plate. The histological changes may be accounted for by a disordered union between the vaginal plate and Müllerian ducts.

DES probably exerts its action by affecting estrogen-responsive tissue during development. This action may be mediated by DES effects or stromal control of epithelial differentiation as has been demonstrated by Cunha.[53] It is unclear whether the damage results from an estrogen effect or from a teratogenic effect of the stilbene molecule.

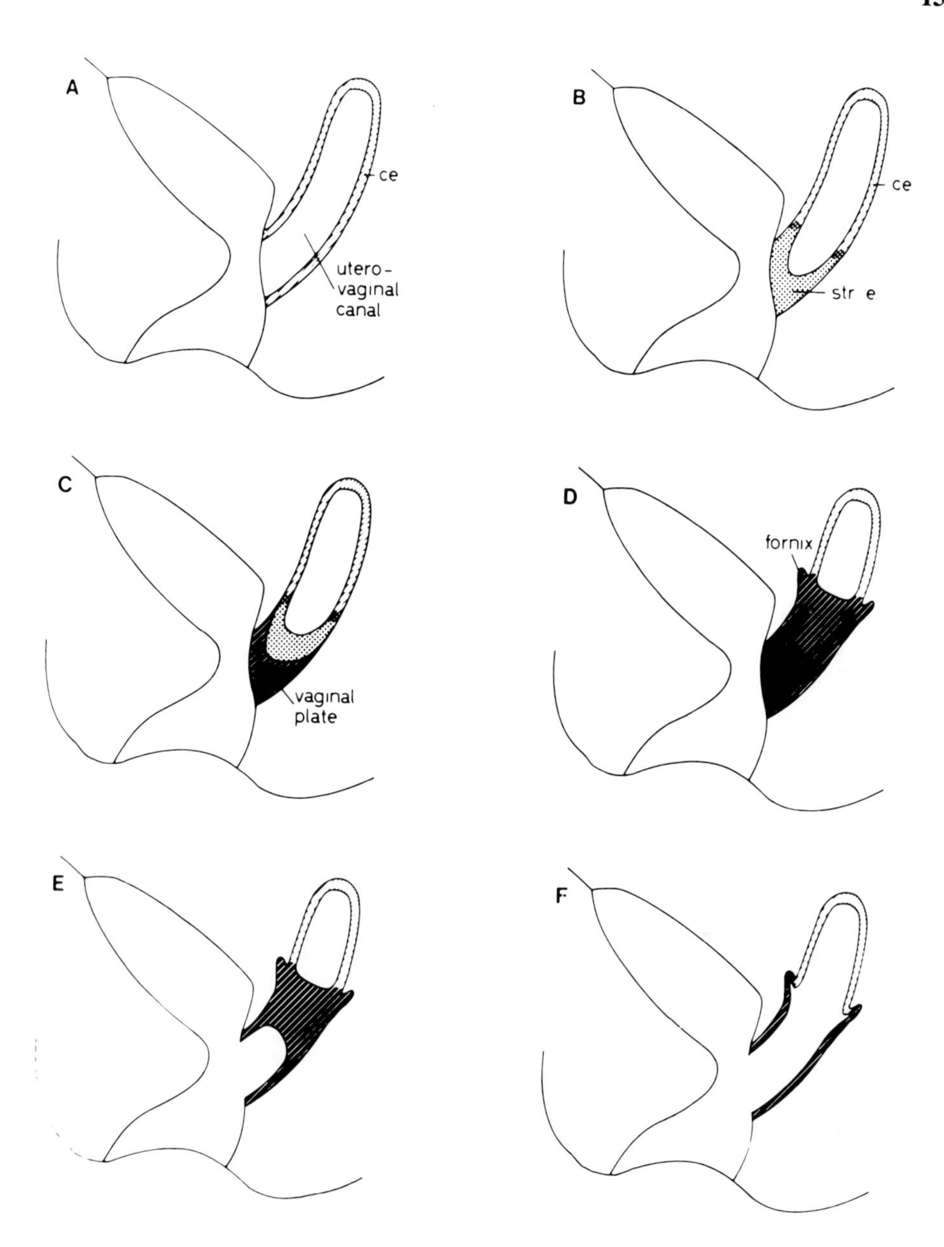

FIGURE 9. Diagrams of successive stages in the development of the human vagina. In (A) the uterovaginal canal is lined with a columnar epithelium which later (B) undergoes a transformation into a stratified squamous type. The transition between these two epithelial types is indefinite. At the same time, as the transformation process takes place, the uterovaginal canal is changed into a solid structure in the same region (B). In (C) the vaginal plate has appeared between the dorsal wall of the urogenital sinus and the region with stratified epithelium in the uterovaginal anlage. The vaginal plate grows cranially, the stratified epithelium is resorbed, and the plate finally is in contact with columnar epithelium (D). Lumen formation in the solid vaginal plate begins caudally (E) and progresses cranially (F). (From Forsberg, J. G. and Kalland, T., in *Developmental Effects of Diethylstilbestrol (DES) in Pregnancy,* Herbst, A. L. and Bern, H. A., Eds., Thieme-Stratton, New York, 1981. With permission.)

REFERENCES

1. **Dodds, E. C., Goldberg, L., Larson, W., et al.,** Oestrogenic activity of certain synthetic compounds, *Nature (London),* 141, 247, 1938.
2. **Smith, O. W., Smith, G. V., and Hurwitz, D.,** Increased excretion of pregnanediol in pregnancy from diethylstilbestrol with special reference to the prevention of late pregnancy accidents, *Am. J. Obstet. Gynecol.,* 51, 411, 1946.
3. **Smith, O. W.,** Diethylstilbestrol in the prevention and treatment of complications of pregnancy, *Am. J. Obstet. Gynecol.,* 56, 821, 1948.
4. **Ferguson, J. H.,** Effect of stilbestrol on pregnancy compared to the effect of a placebo, *Am. J. Obstet. Gynecol.,* 65, 592, 1953.
5. **Dieckmann, W. J., Davis, M. E., Rynkiewicz, S. M., et al.,** Does the administration of diethylstilbestrol during pregnancy have therapeutic value?, *Am. J. Obstet. Gynecol.,* 66, 1062, 1953.
6. **Herbst, A. L. and Scully, R. E.,** Adenocarcinoma of the vagina in adolescence: a report of 7 cases including 6 clear cell carcinomas (so-called mesonephromas), *Cancer,* 25, 745, 1970.
7. **Herbst, A. L., Ulfelder, H., and Poskanzer, D. C.,** Adenocarcinoma of the vagina: association of maternal stilbestrol therapy with tumor appearance in young women, *N. Engl. J. Med.,* 284, 878, 1971.
8. **Greenwald, P., Barlow, J. J., Nasca, P. C., et al.,** Vaginal cancer after maternal treatment with synthetic estrogens, *N. Engl. J. Med.,* 285, 390, 1971.
9. Food and Drug Administration Drug Bulletin, Diethylstilbestrol Contraindicated in Pregnancy, U.S. Department of Health, Education and Welfare, Washington, D.C., 1971.
10. **Lanier, A. P., Noller, K. L., Decker, D. G., et al.,** Cancer and stilbestrol: a follow-up of 1,719 persons exposed to estrogens in utero and born 1943-1959, *Mayo Clin. Proc.,* 48, 793, 1973.
11. **Heinonen, O. P.,** Diethylstilbestrol in pregnancy: frequency of exposure and usage pattern, *Cancer,* 31, 573, 1973.
12. **Melnick, S., Cole, P., Anderson, D., et al.,** Rates and risks of diethylstilbestrol-related clear cell adenocarcinoma of the vagina and cervix: an update, *N. Engl. J. Med.,* in press.
13. **Herbst, A. L., Cole, P., Colton, T., et al.,** Age-incidence and risk of diethylstilbestrol-related clear cell adenocarcinoma of the vagina and cervix, *Am. J. Obstet. Gynecol.,* 128, 43, 1977.
14. **Herbst, A. L., Anderson, S., Hubby, M., et al.,** Risk factors for the development of diethylstilbestrol-associated clear cell adenocarcinoma: a case-control study, *Am. J. Obstet. Gynecol.,* 154, 814, 1986.
15. **Senekjian, E. K., Frey, K. W., Anderson, D., et al.,** Local therapy in stage I clear cell adenocarcinoma (CCA) of the vagina, submitted for publication.
16. **Herbst, A. L. and Anderson, D.,** Clinical correlations and management of vaginal and cervical clear cell adenocarcinoma, in *The Developmental Effects of Diethylstilbestrol (DES) in Pregnancy,* Herbst, A. L. and Bern, H. A., Eds., Thieme-Stratton, New York, 1981.
17. **Scully, R. E. and Welch, W. R.,** Pathology of the female genital tract after prenatal exposure to diethylstilbestrol, in *The Developmental Effects of Diethylstilbestrol (DES) in Pregnancy,* Herbst, A. L. and Bern, H. A., Eds., Thieme-Stratton, New York, 1981.
18. **Holt, L. and Herbst, A. L.,** DES-related female genital changes, *Semin. Oncol.,* 9, 341, 1982.
19. **Stafl, A., Mattingly, R. F., Foley, D. V., et al.,** Clinical diagnosis of vaginal adenosis, *Obstet. Gynecol.,* 43, 118, 1974.
20. **Herbst, A. L., Poskanzer, D. C., Robboy, S. J., et al.,** Prenatal exposure to stilbestrol. A prospective comparison of exposed female offspring with unexposed controls, *N. Engl. J. Med.,* 292, 334, 1975.
21. **O'Brien, P. C., Noller, K. L., and Robboy, S. J.,** Vaginal epithelial changes in women enrolled in the National Cooperative Diethylstilbestrol Adenosis (DESAD) Project, *Obstet. Gynecol.,* 53, 300, 1979.
22. **Noller, K. L., Townsend, D. E., Kaufman, R. H., et al.,** Maturation of vaginal and cervical epithelium in women exposed in utero to diethylstilbestrol (DESAD Project), *Am. J. Obstet. Gynecol.,* 146, 279, 1983.
23. **Robboy, S. J., Welch, W. R., Young, R. H., et al.,** Topographic relation of cervical ectropion and vaginal adenosis to clear cell adenocarcinoma, *Obstet. Gynecol.,* 60, 546, 1982.
24. **Sander, R., Nuss, R. C., and Rhatigan, R. M.,** Diethylstilbestrol-associated vaginal adenosis followed by clear cell adenocarcinoma, *Int. J. Gynecol. Pathol.,* 5, 362, 1986.
25. **Jeffries, J. A., Robboy, S. J., O'Brien, P. C., et al.,** Structural anomalies of the cervix and vagina in women enrolled in the Diethylstilbestrol Adenosis (DESAD) Project, *Am. J. Obstet. Gynecol.,* 148, 59, 1984.
26. **Kaufman, R. H., Adam, E., Binder, G. L., et al.,** Upper genital tract changes and pregnancy outcome in offspring exposed in utero to diethylstilbestrol, *Am. J. Obstet. Gynecol.,* 137, 299, 1980.
27. **Haney, A. F., Hammond, C. B., Soules, M. R., et al.,** Diethylstilbestrol-induced upper genital tract abnormalities, *Fertil. Steril.,* 31, 142, 1979.
28. **DeCherney, A. H., Cholst, I., and Naftolin, A.,** Structure and function of the fallopian tubes following exposure to diethylstilbestrol (DES) during gestation, *Fertil. Steril.,* 36, 741, 1981.

29. **Antonioli, D. A., Burke, L., Friedman, E. A., et al.,** Natural history of diethylstilbestrol-associated genital tract lesions: cervical ectopy and cervicovaginal hood, *Am. J. Obstet. Gynecol.*, 137, 847, 1980.
30. **Burke, L., Antonioli, D., and Friedman, E. A.,** Evolution of diethylstilbestrol-associated genital tract lesions, *Obstet. Gynecol.*, 57, 79, 1981.
31. **Taft, P. D., Robboy, S. J., Herbst, A. L., et al.,** Cytology of clear-cell adenocarcinoma of genital tract in young females: review of 95 cases from the registry, *Acta Cytol.*, 18, 279, 1974.
32. **Welch, W. R., Robboy, S. J., Townsend, D. E., et al.,** Comparison of histologic and colposcopic finding in DES-exposed females, *Obstet. Gynecol.*, 52, 457, 1978.
33. **Robboy, S. J., Keh, P. C., Nickerson, R. J., et al.,** Squamous cell dysplasia and carcinoma in situ of the cervix and vagina after prenatal exposure to diethylstilbestrol, *Obstet. Gynecol.*, 51, 528, 1978.
34. **Robboy, S. J., Noller, K. L., O'Brien, P., et al.,** Increased incidence of cervical and vaginal dysplasia in 3,980 diethylstilbestrol-exposed young women, *JAMA*, 252, 2979, 1984.
35. **Robboy, S. J., Szyfelbein, W. M., Goellner, J. R., et al.,** Dysplasia and cytologic findings in 4,589 young women enrolled in the Diethylstilbestrol Adenosis (DESAD) Project, *Am. J. Obstet. Gynecol.*, 140, 579, 1981.
36. **Bibbo, M., Gill, W. B., Azizi, F., et al.,** Follow-up study of male and female offspring of DES-exposed mothers, *Obstet. Gynecol.*, 49, 1, 1977.
37. **Herbst, A. L., Hubby, M., Blough, R. R., et al.,** A comparison of pregnancy experience in DES exposed and DES unexposed daughters, *J. Reprod. Med.*, 24, 62, 1980.
38. **Barnes, A. B.,** Menstrual history of young women exposed in utero to diethylstilbestrol, *Fertil. Steril.*, 32, 148, 1979.
39. **Herbst, A. L., Hubby, M. M., Azizi, F., et al.,** Reproductive and gynecologic surgical experience in diethylstilbestrol-exposed daughters, *Am. J. Obstet. Gynecol.*, 141, 1019, 1981.
40. **Barnes, A. B., Colton, T., and Gunderson, J.,** Fertility and outcome of pregnancy of women exposed in utero to diethylstilbestrol, *N. Engl. J. Med.*, 302, 609, 1980.
41. **Cousins, L., Karp, W., Lacey, C., et al.,** Reproductive outcome of women exposed to diethylstilbestrol in utero, *Obstet. Gynecol.*, 56, 70, 1980.
42. **Mangan, C. E., Borow, L., Burtnett-Rubin, M. M., et al.,** Pregnancy outcome in 98 women exposed to diethylstilbestrol in utero, their mothers, and unexposed siblings, *Obstet. Gynecol.*, 59, 315, 1982.
43. **Goldstein, D. P.,** Incompetent cervix in offspring exposed to diethylstilbestrol in utero, *Obstet. Gynecol.*, Suppl. 52, 73S, 1978.
44. **Hadjimichael, O. C., Meigs, J. W., Falcier, F. W., et al.,** Cancer risk among women exposed to exogenous estrogens during pregnancy, *JNCI*, 73, 831, 1984.
45. **Greenberg, E. R., Barnes, A. B., Resseguie, L., et al.,** Follow-up study of mothers exposed to diethylstilbestrol in pregnancy, *N. Engl. J. Med.*, 311, 1393, 1984.
46. **Brian, D. D., Tilley, B. C., Labarthe, D. R., et al.,** Breast cancer in DES-exposed mothers. Absence of association, *Mayo Clin. Proc.*, 55, 89, 1980.
47. **Hubby, M., Haenszel, W. M., and Herbst, A. L.,** Effects on the mother following exposure to diethylstilbestrol in pregnancy, in *The Developmental Effects of Diethylstilbestrol (DES) in Pregnancy*, Herbst, A. L. and Bern, H. A., Eds., Thieme-Stratton, New York, 1981.
48. **Forsberg, J. G.,** Cervicovaginal epithelium: its origin and development, *Am. J. Obstet. Gynecol.*, 115, 1025, 1973.
49. **Ulfelder, H. and Robboy, S. J.,** The embryologic development of the human vagina, *Am. J. Obstet. Gynecol.*, 126, 769, 1976.
50. **Forsberg, J. G.,** Estrogen, vaginal cancer, and vaginal development, *Am. J. Obstet. Gynecol.*, 113, 83, 1972.
51. **Forsberg, J. G.,** Late effects in the vaginal and cervical epithelia after injections of diethylstilbestrol into neonatal mice, *Am. J. Obstet. Gynecol.*, 121, 101, 1975.
52. **Forsberg, J. G.,** Developmental mechanism of estrogen-induced irreversible changes in the mouse cervicovaginal epithelium, *Natl. Cancer Inst. Monogr.*, 51, 41, 1979.
53. **Cunha, G. R.,** Epithelial-stromal interactions in development of the urogenital tract, *Int. Rev. Cytol.*, 47, 137, 1976.

Chapter 11

EFFECTS ON HUMAN MALES OF *IN UTERO* EXPOSURE TO EXOGENOUS SEX HORMONES

William B. Gill

TABLE OF CONTENTS

I. Introduction 162

II. Epididymal (Spermatocele) Cysts 162

III. Testicular Abnormalities 162
- A. Testicular Hypoplasia 162
- B. Leydig Cell Hyperplasia 165
- C. Cryptorchidism 165

IV. Hypospadias 166

V. Microphallus (Micropenis) 166

VI. Meatal Stenosis 166

VII. Male Pseudohermaphroditism 167

VIII. Psychosexual Abnormalities 167

IX. Infertility 167

X. Testis Carcinoma 168

XI. Mechanisms of DES Pathogenesis on the Fetal Male Genital Tract (Sex Hormone Compartmentation in Pregnancy Between Mother/Placenta/Fetus) 171

XII. Conclusion 174

References 174

I. INTRODUCTION

Diethylstilbestrol (DES) has been the most extensively used exogenous estrogenic hormone in human pregnancy.[1] DES exposure *in utero* has been linked to a variety of structural and functional alterations in the human male genital tract.[2] These male genital abnormalities are qualitatively comparable to those reported in a variety of animal species following similar exposure to DES during pregnancy.[3,4] The lesions have ranged from relatively minor structural alterations such as epididymal or spermatocele cysts without known functional problems to more major anatomical changes of testicular hypoplasia, cryptorchidism, hypospadias, and microphallus with potential disease dysfunctions of ambiguous genitalia, male pseudohermaphroditism, infertility and testicular carcinoma.

In this chapter the published articles with both positive and negative findings with respect to DES exposure *in utero* on the human male genital tract are reviewed. A tabulation of these reports is given in Table 1. In addition, the mechanisms for potential pathogenic action of DES on the developing male genital tract of the fetus are developed with respect to the steroid hormone compartmentation during pregnancy between the mother, the placenta, and the male fetus (Figure 2).

II. EPIDIDYMAL (SPERMATOCELE) CYSTS

Epididymal cysts were the most frequently found anatomical abnormalities in young adult men 25 years after exposure to DES *in utero* in both the Chicago[5] and Beth Israel[6] studies. In the double-blind Chicago study, epididymal (spermatocele) cysts were found in 20.8% of 308 DES-exposed patients vs. 4.9% of 307 placebo-exposed control ($p < 0.005$). Epididymal cysts were located in the efferent ductule/superior epididymis area. The gross and microscopic appearance of an epididymal cyst from an *in utero* DES-exposed 25-year-old male is shown in Figure 1. Although the cyst was lined by columnar epithelium, the secretory cells characterizing the normal epididymis were absent. Because no spermatozoa were found in the fluid, the criterion for a spermatocele was absent. Aspirates of epididymal (spermatocele) cysts from nine DES-exposed men revealed straw-colored fluid without spermatozoa in six cases and a slightly milky fluid containing sperm in three cases. Aspirates of epididymal cysts from two control patients revealed sperm in one case. Cytologic examinations of all of these aspirated cysts revealed only spermatozoa and/or epithelial cells and/or amorphous precipitates with no suggestions of malignancy.

A Beth Israel study[6] found epididymal cysts in 13% of 48 adult males exposed to DES *in utero*. This study did not have direct examinations of matched normal males but relied on historical controls reported in the literature. A Stanford study[7] found epididymal cysts in 4% of 24 DES-exposed males answering a newspaper request vs. no epididymal cysts in 24 age-matched controls. A Mayo Clinic study[8] based on physical exams of 32% of all males exposed to DES *in utero* between 1939 and 1962 and living within 100 miles of the Mayo Clinic found epididymal cysts in 6.9% of these 265 men vs. 5.1% of matched controls (274 men examined which represented 40% of controls selected). The differences in the incidences of epididymal cysts may be explained by differences in DES dosages (Mayo Clinic dose approximately 1/10 the Chicago dose), lack of patient compliance monitoring in most studies (and thus uncertainty as to DES dosage and timing),[9] the difficulties inherent in examining male genitalia, and the tendency to miss smaller lesions.

III. TESTICULAR ABNORMALITIES

A. Testicular Hypoplasia

Abnormally small testes have been variously designated as testicular atrophy, hypotrophy, or hypoplasia. Our Chicago study[5] found testicular hypoplasia, defined as testis length <3.6

Table 1
MALE GENITAL TRACT ABNORMALITIES REPORTED IN HUMANS AFTER DES EXPOSURE *IN UTERO*

	Positive findings				Negative findings			
	DES-exposed	Control	Remarks	Ref.	DES-exposed	Control	Remarks	Ref.
Epididymal cysts	20.8%	4.9%		5	6.9%	5.1%		8
	4%	0%		7				
	13%	—		6				
Testis hypoplasia	8.4%	1.9%		5				
	4%	0%		7				
	10%	—		6				
	87%	40%		11				
Leydig cell hyperplasia	0.49 *	0.15*	*ISC/SP cell ratio	12				
	67%	—	In testicular cancer	15				
Cryptorchidism	65%	17%	With testicular hypoplasia	5	27%	25%		16
	8%	—		6	8%	6.6%		17
Hypospadias	9%	0%		16				
	8.3%**	1.8%	**Progestins also	21				
Microphallus	1.3%	0%	Severe	25				
	31.5%	8.8%	Mild to moderate	25				
	2%	—		6				
Urethral meatal stenosis	36%	0%		16				
Male pseudohermaphroditism (anecdotal cases)				27, 32 30,29 28				
Psychosexual abnormalities				35				
Infertility and/or abnormal semen	27%	15%	Eliasson score >5	5	30%	25%	Eliasson score >5	7

Table 1 (continued)
MALE GENITAL TRACT ABNORMALITIES REPORTED IN HUMANS AFTER DES EXPOSURE *IN UTERO*

	Positive findings				Negative findings			
	DES-exposed	Control	Remarks	Ref.	DES-exposed	Control	Remarks	Ref.
	9.1%	1.4%	Infertility	5				
	16.9%	5%	Infertility (see Table 3)					8
	50%	—	Eliasson score >5	6				
	82%	17%	Abnormal sperm penetration ova					
			Hembree-abnormal meiotic cytogenetics	86				
Testis carcinoma	9.3%	1.9%	Exogenous estrogens	43				45, 46
	Testicular cancer		Maternal hormones	42				47
	4.33 R.R. vs. controls		+ nausea		1%	0%		8
			Anecdotal cases	15, 44, 5				
	Estrogens in 5.8% testicular cancer vs. 2.1% controls (P <0.17) (medications during pregnancy unknown in 45% of testicular cancer cases and in 63% of controls)			41				

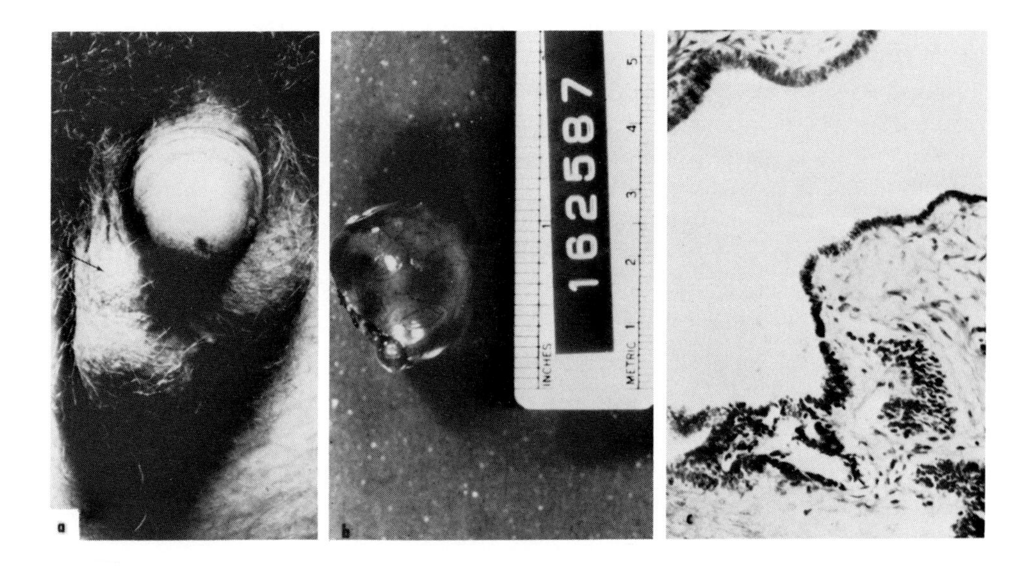

FIGURE 1. Right epididymal cyst. (a) Typical appearance on physical examination (arrow) superior and slightly posterior to main body of testes. (b) Gross appearance of a surgically excised epididymal cyst. (c) Photomicrograph of an epididymal cyst, which is lined by columnar epithelium without secretory cells. (H & E, magnification × 250.)

cm, in 8.4% of 308 men exposed to DES *in utero* vs. 1.9% in 307 placebo-exposed controls (<0.005). A Stanford study[7] found 1 of 24 DES-exposed patients with testicular hypoplasia vs. 0 of 24 controls. Beth Israel[6] reported 5 of 48 DES-exposed men to have hypoplastic testes. A University of Southern California study[11] found that, "The post pubertal boys in the DES exposed group had significantly smaller testes ($p < 0.05$): in 7 of the 8 exposed boys compared with only 6 of the 15 control subjects, the testes were less than 3,000 mm^2 in area". A Mayo Clinic study[8] reported no differences in the mean testicular dimensions between 205 DES-exposed males and 274 controls, but no information was given as to the frequency of occurrence of small testes in the two groups.

B. Leydig Cell Hyperplasia

Leydig cell hyperplasia, defined as an increase in the ratio of interstitial to spermatogenic cells, was found by Driscoll and Taylor in a perinatal autopsy study[12] between 1951 and 1976 in 38 males exposed to DES *in utero* (I/SP = 0.49 DES vs. 0.15 for controls, $p < 0.001$). Since the mean spermatogenic cell count was constant in the exposed and control groups, this was true hyperplasia. The authors pointed to the abundant literature dealing with animals exposed to DES *in utero* and Leydig cell hyperplasia and interstitial cell tumors.[13,14]

Leydig cell hyperplasia was found by Sant et al.[15] in 8 of 12 men with testicular cancer who had been exposed to DES *in utero*. Whether this Leydig cell hyperplasia reflects a persistent stigmata of DES exposure *in utero* or merely a reflection of stimulation from human chorionic gonadotropin being produced by the testis carcinomas is not yet clear.

C. Cryptorchidism

Cryptorchidism has long been known to occur in animals exposed to estrogens,[3] including DES, estradiol, and estradiol esters. In our Chicago study[5] of adult human males, 65% of the DES-exposed males who had testicular hypoplasia (26 of 308) gave a history of cryptorchidism, whereas only 16% of placebo-exposed controls with testicular hypoplasia (6 of 307) reported cryptorchidism. Beth Israel[6] reported 8% of 48 DES-exposed patients had a history of cryptorchidism.

A small study from Southern California[16] did not find differences in cryptorchidism among 3 of 11 DES-exposed and 1 of 4 controls giving a positive history. In a much larger but retrospective study[17] of cryptorchidism diagnosed on infant medical records for the years 1943 to 1973, no significant differences in estrogen exposure for cryptochids (8%) from controls (6.6%) was found. Retrospective analysis[9] of patient records poses problems with respect to reliability of patient compliance (and hence true DES exposure) and reliability of diagnosing cryptorchidism from infant records because it leaves doubt about the reliability of the data so obtained.[9,10]

IV. HYPOSPADIAS

Urethral fusion in the human male fetus occurs during the 8th to 14th week of gestation[18] and is under the control of the fetal testis production of testosterone and conversion in the end organ by $\Delta^4 5\alpha$-reductase to 5α-dihydrotestosterone.[19] Since both of these processes (testis production of testosterone and end-organ conversion to 5α-dihydrotestosterone) can be inhibited by DES or other estrogens in experimental animals,[20] one might expect qualitatively similar results in humans.

Surprisingly little attention has been paid to hypospadias in human males following DES exposure *in utero*. A Southern California study[16] found hypospadias in 1 of 11 DES-exposed men vs. 0 of 11 controls. Severe hypospadias as a component of ambiguous genitalia and intersex problems will be discussed under male pseudohermaphroditism.

Progestin plus estrogen exposure *in utero* was found in 5 of 130 hypospadias patients by Aarskog[21] and prosgestins alone in an additional 6 of the 130 hypospadias, resulting in a total incidence of 8.3% of hypospadias patients being exposed to exogenous progestational agents in early pregnancy vs. 1.8% of controls without hypospadias being exposed to progestins. Interestingly, the severity of the hypospadias correlated with the earlier use of progestins in pregnancy.

V. MICROPHALLUS (MICROPENIS)

The morphogenesis and growth of the fetal penis is under the control of testosterone from the fetal testis.[19] Testosterone production from this fetal testis commences during the 8th week of gestation with linear penile growth of approximately 0.7 mm/week from the 10th week of gestation onward until the 3.5 cm of a full-term birth is reached.[22] The normal postpubertal male size of 12 cm (9 to 15 cm, 10th to 90th percentile) is attained at the age of 16.[23]

A variety of conditions which interfere with the normal male fetal production of testosterone or its metabolism have been described[24] as the cause of microphallus. With respect to exposure to DES *in utero,* we found in the Chicago study[25] 4 DES-exposed patients (1.3%) with severe microphallus (stretched penile length $<$4 cm) vs. none in control patients (0 to 307). Mild microphallus of 7 cm or less penile length was found in 31.5% of the DES-exposed patients vs. 8.8% of the controls ($P < 0.005$).

The Beth Israel study of Whitehead and Leiter[6] described a "hypoplastic penis" in 2% of 48 DES-exposed patients. A Stanford study[7] of relatively small numbers found an 8% incidence of micropenis in both the DES-exposed and control groups. Definitions of microphallus or hypoplastic penis were not given in these later papers. Further discussion of microphallus as a component of ambiguous genitalia will be given under the heading of male pseudohermaphroditism.

VI. MEATAL STENOSIS

External urethral meatal stenosis was reported in 4 of 11 males exposed to DES *in utero* vs. 0 of 4 controls by Cosgrove et al. at the University of Southern California.[16] Since other studies of DES-exposed patients have not mentioned either the presence or absence of meatal stenosis, this abnormality has undoubtedly not been systematically evaluated.

VII. MALE PSEUDOHERMAPHRODITISM

Experimental intersexuality dates back to the 1930s with the production of male pseudohermaphroditism in animals by *in utero* exposure to estrogens. Although Greene in 1944[26] predicted on the basis of animal studies that human pseudohermaphroditism could be caused by excess maternal estrogens or androgens, it was not until 1959 that Kaplan[27] first reported on human male pseudohermaphroditism following DES exposure *in utero*. In this case DES had been started at 50 mg/day during the 6th week of gestation to prevent another abortion. The DES dosage was increased ''every few days'' until 200 mg/day was being taken by the 8th week of gestation and was continued throughout the pregnancy. Progesterone was also given at a dose of 25 mg biweekly during the 7th through 10th weeks of gestation. Microphallus (15 mm) and ''complete hypospadias and cryptorchidism with wrinkled and flabby labioscrotal folds'' were described by Kaplan.[27] At 10 $^1/_2$ months a biopsy of ''labioscrotal masses'' revealed hypoplastic testicular tissue with ''tubules made up exclusively of undifferentiated Sertoli cells''.

Other cases of ambiguous genitalia following exposure to DES *in utero* have been reported. Cleveland and Chang[28] reported a problem of intersex with an XX karyotype, but no uterus or Fallopian tubes, and with glands only being in scrotal folds and containing aplastic testicular tissue. Hoefnagel[29] reported prenatal DES exposure and ''typical clinical and laboratory findings of hypogonadotropic hypogonadism with anosmia''. (Histologic description of the gonads was not given.) Beral and Colwell[30] reported a case of hermaphroditism and a case of ''underdeveloped external genitalia'' following *in utero* exposure to DES and ethisterone.

Specific enzymatic defects or decreases in the testicular steroidogenesis of testosterone have been induced in animals (rats, mice, monkeys) by DES exposure *in utero*[31] (17α-hydroxylase, C 17-20 lyase, and 17β-hydroxysteroid dehydrogenase). Imperato-McGinley et al.[32] reported a case of male pseudohermaphroditism with 17β-hydroxysteroid dehydrogenase deficiency who had bilateral cryptorchidism, hypospadias, and microphallus corrected surgically at age of 14 and a testicular carcinoma, ''teratocarcinoma seminoma'', at age 30. This patient was subsequently discovered to have been exposed to DES during the 9th and 10th week of gestation.[33]

VIII. PSYCHOSEXUAL ABNORMALITIES

The possible occurrence of psychosexual developmental abnormalities in humans has been of concern, since animal studies[34] have shown that *in utero* exposure of the developing fetus to exogenous steroid hormones of the opposite sex can produce reversed sex role preferences and other gender identity disorders. Yalom et al.[35] studied the male offspring of diabetic mothers. They reported that 6- and 16-year-old boys exposed to exogenous estrogens and progestins during fetal life had a lowering of general masculinity, assertiveness, and athletic interest, compared with a sample of volunteer unexposed males, some of whose mothers were not diabetic. An adequate controlled study of the psychosexual development of human males exposed to DES *in utero* has not been reported.

IX. INFERTILITY

Infertility in human males following DES exposure *in utero* has been of concern, because of high rates of infertility in male animals following DES exposure *in utero* and the increased incidence of structural abnormalities of the human male genital tract which have been discussed in the preceding sections. In 1982, Stillman[36] reviewed the literature about adverse effects of DES exposure *in utero* on reproductive performance in both human females and males.

Infertility data are more voluminous with respect to females exposed to DES *in utero*, but a number of studies suggest that human male fertility may also be affected by DES exposure *in utero*.[36] The Beth Israel group[6] found highly abnormal semen analyses (Eliasson score ≥5, pathologic semen) in 50% of 20 males exposed to DES *in utero*. A Stanford study reported a high frequency of pathologic semen in both control and DES groups: 30% of 23 DES males vs. 25% of 24 controls. Stenchever et al.[37] found that 82% of 17 DES-exposed males should be infertile by sperm penetration assays (SPA) with hamster ova, whereas only 17% of 12 non-DES-exposed volunteers met the infertility criterion of less than 14% sperm penetration with all 11 fertile controls having normal SPAs.

Semen analysis on the follow-up on male offspring of the study of Dieckmann et al.[38] at the University of Chicago was previously reported in two groups.[39] This was because of possible selection bias of recall in some patients after their first visit for physical exams and in others having semen analysis at the time of their physical exam when they still did not know whether they had been exposed to DES *in utero* or had been placebo-exposed controls in the original randomized double-blind study. Because of financial curtailments further studies with increased numbers of individuals from each group have not been possible. Recognizing the relatively small percentages of recall from the total of both DES-exposed (37.6% of 420 individuals exposed to DES) and the controls (27.3% of 403 total control individuals), we herein report our final total for all of the men with semen analysis whether given on the first visit when double-blind conditions were in effect or on subsequent visits.

Results of first semen analyses in Table 2 show several parameters that are more frequently abnormal in the DES-exposed group than the controls. Mean sperm concentrations were 93.4 million/mℓ in the DES-exposed group as contrasted with 126.6 million/mℓ in the controls. Percent motile sperm was over the normal limit of 60% in 70.9% of the controls, but only in 51.9% of DES-exposed. Lower motility grades 2 to 2.5 were significantly greater in the DES group. Most importantly the Eliasson scores, which reflect the combination of abnormalities of decreased numbers, decreased motility, and decreased normal morphology, were greater in the DES-exposed.

Although semen analyses give insight into the probability of infertility,[40] many years of follow-up are required before the fertility rate of DES-exposed males can be accurately ascertained. An assessment of infertility in 142 DES-exposed married males showed a statistically significant increase in primary infertility and total infertility (Table 3): 9.1% of 142 DES males had primary infertility vs. 1.4% in controls. In Table 3 it is evident that almost half of the primarily infertile DES males had a demonstrable abnormality such as testicular hypoplasia, cryptorchidism, and/or sperm count $\leq 30 \times 10^6$/mℓ. Somewhat surprisingly, secondary infertility also was higher in the DES-exposed group. Total infertility (primary and secondary) was highly statistically different with a *P* value of <0.005.

X. TESTIS CARCINOMA

An increase in testis carcinoma in adult males, following DES-exposure *in utero*, had been predicted.[41] To date both positive (increased incidence) and negative (incidence equals general population) reports have appeared in the literature. Risk factors for cancer of the testis in young men were studied in 1979 by Henderson et al.[42] with the formulation of the hypothesis ''that a major risk factor for testis cancer is relative excess of certain hormones (in particular estrogen) at the time of differentiation of the testis''. In their study, a questionnaire evaluation of 131 men with testis cancer, their mothers, and their matched controls

Table 2
RESULTS OF FIRST SEMEN ANALYSIS IN CHICAGO STUDY OF ADULT MALES EXPOSED TO EITHER PLACEBO (CONTROLS) OR DES *IN UTERO*

	Controls	DES-exposed	Significant P values[a]
Number of adult males	110	158	
Sperm density, million/mℓ	7.2%	12 %[b]	
≦20			
>60	66.4%	61.4%	
Mean	126.6 ± 17.6[c]	93.4 ± 8.1[c]	$P < 0.1$
% motile sperm			
≦30	2.7%	5.1%	
>60	70.9%	51.9%	$P < 0.005$
Motility grade			
2—2.5	2.7%	17.7%	$P < 0.001$
4	50.9%	43 %	
% oral (normal) forms			
≦40	0.9%	3.2%	
>60	80 %	70.9%	
% eosin — Y negative			
≦40	2.7%	2.5%	
>70	40.9%	41.4%	
Ejaculated volume			
≦1.4 mℓ	9.1%	12.7%	
>5.0 mℓ	11.8%	7 %	
pH of semen			
≦7.9	12.7%	5.7%	
>8.5	3.6%	5.1%	
Eliasson semen quality score			
≦4 (normal)	84.5%	72.8%	
5—10 (pathological)	8.2%	13.3%	
>10 (highly pathological)	7.3%	13.9%	
≧5 (pathological + highly pathological)	15.4%	27.2%	$P < 0.05$

[a] P values were calculated by use of either Yates continuity correction for chi-square test or students two-sample *t* test with unequal variances.
[b] Including three cases of azospermia.
[c] Standard error of the mean.

was completed in Los Angeles county. Several potential factors were evaluated and the following relative risks (RR) with one-sided *P* values in parentheses were found: a history of undescended testis (a known risk factor for testis cancer), RR 5.00 ($P = 0.02$); mother's report of excessive nausea in the index pregnancy (hypothesized to be due to relative excess of some endogenous hormone, probably estrogen), RR 4.00 ($P = 0.06$); and hormones (mostly unidentified) taken during the index pregnancy, RR 5.99 ($P = 0.11$). Although the last item is not significant, combining the information on the maternal hormone treatment with excessive nausea resulted in a relative risk of 4.33 ($P = 0.01$).

Depue et al.[43] in a case control study of 108 cases of testicular cancer in men under thirty years of age found the following statistically significant relative risks and *P* values: (1) cryptorchidism, RR 9.00 ($P = 0.01$); (2) exogenous hormone exposure during first trimester, RR 8.00 ($P = 0.02$) (hormones used in 9 cases of testicular carcinoma, DES in two, estrogen in one, progestin in one and estrogen-progestin in five for pregnancy tests); (3) birth weight under six pounds, RR 3.20 ($P = 0.01$); and (4) patient treated for acne, RR 0.37 ($P =$

Table 3
INFERTILITY PROBLEMS NOTED AS OF 1981 IN MALE OFFSPRING (CHICAGO STUDY)

	Control	DES-exposed
Total number contacted	222	212
Total ever married	140	142
Never had a pregnancy, tried 1 year or more		
Demonstrable male factors	0	6
Hypoplastic testes, cryptorchid or sperm count $<30 \times 10^6$/mℓ		
Unknown factors	2	7
Total "primary" infertility	2	13
% of married men	1.4%[a]	9.1%[a]
Had a pregnancy, waited over 1 year for it or for another one		
Demonstrable male factors		
Hypoplastic testes or sperm count $<30 \times 10^6$/mℓ	2	5
Unknown factors	3	6
Total "secondary" infertility	5	11
% married men	3.6%	7.8%
Total primary and secondary	7	24
% married men	5%[b]	16.9%[b]

[a] $P < 0.01$.
[b] $P < 0.005$.

0.01). Interestingly, the effects of exogenous hormones on the development of testicular cancer was not an indirect one through cryptorchidism in that "none of the mothers of cryptorchid subjects were exposed to exogenous hormones during the index pregnancy". The authors concluded that "a modified hormonal milieu in the mother appears to be important in the later development of testicular cancer in her sons".

Anecdotal cases of testicular carcinoma following DES exposure *in utero* were reported by our group in 1979.[5] In 1983 Conley et al.[44] reported two cases of seminoma and epididymal cysts in men exposed to DES *in utero*. Sant et al.[15] subsequently reported in 1985 on his registry of twelve testis tumors in adult males following DES exposure *in utero*. Of these twelve tumors six were seminomas, two were embryonal tumors, and four were mixed histology. Several additional histological features were seen: *in situ* tumor in five of the tumors (four seminomas and one embryonal tumor), involvement of the rete testis in three, and microlithiasis in two tumors. Leydig cell hyperplasia was found in eight for a 67% incidence. Although data for human chorionic gonadotropin (hCG) hormone secretion by tumors were not given, it should be remembered that Driscoll and Taylor[12] found Leydig cell hyperplasia to be a prominent feature of a perinatal autopsy study of humans exposed to DES *in utero*, but not in nonexposed perinatal autopsies. Therefore, one should consider Leydig cell hyperplasia as a suggestive histologic marker for DES exposure *in utero*.

Negative studies on DES exposure *in utero* and an increased incidence of testis carcinoma have been reported by Greenwald et al.[45] on the New York State Cancer Registry (1973), by Kinlen et al.[46] on a Registry in England (1974), by Carstens and Clemmesen in Denmark[47] (1972), and from the Mayo Clinic by Lanier et al.[48] (1973), Beard et al.[17] (1981), and Leary et al.[8] (1982).

The reliability of retrospective epidemiological studies has been seriously questioned as to the accuracy of the data with respect to actual DES exposure or the lack of exposure in controls. Shottenfeld et al.[41] reported on the epidemiology of various risk factors for testis cancer at Memorial Sloan-Kettering Cancer Center and found to their dismay that "the specific type of medication taken for bleeding or spotting or a threatened miscarriage is unknown in 45% of the cases (testis cancer), 40% of the hospital controls, and 63% of the neighborhood controls''. Shottenfeld enumerated the following reason for the ''little success'' in specifying and validating drug use during pregnancy: (1) physicians managing the pregnancies 20 to 40 years earlier have usually died, retired, or not retained the old records, (2) hospital obstetrical records usually contain only brief and limited summaries of the prenatal period, and (3) ''pharmacy records are uniformly unavailable, most pharmacists indicating that records by name are kept only for five years''.

Last year the DES-adenosis (DESAD) project, a cooperative multi-institutional study of the effects of DES-exposure during fetal life, reported[9] on ''a comparison of recall and medical records''. There was good agreement for mother's recall with their medical records for past miscarriages and pregnancies. Agreement was poor, however, for recall of medical interventions such as drugs and X-rays. Of the DES-exposed mothers identified through review of their prenatal records, 29% could not remember whether they took DES or not. An additional 8% said that they did not take DES when it was recorded in their chart that they did.

Moss et al. in a 1986 report[49] on hormonal risk factors in testicular cancer made an attempt to document all positive reports of hormone use during pregnancy in testicular cancer patients and ''peer control'', but was able to find confirmation in only 21%. These authors concluded that ''given the low prevalence of DES exposures in controls and the uncertainties associated with verification of exposure, a much larger study than ours would be necessary to resolve this issue''. Leary et al. in another report[8] from the Mayo Clinic stated that ''the relative paucity of case reports of testicular cancer in men who were exposed to DES *in utero* undoubtedly reflects ignorance of DES exposure among young men rather than a reduction in cancer occurrence''.

The question of male genital tract carcinogenesis in human males exposed to DES *in utero* has been raised for several reasons:

1. Oncogenic biochemical effects of DES on human cell cultures (DNA clastogenesis,[50] inhibition of microtubule assembly with resultant metaphase arrest)[51]
2. Causation of testicular[14] and prostatic tumors[52] in animals by DES exposure *in utero*
3. DES induction of tumors in human female sex organs (endometrial, breast, and vaginal carcinomas)[1,53]
4. Production of cryptorchidism and/or testicular hypoplasia in humans[5] and animals[4] which are known predisposing factors for testicular carcinomas[54,55]
5. Occurrence in older men of endometrial-like carcinomas from the prostatic utricle,[56] the Müllerian duct remnant homologous to the upper vagina in females
6. The fact that prostatic adenocarcinomas in human males are sex steroid hormone sensitive[57]
7. Reports on human male testicular carcinomas reviewed in the preceding paragraphs

XI. MECHANISMS OF DES PATHOGENESIS ON THE FETAL MALE GENITAL TRACT (SEX HORMONE COMPARTMENTATION IN PREGNANCY BETWEEN MOTHER/PLACENTA/FETUS)

The production, transport, metabolism, and target organ binding of natural and exogenous sex hormones during pregnancy are diagrammed in Figure 2. The key feature in DES pathogenesis is the lack of DES binding to maternal steroid hormone-binding protein[58] so that the free transplacental transportable fraction of DES is significantly higher in the fetus[59] than the free fraction of natural estrogens from both the maternal ovary and the placenta, with a resulting interference of fetal genital organogenesis mediated by fetal testicular testosterone.[60,61]

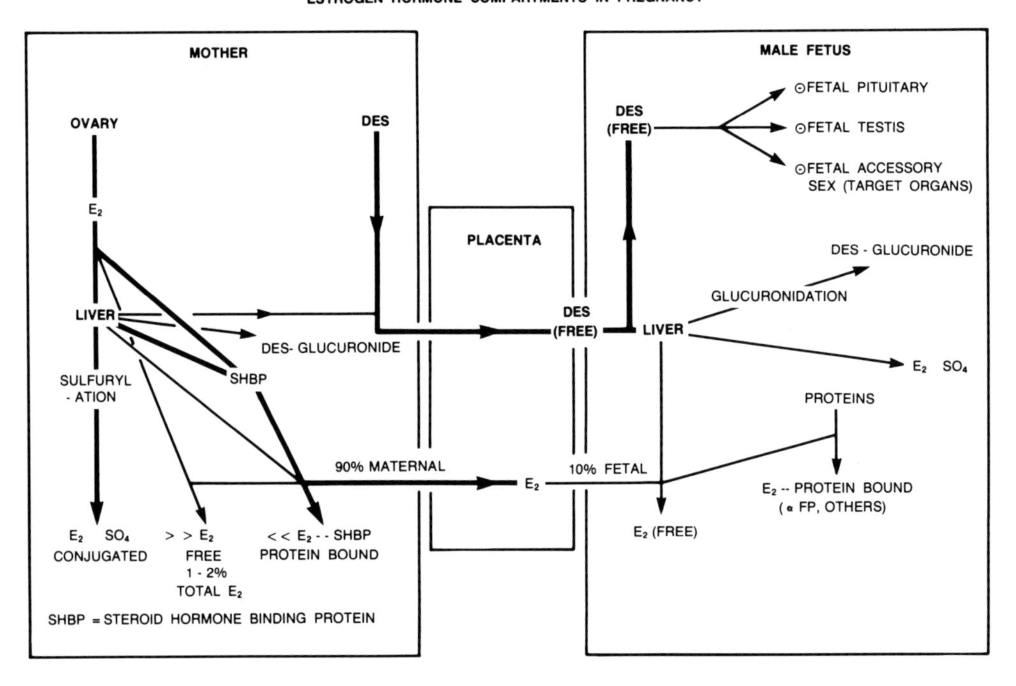

FIGURE 2. Production, transport, metabolism, and target organ binding of natural and endogenous sex hormones during pregnancy.

Contrary to a once widely held misconception that the human male fetus is bathed in a sea of estrogens, the level of free (hormonally active and transmembrane transportable) estradiol (E_2) is quite low in the fetus.[62,63,85] The free fraction of maternal E_2 is only a few percent of the total E_2 with the majority of E_2 being bound to plasma proteins (sex hormone-binding globulin and albumin). In addition, the main source of endogenous estrogen during pregnancy (E_2) is the placenta which shunts 90% of E_2 production to the mother's side[64] and only 10% to the fetus where further reduction in the free "active" form occurs via plasma protein binding and liver conjugation to sulfates and glucuronides.[65] Confusion with rodent models of estrogen pharmacokinetics formerly existed because total estrogen levels in rodent maternal plasma were quite low as compared to humans. Rodents, which do not have sex hormone-binding globulin,[58] have comparably free maternal estrogens.

Evidence that DES crosses the placenta and arrives at target organs in the fetus is derived from quantitative animal data[59,62,66] and the observation in humans by Karnacky[67] that "all the babies exhibited a darkening of the areolae around their nipples, labia, and linea alba, similar in intensity to that of their mothers, indicating that this effect of DES also is shared by the fetus".

McLachlan and Shah and McLachlan[62] have quantitated ^{3}H-DES distribution in pregnant mice and found that free and conjugated DES were of approximately equal concentrations in fetal plasma 30 min after injection of the pregnant mothers and that the levels in the fetal plasma exceeded that in the maternal plasma. Most importantly, free DES was ten times greater than conjugated DES in the fetal genital tract and the level of free DES was two times greater than in the fetal liver. The β-dienestrol metabolites of DES found in mouse, rat, monkey, and human urine have been found to have estrogenic activity in both the in vivo mouse uterine weight bioassay and the in vitro with E_2 receptor binding assay.[68]

A comparison of the transplacental pharmacokinetics of E_2 and DES in the rhesus monkey was made by Slikker, Jr. and Hill.[66] "With DES, the hormonally active parent compound was found in significantly greater concentration in the fetal compartment as compared to the maternal compartment: after the estrogens entered the fetal circulation, conjugation of endogenous estrogens were primarily with sulfates, whereas DES was glucuronoidated." The percentage of nonconjugated DES was greater than the percentage of nonconjugated estrone (E_1), the less potent metabolite of E_2 to which the bulk of E_2 had been converted. "Therefore, the presence of E_1 rather than E_2 reduces the estrogenic potency and/or toxicity of the endogenous estrogen whereas the potency of the synthetic estrogen, DES, remains undiminished." In conclusion these authors stated, "These findings support the hypothesis that pharmacokinetic differences between the synthetic estrogen DES and the endogenous hormone E_2 result in a greater potential for DES-mediated toxicity."

From the preceding descriptions, it is apparent that the synthetic estrogen DES differs considerably from natural steroidal estrogen E_2 with respect to lack of DES protein binding in maternal plasma. This results in greater transplacental transfer of DES to the fetal circulation where its slower conjugation by the fetal liver results in greater free DES. This is the hormonally active form found bound at the target organs of the fetal male genital tract. In the fetus, free DES can bind to target organ sites with resultant negative or inhibitory effects on the pituitary, testis, and accessory male sexual organs (i.e., prostate, seminal vesicles, epididymis, penis, etc.). Any or all of these mechanisms can interfere with the normal development of the fetal male genital tract which is dependent on the normal production and distribution of testosterone from the fetal testis.[60,61]

Fetal pituitary-gonadal axis studies have demonstrated that initiation of testosterone production by fetal testis Leydig cells is from placenta-derived hCG stimulation and probably remains the dominant gonadotropin during the first trimester.[69] Immunoreactive gonadotropin-releasing hormones can be detected in fetal hypothalami by 8 weeks of age.[70] Fetal pituitary gonadotropin can be detected as early as 10 weeks of fetal age and continues to rise during the second trimester,[71] which is the reverse of the falling hCG levels during this period. Speroff et al.[71] have concluded that "Initial testosterone production and sexual differentiation are in response to the high fetal levels of hCG, whereas further testosterone production and differentiation appear to be maintained by the fetal pituitary gonadotropins". The negative feedback of estrogens and DES on the hypothalamic pituitary-gonadal axis has been extensively described.[72] Thus it is apparent that during the second trimester and perhaps as early as the 8th to 10th week of fetal age, DES can be deleterious to the normal development of the male genital tract.

Direct depressive effects of DES on several testicular enzymes involved in testosterone steroidogenesis have been demonstrated in several species. Samuels et al.[31] found both in vivo and in vitro suppression of 17α-hydroxylase, 17-20 lyase, and 17β-hydroxysteroid dehydrogenase in rodents (mice and rats) and monkeys. Following 2 weeks of in vivo DES administration to neonatal mice, the activities of these three enzymes were markedly decreased in in vitro assays and the enzyme suppressions persisted for at least 4 months which would be approximately equivalent to 5 years in humans. Hsueh et al.[20] and Bartke et al.[73] also demonstrated direct inhibitory effects on rodent enzymatic testicular steroidogenesis by DES and E_2, with DES being an order of magnitude more potent as an inhibitor.

In humans with prostatic carcinoma, DES promptly reduced plasma testosterone, but it was 2 weeks before luteinizing hormone (LH) levels fell (Sholiton et al.).[74] Jones et al.[75] showed that in normal human males plasma testosterone was decreased within 3 hours of administration of estrogens while plasma LH levels remained normal.

Direct depressive effects of estrogens on testicular LH/hCG receptors is another subset of direct testicular actions of estrogens. The effects are such that they produce inhibition of testicular androgen steroidogenesis.[72,76,84]

The third mechanism of DES antagonism of testicular androgens in the developing male fetus is directly on the target organs. The target organs are the accessory or secondary sexual organs which are the epididymis, vas deferens, seminal vesicles, prostate, urethra, penis, and scrotum. Liao and Fang[77] reported that DES competed with 5α-dihydrotestosterone for 5α-dihydrotestosterone nuclear receptor at concentrations of 10 to 50 μ *M/ℓ* rodent prostates. Rennie and Bruchovsky[78] and Bellis et al.[79] have extensively studied the competitive binding of androgen and estrogen binding to target organ epithelium and muscle in both in vivo and in vitro studies. Lasnitzki[80] demonstrated estrogen antagonism of androgens on prostatic epithelium in organ culture. Several groups (Baulieu and Robel,[81] Shimazaki et al.,[82] and Lee et al.[83]) have studied the direct inhibition of estrogens on the enzyme 5α-reductase which is necessary to convert testosterone to the more active dihydrotestosterone in certain target organs (prostate, urethra, penis).

XII. CONCLUSION

In conclusion, DES exposure of human males *in utero* has resulted in an increased incidence of structural and functional abnormalities of the male genital tract. The human male, therefore, sustains damage that is qualitatively similar to experimental animals exposed to DES *in utero*. The pathogenetic mechanisms of DES on the male embryo/fetus follow from the unique pharmacokinetics of this synthetic, nonsteroidal estrogen, which in large part escapes the usual protein binding and conjugation mechanisms that protects the developing male genital tract from damage by natural estrogens.

REFERENCES

1. **Herbst, A. L. and Cole, P.,** Epidemiologic and clinical aspects of clear cell adenocarcinoma in young men, in *Intrauterine Exposure to Diethylstilbestrol in the Human,* Herbst, A. L., Ed., American College of Obstetrics and Gynecology, Chicago, 1978, chap. 1.
2. **Gill, W. B., Schumacher, F. B., and Bibbo, M.,** Structural and functional abnormalities in the sex organs of male offspring of mothers treated with diethylstilbestrol, *J. Reprod. Med.,* 16, 147,1976.
3. **Greene, R. R., Burrill, M. W., and Ivy, A. C.,** Experimental intersexuality: effects of estrogens on antenatal sexual development of the rat, *Am. J. Anat.,* 67, 305, 1940.
4. **McLachlan, J. A., Newbold, R. R., and Bullock, B.,** Reproductive tract lesions in male mice exposed prenatally to diethylstilbestrol, *Science,* 190, 991, 1975.
5. **Gill, W. B., Schumacher, G. F. B., Bibbo, M., Straus, F. H., II, and Schoenberg, H. W.,** Association of diethylstilbestrol exposure *in utero* with cryptorchidism, testicular hypoplasia and semen abnormalities, *J. Urol.,* 122, 36, 1979.
6. **Whitehead, E. D. and Leiter, E.,** Genital abnormalities and abnormal semen analyses in males exposed to diethylstilbestrol (DES) *in utero, J. Urol.,* 125, 47, 1981.
7. **Andonian, R. W. and Kessler, R.,** Transplacental exposure to diethylstilbestrol in men, *Urology,* 13, 276, 1979.
8. **Leary, F. J., Resseguie, L. J., Kurland, L. T., O'Brien, P. C., Emslander, R. F., and Noller, K. L.,** Males exposed *in utero* to diethylstilbestrol, *JAMA,* 252, 2984, 1984.
9. **Tilley, B. C., Barnes, A. B., Bergstralh, E., Labarthe, D., Noller, K. L., Colton, T., and Adam, E.,** A comparison of maternal history recall and medical records: implications for retrospective studies, *Am. J. Epidemiol.,* 12, 269, 1985.
10. **Vessey, M. P., Fairweather, D. V. I., and Norman-Smith, B.,** A randomized double blind controlled trial of the value of stilbestrol therapy in pregnancy: long term follow-up of mothers and their offspring, *Br. J. Obstet. Gynecol.,* 90, 1007, 1983.
11. **Ross, R. K., Garbeff, P., Paganini-Hill, A., and Henderson, B. E.,** Effect of *in-utero* exposure to diethylstilbestrol on age at onset of puberty and on postpubertal hormone levels in boys, *Can. Med. Assoc. J.,* 128, 1197, 1983.
12. **Driscoll, S. G. and Taylor, S. H.,** Effects of prenatal maternal estrogen on the male urogenital system, *Obstet. Gynecol.,* 56, 537, 1980.
13. **Hooker, C., Gardner, W., and Pfeiffer, C.,** Testicular tumors in mice receiving estrogens, *JAMA,* 115, 443, 1940.

14. **Shimkin, M. B., Grady, H. G., and Andervont, H. B.,** Induction of testicular tumors and other effects of diethylstilbestrol cholesterol pellets in strain C mice, *J. Natl. Cancer Inst.*, 2, 65, 1941.
15. **Sant, G. R., Mitcheson, H. D., and Ucci, A. A., Jr.,** Sons and testicular cancer — a histopathologic study of testis tumor occurring in DES sons, in *Summary of Conference on Estrogens and Cancer: A Multidisciplinary Focus on Diethylstilbestrol (DES)*, Tunel, J. and Wingard, D., Eds., University of California, Berkeley, 1985, 16.
16. **Cosgrove, M. D., Benton, B., and Henderson, B. E.,** Male genitourinary abnormalities and maternal diethylstilbestrol, *J. Urol.*, 117, 220, 1977.
17. **Beard, C. M., Melton, L. J., III, O'Fallon, W. M., Noller, K. L., and Benson, R. C.,** Cryptorchidism and maternal estrogen exposure, *Am. J. Epidemiol.*, 120, 707, 1984.
18. **Campbell, M. F.,** Embryology of the urogenital tract, in *Campbells' Urology*, 2nd ed., W. B. Saunders, Philadelphia, 1963, chap. 33.
19. **Rajfer, J., Namkung, P. C., and Petra, P. P.,** Identification, characterization and age-related changes of a cytoplasmic androgen receptor in the rat penis, *J. Steroid Biochem.*, 13, 1489, 1980.
20. **Hsueh, A. J. W., Dufau, M. L., and Catt, K. J.,** Direct inhibitory effect of estrogen on Leydig cell function of hypophysectomized rats, *Endocrinology*, 103, 1096, 1978.
21. **Aarskog, D.,** Maternal progestins as a possible cause of hypospadias, *N. Engl. J. Med.*, 300, 75, 1979.
22. **Smith, D. W.,** Micropenis and its management, in *Birth Defects: Original Article Series*, Vol. 8, The National Foundation, 1977, 147.
23. **Hinman, F., Jr.,** Microphallus: distinction between anomalous and endocrine types, *J. Urol.*, 123, 412, 1980.
24. **Wilson, J. D. and Griffin, J. E.,** Disorders of sexual differentiation, in *Harrison's Principles of Internal Medicine*, 9th ed., McGraw-Hill, New York, 1980, chap. 344.
25. **Gill, W. B., Schumacher, G. F. B., Hubby, M. M., and Blough, R. R.,** Male genital tract changes in humans following intrauterine exposure to diethylstilbestrol (DES), in *Developmental Effects of Diethylstilbestrol (DES) in Pregnancy*, Herbst, A. L. and Bern, H. A., Eds., Thieme Stratton, New York, 1981, chap. 8.
26. **Greene, R. R.,** Embryology of sexual structure and hermaphroditism, *J. Clin. Endocrinol.*, 4, 335, 1944.
27. **Kaplan, N. M.,** Male pseudohermaphroditism, *N. Engl. J. Med.*, 261, 641, 1959.
28. **Cleveland, W. and Chang, G.,** Male pseudohermaphroditism with female chromosomal constitution, *Pediatrics*, 36, 892, 1985.
29. **Hoefnagel, P.,** Prenatal diethylstilbestrol exposure and male hypogonadism, *Lancet*, 1, 152, 1976.
30. **Beral, V. and Colwell, L.,** Randomized trial of high doses of stilboestrol and ethisterone therapy in pregnancy: long term follow-up of the children, *J. Epidemiol. Community Health*, 35, 115, 1981.
31. **Samuels, L. T., Uchikawa, T., and Huseby, R. A.,** Direct and indirect effects of oestrogens on the enzymes of the testis, in *Ciba*, Vol. 16, Little, Brown, Boston, 1967, 211.
32. **Imperato-McGinley, J., Peterson, R. E., Stoller, R., and Goodwin, W. E.,** Male pseudohermaphroditism secondary to 17 β-hydroxysteroid dehydrogenase deficiency: gender role change with puberty, *J. Clin. Endocrinol. Metabl.*, 49, 391, 1979.
33. **Mix, T.,** private communications.
34. **Maclusky, N. J. and Naftolin, F.,** Sexual differentiation of the central nervous system, *Science*, 211, 1294, 1981.
35. **Yalom, I. D., Green, R., and Fisk, N.,** Prenatal exposure to female hormones: effect on psychosexual development in boys, *Arch. Gen. Psychiatry*, 28, 554, 1973.
36. **Stillman R. J.,** *In utero* exposure to diethylstilbestrol: adverse effects on the reproductive performance in male and female offspring, *Am. J. Obstet. Gynecol.*, 142, 905, 1982.
37. **Stenchever, M. A., Williamson, R. A., Leonard, J., Karp, L. E., Ley, B., Shy, K., and Smith, D.,** Possible relationship between *in utero* diethylstilbestrol exposure and male fertility, *Am. J. Obstet. Gynecol.*, 140, 186, 1981.
38. **Dieckmann, W. J., Davis, M. E., Rynkiewicz, L. M., and Pottinger, R. E.,** Does the administration of diethylstilbestrol during pregnancy have therapeutic value?, *Am. J. Obstet. Gynecol.*, 66, 1062, 1953.
39. **Schumacher, G. F. B., Gill, W. B., Hubby, M. M., and Blough, R. R.,** Semen analysis in males exposed in utero to diethylstilbestrol (DES) or placebo, *IRCS Med. Sci.*, 9, 100, 1981.
40. **Eliasson, R.,** Analyses of semen, in *Progress in Infertility*, 2nd ed., Behrman, S. J., and Kistner, R. W., Eds., Little, Brown, Boston, 1975, 691.
41. **Schottenfeld, D., Warshauer, M. E., Sherlock, S., Zauber, A. G., Leder, M., and Payne, R.,** The epidemiology of testicular cancer in young adults, *Am. J. Epidemiol.*, 112, 232, 1980.
42. **Henderson, B. E., Benton, B., Jing, J., Yu, M. C., and Pike, M. C.,** Risk factors for cancer of the testis in young men, *Int. J. Cancer*, 23, 598, 1979.
43. **Depue, R. H., Pike, M. C., and Henderson, B. E.,** Estrogen exposure during gestation and risk of testicular cancer, *J. Natl. Cancer Inst.*, 71, 1151, 1983.

44. **Conley, G. R., Sant, G. R., Ucci, A. A., and Mitcheson, H. D.,** Seminoma and epididymal cysts in a young man with known diethylstilbestrol exposure *in utero*, *J. Am. Med. Assoc.*, 249, 1325, 1983.
45. **Greenwald, P., Nasca, P. C., Burnett, W. S., and Polan, A.,** Prenatal stilbestrol experience of mothers of young cancer patients, *Cancer (Philadelphia)*, 31, 568, 1973.
46. **Kinlen, L. J., Badaracco, M. A., Moffett, J., and Vessey, M. P.,** A survey of the use of estrogens during pregnancy in the United Kingdom and of the genitourinary cancer mortality and incidence rates in young people in England and Wales, *J. Obstet. Gynecol. Br. Commonw.*, 81, 840, 1974.
47. **Carstens, P. H. B. and Clemmesen, J.,** Genital tract cancer in Danish adolescents, *N. Engl. J. Med.*, 287, 198, 1972.
48. **Lanier, A. P., Noller, K. L., Decker, D. G., Elveback, L. R., and Kurland, L. T.,** Cancer and stilbestrol: a follow-up of 1719 persons exposed to estrogens *in utero* and born 1943—1959, *Mayo Clin. Proc.*, 48, 793, 1973.
49. **Moss, A. R., Osmond, D., Bacchetti, P., Torti, F. M., and Gurgin, V.,** Hormonal risk factors in testicular cancer: a case control study, *Am. J. Epidemiol.*, 124, 39, 1986.
50. **Birnboim, H. C.,** DNA Clastogenic activity of diethylstilbestrol, *Biochem. Pharmacol.*, 34, 3251, 1985.
51. **Hartley-Asp, B., Deinum, J., and Wallin, M.,** Diethylstilbestrol induces meaphase arrest and inhibits microtubule assembly, *Mutat. Res.*, 143, 231, 1985.
52. **Arai, Y., Chen, C., and Nishizuka, Y.,** Cancer Development in male reproductive tract in rats given diethylstilbestrol at neonatal age, *Gann*, 69, 861, 1978.
53. **Greenberg, E. R., Barnes, A. B., and Resseguie, L., et al.,** A follow-up study of mothers exposed to diethylstilbestrol in pregnancy, *N. Engl. J. Med.*, 311, 1393, 1984.
54. **Batata, M. A., Whitmore, W. F., Chu, F. C. H., Hilary, B. S., Loh, J., Grabstald, H., and Golbey, R.,** Cryptorchidism and testicular cancer, *J. Urol.*, 124, 382, 1980.
55. **Haines, J. S. and Grabstald, H.,** Tumor formation in atrophic testes, *Arch. Surg. Chicago*, 60, 857, 1950.
56. **Melicow, M. M. and Tannenbaum, M.,** Endometrial carcinoma of uterus masculinus (prostatic utricle): report of 6 cases, *J. Urol.*, 106, 892, 1971.
57. **Huggins, C. B. and Hodges, C. V.,** Studies on prostatic cancer: effect of castration, of estrogens and of androgen injection on serum phosphatases in metastatic carcinoma of the prostate, *Cancer Res.*, 1, 293, 1941.
58. **Renoir, J., Mercier-Bodard, C., and Baulieu, E. E.,** Hormonal and immunological aspects of the phylogeny of sex steroid binding plasma protein, *Proc. Natl. Acad. Sci. U.S.A.*, 77, 4578, 1980.
59. **McLachlan, J. A.,** Prenatal exposure to diethylstilbestrol in mice: toxicological studies, *J. Toxicol. Environ. Health*, 2, 527, 1977.
60. **Speroff, L., Glass, R. H., and Kase, N. G.,** Normal and abnormal sexual development, in *Clinical Gynecologic Endocrinology and Infertility*, 3rd ed., Williams & Wilkins, Baltimore, 1983, chap. 12.
61. **Bardin, C. W. and Catterall, J. F.,** Testosterone: a major determinant of extragenital sexual dimorphism, *Science*, 211, 1285, 1981.
62. **Shah, H. C. and McLachlan, J. A.,** The fate of diethylstilbestrol in the pregnant mouse, *J. Pharmacol. Exp. Ther.*, 197, 687, 1976.
63. **Speroff, L., Glass, R. H., and Kase, N. G.,** Hormone biosynthesis, metabolism and mechanism of action, in *Clinical Gynecologic Endocrinology and Infertility*, 3rd ed., Williams & Wilkins, Baltimore, 1983, chap. 1.
64. **Pritchard, J. A., Macdonald, P. C., and Gant, N. F.,** The placental hormones, in *Williams Obstetrics*, 17th ed., Appleton-Century-Crafts, Norwalk, Conn., 1985, 127.
65. **Speroff, L., Glass, R. H., and Kase, N. G.,** The endocrinology of pregnancy, in *Clinical Gynecologic Endocrinology and Infertility*, 3rd ed., Williams & Wilkins, Baltimore, 1983, 276.
66. **Slikker, W., Jr. and Hill, D. E.,** Comparison of the transplacental pharmacokinetics of estradiol-17β and diethylstilbestrol in the rhesus monkey, in *NCTR Final Report Exp. 225*, National Technical Information Service, Springfield, Va., 1981.
67. **Karnacky, K. J.,** Estrogenic tolerance in pregnant women, *Am. J. Obstet. Gynecol.*, 53, 312, 1947.
68. **Korach, K. S., Metzler, M., and McLachlan, J. A.,** Estrogenic activity *in vivo* and *in vitro* of some diethylstilbestrol metabolites and analogs, *Proc. Natl. Acad. Sci., U.S.A.*, 75, 468, 1978.
69. **Winter, J. S. D., Faiman, C., and Reyes, F. I.,** Sex steroid production by the human fetus: its role in morphogenesis and control by gonadotropins, in *Birth defects: Original Article Series*, Vol. 8, The National Foundation, 1977, 147.
70. **Kaplan, S. L., Grumbach, M. M., and Aubert, M. L.,** The ontogenesis of pituitary hormones and hypothalamic factors in the human fetus: maturation of central nervous system regulation of anterior pituitary function, *Res. Prog. Horm. Res.*, 32, 161, 1976.
71. **Speroff, L., Glass, R. H., and Kase, N. G.,** Neuroendocrinology, in *Clinical Gynecologic Endocrinology and Infertility*, 3rd ed., Williams & Wilkins, Baltimore, 1983, 68.

72. **Vigersky, R. A.,** Pituitary-testicular axis, in *Infertility in the Male,* Lipschultz, L. D. and Howards, S. S., Eds., Churchill Livingstone, New York, 1983, chap. 2.
73. **Bartke, A., Williams, K. I. H., and Dalterio, S.,** Effects of estrogens on testicular testosterone production in vitro, *Biol. Reprod.,* 17, 645, 1977.
74. **Sholiton, L. J., Srivastava, L., and Taylor, B. B.,** The *in-vitro* and *in-vivo* effects of diethylstilbestrol on testicular synthesis of testosterone, *Steroids,* 26, 797, 1975.
75. **Jones, T. M., Fang, V. S., Landau, R. L., and Rosenfield, B.,** Direct inhibition of Leydig cell function by estradiol, *J. Clin. Endocrinol. Metab.,* 47, 1368, 1978.
76. **Murphy, J. B., Emmott, R. C., Hicks, L. L., and Walsh, P. C.,** Estrogen receptors in the human prostate, skin, seminal vesicle, epididymis, testis and genital skin: a marker for estrogen-responsive tissues?, *J. Clin. Endocrinol. Metab.,* 50, 938, 1980.
77. **Liao, S. and Fang, S.,** Factors and specificities involved in the formation of 5α-dihydrotestosterone — nuclear receptor protein complex in rat ventral prostate, in *Some Aspects of the Aetiology and Biochemistry of Prostate Cancer,* 3rd Tenovus Workshop, Griffith, K. and Pierrepont, C. G., eds., Cardiff, 1970, 105.
78. **Rennie, P. and Bruchovsky, N.,** Studies on the relationship between androgen receptors and the transport of androgens in the prostate, *J. Biol. Chem.,* 248, 3288, 2973.
79. **Bellis, J. A., Blume, C. D., and Mawhinney, M. G.,** Androgen and estrogen binding in male guinea pig accessory sex organs, *Endocrinology,* 101, 726, 1977.
80. **Lasnitzki, I.,** The effect of hormones on rat prostatic epithelium in organ culture, in *Normal and Abnormal Growth of the Prostate,* Goland, M., Ed., Charles C Thomas, Springfield, Ill., 1974, 29.
81. **Baulieu, E. E. and Robel, P.,** Testosterone metabolites: their receptors, metabolism and action in the rat ventral prostate, in *Some Aspects of the Aetiology and Biochemistry of Prostatic Cancer,* 3rd Tenovus Workshop, Griffith, K. and Pierrepont, C. G., Eds., Cardiff, 1970, 74.
82. **Shimazaki, J., Furuya, N., Yamanaka, H., and Shida, K.,** Effects of estrogen administration on testosterone-induced growth and increase in enzyme activities in the ventral prostate of castrated rats, *Endocrinol. Jpn.,* 16, 163, 1969.
83. **Lee, K. H., Bird, C. E., and Clark, A. F.,** *In vitro* effects of estrogens on rat prostatic 5α reduction of testosterone, *Steroids,* 22, 677, 1973.
84. **Hultaniemi, I. T., Leinonen, P., Hammond, G. L., and Vihko, R.,** Effect of oestrogen treatment on testicular LH/hCG receptors and endogenous steroids in prostatic cancer patients, *Clin. Endocrinol.,* 13, 561, 1980.
85. **Soloff, M. S., Morrison, M. J., and Swartz, T. L.,** A comparison of the estrone-estradiol-binding proteins in the plasmas of prepubertal and pregnant rats, *Steroids,* 20, 597, 1972.
86. **Hembree, W. C., Nagler, H. M., Fang, J., Myles, E. L., and Jagiello, M.,** Infertility in a patient with abnormal spermatogenesis and *in utero* DES exposure, *J. Urol.,* in press.

INDEX

A

A cells, 43—45, see also specific types
Adenocarcinomas, 67, see also specific types
  cervical, 17, 49
  clear cell, see Clear cell adenocarcinoma (CCA)
  DES-associated, 145—146
  rete testicular, 90, 92
  uterine, 53, 70
  vaginal, 17, 49
Adenocarcinomatous lesions, 3
Adenomyosis, 52, 70
Adenosis, 2, 46, 49—51, 65, 73, 151, see also specific types
Adrenal gland changes, 72—73
Androgen receptors, 23—25, see also specific types
Androgens, see also specific types
  adult anatomy and, 134—135
  anatomical change timing and, 133—134
  aromatizable, 66, 71
  behavioral consequences of, 128—129
  epithelial proliferation and, 24
  genital tract effects of, 133
  hormonal action principles and, 133
  insensitivity to, 24
  juvenile behavior and, 128—129
  mammary tumorigenesis and, 82
  mammotropic hormone secretion and, 84
  menarche and, 129—132
  nonaromatizable, 66
  prostatic glands and, 139
  receptor-mediated action of, 24
  sexual behavior and, 129—132
  squamous vaginal epithelial transformation and, 45—46
  urogenital organs and, 135—139
Aneoploidy, 54
Anovulatory syndrome, 64, 71, 82
Antiandrogens, 85, see also specific types
Antibody responses, 113
Antiestrogens, 65, see also specific types
Antigens, 23, see also specific types
Arcuate nucleus (ARCN), 12, 16, 17
Aromatase, 66
Aromatizable androgens, 66, 71, see also specific types
Arthritis, 115—116
Autoantibodies, 115, see also specific types
Autoimmunity, 115—116, see also Immunity
Axonal growth, 16

B

Basal lamina, 45, 64, 69
B cells, 43—45, 112—115, 118, see also specific types
Bed nucleus of stria terminalis, 11
Behavior, 128—132, see also specific types
Birth control pills, 74, 115
Bisexuality, 17
Bladder, 23, 25
Brain
  neuroendocrine, 11, 16—18
  sexual differentiation in, 10—11
  sexually dimorphic, 11—16
Breast cancer, 74, 155—156

C

Cancerous changes, 45—46
Cancers, 69, 74, 82, 155—156, see also specific types
Carcinogens, 83—84, see also specific types
Carcinomas, 73, see also specific types
  cervical, 150
  epidermoid, 45
  in situ, 51
  squamous-cell, 64, 65
  testicular, 168—171
  vaginal, 150
Castration, 113
Catechol estrogen, 55
CB-154, 83
CCA, see Clear cell adenocarcinoma
CCC, see Common cervical canal
C cells, 45, see also specific types
Cell division control, 44
Ceroid deposition, 71, 73
Ceroidogenesis, 72
Cervicovaginal epithelium, 50
Cervicovaginal tumorigenesis, 3
Cervix, 17, 46—49, 51, 82, 135, 139, 150
Chromosomes, 48, 112, 115, see also specific types
Clear cell adenocarcinoma (CCA), 4, 17, 51, 65, 144—149
Clitoris, 69, 135
Cloniphene, 50
Colchicine-like effects, 54
Collagen arthritis, 116
Colposcopy, 152, 154
Columnar epithelium, 46—51
Common cervical canal (CCC), 40, 50
Cornification, 27, 30, 42, 64, 74
Corpora lutea, 53
Corpus testicular lesions, 100—105
Coumestrol, 5, 67, 72, 83
Critical periods, 1
Cryptorchidism, 92, 105, 165—166
Cyclic AMP, 47
Cyclicity of hormone secretion, 84
Cyproterone acetate, 85
Cystic glandular hyperplasia, 52, 53
Cysts, 97, 106, 162, see also specific types
Cytodifferentiation, 24, 32

Cytosolic estrogen binding sites, 67
Cytosolic estrogen receptor sites, 70
Cytotoxic effects, 50, see also specific types

D

Defeminization, 10
Dendritic growth, 16
DES, see Diethylstilbestrol
Desmosomes, 45, 69
5α-DHT, see 5α-Dihydrotestosterone
Dietary estrogen, 67
Diethylstilbestrol (DES), 2—5
  adenocarcinomas and, 145—146
  B cells and, 118
  bisexuality and, 17
  breast cancer and, 155—156
  clear cell adenocarcinoma and, see Clear cell adenocarcinoma (CCA)
  colchicine-like effects of, 54
  columnar epithelium and, 46—48
  cystic glandular hyperplasia and, 53
  cytotoxic effects of, 50
  efficacy of, 144
  epidemiology and, 144—145
  female lower genital tract changes and, 149—152
  fertility and, 167—168
  genotoxic effects of, 47, 49
  history of, 144
  homosexuality and, 17
  immune system and, 117—120
  in utero, 96
  male exposure to, 90, 162, 167, 171—174
  male genital tract abnormalities and, 163—164
  mammary glands and, 82, 83
  mammotropic hormone secretion and, 84
  management of females exposed to, 152—154
  metabolism of, 55
  pregnancy and, 155
  primate genital tract and, 50
  protein synthesis and, 54
  surgery and, 154—155
  testicular lesions and, 104
  transplacental exposure to, 98
  uterine adenocarcinoma and, 70
  vaginal adenosis and, 51
  vaginal cornification and, 31
  vaginal epithelium and, 50—51, 64
  vaginal hyperplasia and, 31
Differentiation, epithelial, see Epithelial differentiation
5α-Dihydrotestosterone, 24, 66, 82, 83, 166
Dihydrotestosterone propionate, 132
7,12-Dimethylbenz(*a*)anthracene, 50
DNA, 45, 55
Dopamine, 84
Dorsomedial POA, 12
Downgrowths, 64
Dysplasia, 51, see also specific types

E

EB, see Estradiol benzoate
Ectopic pregnancy, 71
Endometrial cancer, 74
Endometrial glandular hyperplasia, 52
Environmental estrogens, 5, see also specific types
Epidermoid carcinomas, 45
Epidermoid tumors of vagina, 49
Epididymal cysts, 97, 162
Epigenetic effects, 54—55, see also specific types
Epithelial cells, 22
  androgen-induced proliferation of, 24
  cervical, 47, 48
  cultures of, 30
  morphogenesis of, 27
  morphology of, 32
  proliferation of, 30
  prostatic, 24
  stroma and, 22
Epithelial differentiation, 21—32
  abnormal, 32
  androgenic effects and, 24—25
  mesenchyme and, see Mesenchymal-epithelial interactions
  stroma and, 25—32
Epithelial hyperplasia of rete testis, 90
Epithelial hypertrophy of uterus, 27
Epithelial projections, 71
Epithelium
  cervical, 47
  cervicovaginal, 50
  columnar, 46—51
  epithelial, 27, 64
  heterotopic columnar, see Heterotopic columnar epithelium (HCE)
  morphogenesis of, 24
  Müllerian, 41, 51
  squamous vaginal, 45—46
  stromal interactions with, 32
  transformation of, 41, 45—46
  uterine, 25, 47, 52
  vaginal, see Vaginal epithelium
Estradiol, 30, 31, 46—48, 53, 82, 113
17β-Estradiol, 1
Estradiol benzoate (EB), 83
Estrogen binding sites, 27, 67, 85, see also specific types
Estrogenic mycotoxins, 67, see also specific types
Estrogen receptors, 28, 30, 31, 52, 54, see also specific types
Estrogen receptor sites, 70, see also specific types
Estrogens, see specific types
  autoimmunity and, 115—116
  binding capacities of, 85
  castration and, 113
  cathechol, 55
  columnar epithelium and, 46—51
  dietary, 67

environmental, 5
genetic effects of, 54—55
genital tract development and, 1
immunity and, 112—116
long-term administration of, 1
mammary tumorigenesis and, 82
mitotic activity and, 44
nonreceptor mediated effects of, 53—55
oviductal morphology and, 53
pituitary response to, 84
plant, 5
proliferative inhibition and, 45
receptor mediated effects of, 53—55
serum levels of, 84
squamous vaginal epithelial transformation and, 45—46
synthetic, see Diethylstilbestrol (DES)
uterine effects of, 51—52
Estrogen-target epithelium, 25—32
Estrogen-target organ, 27
Estrous cycles, 27, 84
Estrous state, 64

F

Female genital tract, 39—55, 65, 149—152, see also specific parts
Feminization, 10, 24
Fertility, 154, 167—168
Follicles, 3, 54, 71, 72, see also specific types
Follicle-stimulating hormone (FSH), 17

G

Genetic effects, 54—55, see also specific types
Genital tract, 1—5, 50, 65, 69, 133, see also specific parts
Genome effects, 55, see also specific types
Genotoxic effects, 47, 49, see also specific types
GH, see Growth hormones
Glandular morphogenesis, 22
Gonadal steroids, 10—11, 16—18, see also specific types
Gonadal agenesis,XY, 18
Grafting, 83, see also specific types
Granulomas, 105, see also specific types
Granulosa cell tumors, 72
Growth factors, 85, see also specific types
Growth hormones (GH), 84, see also specific types
Growth regulators, 30, see also specific types

H

HCE, see Heterotopic columnar epithelium
HeLa cells, 54
Hemidesmosomes, 45
Herpesvirus, 51
Heterotopic columnar epithelium (HCE), 46—50
Histogenesis, 156
Homosexuality, 17
17α-Hydroxyprogesterone caproate, 66, 84
Hyperplasia, 30, 31, 52—53, 165
Hyperplastic lesions, 45, 74
Hypoplasia, 51, 69, 162—165
Hypospadias, 46, 69, 166, see also specific types
Hypothalamo-hypophysial system, 1, 64, see also specific types
Hypothalamo-hypophysial-adrenal system, 73
Hypothalamo-hypophysial-ovarian system, 64, 70, 72
Hypothalamus, 84

I

Immunity
B cells and, 114—115, 118
DES and, 117—120
estrogens and, 112—116
nonspecific, 113
physiological, 112—115
pregnancy and, 112
sexual dimorphism and, 112—113
T cells and, 113—114, 117—118
Infertility, 167—168
Interstitial cell tumors, 105
Intraepithelial neoplasia, 51

J

Juvenile behavior, 128—129

K

Kidney cells, 48

L

Labeling index, 47
Labia majora, 135
Labia minora, 135
Leiomyomas, 53, see also specific types
Lesions, see also specific types
adenocarcinomatous, 3
corpus testicular, 100—105
epithelium, 74
genital tract, 65
hyperplastic, 45, 74
Müllerian duct remnant, 94—99
permanent, 74
precancerous, 64
rete testicular, 90—93
vaginal, 64, 65, 74
Leydig cell hyperplasia, 165
LH, see Luteinizing hormone
Lordosis facilitation center, 10
Lordosis-inhibiting system, 10
Lupus, 115
Luteinizing hormone (LH), 17, 18, 53
Lymphocytes, 50, 114, 117—118, see also specific types

Lymphoid organ pathotoxicology, 117

M

Macrophages, 113, see also specific types
Male genital tract, 89—106, see also specific parts
  corpus testis lesions, 100—105
  DES and, 163—164, 171—174
  mesenchymal-epithelial interactions and, 22—24
  Müllerian duct remnant lesions, 94—99
  rete testicular lesions, 90—93
Males
  DES exposure in, 90, 162, 167, 171—174
  homosexual, 17
  pseudohermaphroditism in, 167
  sex hormone exposure in, 161—174, see also specific effects
Malignancy, 49—50, see also specific types
Mammary glands, 24, 25, 81—85
Mammary tumor virus (MTV), 82, 83
Mammotropic hormone secretion, 84
MAN, see Medial amygdaloid nucleus
Masculine behavior, 129, 132
Masculinity, 167
Masculinization, 10, 18
MCF-7 cells, 30
Meatal stenosis, 166—167
Medial amygdaloid nucleus (MAN), 11, 13
Medial preoptic nucleus (MPN), 11, 17
Menarche, 129—132
Mesenchymal cells, 27, 32
Mesenchymal-epithelial interactions, 22—25
Mesenchymal estrogen receptors, 28
Mesenchyme, 22, 25
Mesonephric duct, 105, 106
Mesonephric remnants, 71
Metaplasia, 48—52, 70, see also specific types
3-Methylcholanthrene, 50
Micropenis, 166
Microphallus, 166
Milk, 84
MIS, see Müllerian-inhibiting substance
Mitochondria, 69
Mitogens, 28, see also specific types
Mitotic activity, 44, 45, 47, see also specific types
Moldy corn syndrome, 67
Monosodium glutamate (MSG), 82
Morphogenesis, 22—24, 40, 41, 166
Morphology, 53, see also specific types
MPN, see Medial preoptic nucleus
MSG, see Monosodium glutamate
MTV, see Mammary tumor virus
Mucification, 66
Müllerian duct, 24, 94—99, 105
Müllerian epithelium, 41, 51, 65
Müllerian-inhibiting substance (MIS), 94
Müllerian vagina, 40, 44, 45, 47
Mycotoxins, 67, see also specific types

N

Nafoxidine, 50
Natural killer cells, 50, 118—120
Neoplasia, 51—53, see also specific types
Neuroendocrine brain, 11, 16—18
Nonaromatizable androgens, 66, see also specific types
Nonspecific immune functions, 113, see also specific types
Nuclear estrogen binding sites, 27, 67
Nuclear estrogen receptors, 54, 70

O

Oral contraceptives, 74, 115
Ovarian anovalatory syndrome, 82
Ovaries, 53—54, 71—72, 84
Ovary-dependent cornification, 42, 64
Ovary-independent cornification, 42—45, 64
Oviduct, 53, 70—71

P

PAGE, see Polyacrylamide gel electrophoresis
Papillomas, 73, see also specific types
Papillomavirus, 51
Paramesonephric duct, 105
Paraovarian cysts, 106
Pathotoxicology, 117, see also specific types
PE, see Persistent estrus
Penis, 135, 166
Peritoneal macrophages, 113
Persistent anovulatory diestrous syndrome, 17
Persistent estrus (PE), 1
Persistent vaginal estrous state, 64
PHA, see Phytohemagglutinin
Phenol red, 30
Photoestrogens, 83, see also specific types
Phytoestrogens, 67, see also specific types
Phytohemagglutinin (PHA), 113
Pituitary gland, 83, 84
Plant estrogens, 5, see also specific types
Plasma growth hormones, 84
POA, 11, 12, 17
Polyacrylamide gel electrophoresis (PAGE), 23
Polyovular follicles, 3, 54, 71, 72
Polyploidy, 54
Precancerous changes, 45—46
Precancerous lesions, 64
Pregnancy, 71, 82, 112, 115, 155
Primate genital tract, 50
Primordia, 134
Progesterone, 4, 45—46, 82—85
Progesterone receptors, 30, see also specific types
Prolactin, 82, 84
Proliferative activity, 46, see also specific types
Prostate, 22—25, 139

Protein kinase, 47
Proteins, 54, see also specific types
Pseudohermaphroditism, 128, 132, 167
Psychological effects, 128—129, see also Behavior
Psychosexual abnormalities, 167, see also specific types

R

5α-Reductase, 66
Reproductive tract, see Genital tract
Rete testis, 90—93
Reticuloendothelial system, 113
Rheumatoid arthritis, 115—116

S

Sarcomas, 50, 73, see also specific types
SCE, see Sister chromatid exchanges
SDN, see Sexually dimorphic nucleus of POA
Seminal vesicle, 22, 23
Seminomas, 100, see also specific types
Sex differences, 10—11, 17, see also Feminization; Males; Masculinization
Sex hormones, 1—5, 112, 167, see also specific types
Sexual behavior, 129—132, 167, see also specific types
Sexual dimorphism, 112—113
Sexually dimorphic brain, 11—16
Sexually dimorphic nucleus of POA (SDN-POA), 11, 17
Sinus vagina, 40, 44
Sister chromatid exchanges (SCE), 48, 55
SLE, see Lupus; Systemic lupus erythematosus
Spermatocele cysts, 162
Sperm granulomas, 105
Spontaneous mammary tumorigenesis, 82—83
Squamous-cell carcinoma, 3, 64, 65
Squamous metaplasia, 49, 51—52, 70
Squamous neoplasia, 51
Squamous stratification, 69
Squamous vaginal epithelium, 45—46
Stenosis, 166—167, see also specific types
Steroids, 82, 83, see also specific types
Stria terminalis, 11
Stroma, 22, 25—32
Stromal cells, 24, 25, 30, 47
Stromal mediators, 30, see also specific types
Structural sex difference, 11
Synaptogenesis, 16
Systemic lupus erythematosus (SLE, lupus), 115

T

Tamoxifen, 50, 65, 69, 70, 72
T cells
  antibody responses and, 113
  autoimmunity and, 115—116
  function of, 112, 114
  hyperactivity of, 115
  immunity and, 113—118
Testes, 139
  abnormalities in, 162—166, see also specific types
  adenocarcinoma of, 90, 92
  carcinoma of, 168—171
  corpus, see Corpus testicular lesions
  degenerative changes in, 100, 101
  hypoplasia of, 162—165
  retained, 100
  rete, see Rete testis
  tumors of, 105
Testicular feminization, 24
Testosterone, 24, 82, 134, 135, 139
Testosterone propionate, 83, 132
Thymus, 113, 117
Triphenylethylene derivatives, 50, see also specific types
Tumorigenesis, 3, 82—84
Tumors, see also specific types
  epidermoid, 49
  granulosa cell, 72
  interstitial cell, 105
  natural killer cells and susceptibility to, 118—120
  ovarian, 84
  testicular, 105
  uterine, 84
  vaginal, 2, 46, 49
Type A urogenital organs, 135, see also specific types
Type B urogenital organs, 135, see also specific types
Type C urogenital organs, 135—139, see also specific types

U

Urethra, 139
Urinary bladder, 23
Urogenital organs, 135—139, see also specific types
Urogenital sinus, 22, 23, 25, 40
Uterine body, 135
Uterine cervix, 46, 47, 51
Uterine epithelium, 25, 27, 47, 52, 69
Uterus, 51—53, 69—70, 84, 139

V

Vagina, 23, 51, 135, 139
  adenocarcinoma of, 17, 49
  adenosis of, 46, 51, 73
  cancers of, 69, 82, see also specific types
  carcinoma of, 160
  changes in, 64—69
  concretions of, 46
  development of, 41
  dysplasia of, 51
  hyperplasia of, 30, 31
  irreversible changes in, 66
  lesions of, 64, 65

mitotic activity in, 44
morphogenesis of, 41
Müllerian, see Müllerian vagina
permanent changes in, 4
ridging of, 73
sinus, 40, 44
stroma of, 25
tumors in, 2, 46
Vaginal cornification, 42—45, 64, 66
Vaginal differentiation, 25
Vaginal epidermoid tumors, 49
Vaginal epithelium, 27, 45—46, 50—51, 64, 74
Vaginal estrous state, 64
Vaginal plate, 41
Ventromedial nucleus (VMN), 11, 15, 16
Vestibule, 135
Viruses, 51, 82, see also specific types
Vitamin A, 43, 82
VMN, see Ventromedial nucleus

W

Wolffian ducts, 22, 24, 105, 135—136

X

X chromosomes, 112, 115
XY gonadal agenesis, 18

Y

Y chromosomes, 112
YMN, 13

Z

Zearalenone, 5, 47, 67, 72, 83